Imaging Biomarkers in Epilepsy

Imaging Biomarkers in Epilepsy

Edited by

Andrea Bernasconi
Montreal Neurological Institute, McGill University

Neda Bernasconi
Montreal Neurological Institute, McGill University

Matthias Koepp
Institute of Neurology, University College London

CAMBRIDGE
UNIVERSITY PRESS

CAMBRIDGE
UNIVERSITY PRESS

University Printing House, Cambridge CB2 8BS, United Kingdom

One Liberty Plaza, 20th Floor, New York, NY 10006, USA

477 Williamstown Road, Port Melbourne, VIC 3207, Australia

314–321, 3rd Floor, Plot 3, Splendor Forum, Jasola District Centre, New Delhi – 110025, India

79 Anson Road, #06–04/06, Singapore 079906

Cambridge University Press is part of the University of Cambridge.

It furthers the University's mission by disseminating knowledge in the pursuit of education, learning, and research at the highest international levels of excellence.

www.cambridge.org
Information on this title: www.cambridge.org/9781107108356
DOI: 10.1017/9781316257951

First published 2019

Printed and bound in Great Britain by Clays Ltd, Elcograf S.p.A.

A catalogue record for this publication is available from the British Library.

ISBN 978-1-107-10835-6 Hardback

To our dear mentors, Frederick Andermann and Dieter Janz

Contents

Preface

Epilepsy is one of the most common and disabling neurological disorders, affecting 50 million people worldwide. Its prevalence appears to be increasing and is double that of multiple sclerosis, Parkinson's disease, and autism spectrum disorders combined. In nearly 30% of patients with epilepsy, seizures remain insufficiently controlled despite the availability of numerous antiseizure drugs. No treatments exist to prevent the development of epilepsy, despite an increasing understanding of its underlying molecular and cellular pathways. Major barriers in these areas are the complex and multifactorial nature of epilepsy and its heterogeneity. No tests exist for measuring the presence or severity at any stage of the disease, other than counting seizures, and no tools are available for predicting response to therapeutic interventions, be it medical or surgical, other than waiting for the next seizure to occur. Thus, the vision of new treatments relies upon the development of biomarkers that would ultimately allow individualized treatments.

Epilepsy may broadly be defined as a state of recurrent spontaneous seizures that arise when balance between neuronal excitation and inhibition is disrupted. Epileptogenesis, the gradual process by which a brain develops epilepsy after diverse pathogenetic insults, can be examined at different levels of the nervous system: ions and membranes, cells and circuits/synapses, and large-scale neuronal networks. Importantly, there have been two recent paradigm shifts in our conceptualization and understanding of these phenomena, which make this book particularly timely. Indeed, epilepsy is now defined as a disorder characterized by an enduring predisposition to generate epileptic seizures and associated cognitive, psychological, and social consequences. This widened approach is mirrored by the generally accepted view that epileptogenesis includes both the development of an epilepsy condition and its progression.

During the last few decades, we have witnessed unprecedented advances in imaging techniques, particularly MRI, which has revolutionized epileptology by shifting the field from prevailing electroclinical correlations to a multidisciplinary practice. Importantly, neuroimaging provides a broad spectrum of research and clinical tools with the capacity to identify biomarkers, namely objectively measurable characteristics of biological processes that identify the development, presence, severity, progression, and localization of epileptogenic abnormalities. In this regard, epilepsy research, with its established animal models that recapitulate many syndromes seen in human subjects, provides a unique opportunity to assess the validity of preclinical biomarkers in clinical settings and vice versa. The ideal biomarker should be sensitive, specific, and reproducible, as well as non-invasive, easily repeatable, and cost-effective. It is fair to say that imaging markers, particularly MRI-derived indicators of the "epileptic state," meet the majority of these criteria.

We wanted this book to be truly translational and provide both clinicians and researchers with a series of chapters that integrate in one medium state-of-the-art science on imaging biomarkers aimed at capturing the development and progression of various epilepsies, as well as their neurobiological and cognitive consequences. The central concept to this work is that clinically relevant topics, such as mapping epileptogenic lesions, tracing disease consequences, or predicting therapy responses, should be linked with multidisciplinary research that takes advances from basic science directly into the medical arena, with the ultimate goal to tailor the right diagnostic and therapeutic strategy for the right person at the right time. We anticipate that the discovery of novel imaging-derived biomarkers for prediction and monitoring of treatment response and outcome (including comorbidities) will have a fundamental impact not only on the way we treat epilepsy but also on associated conditions, such as stroke and neurodegeneration.

The book is divided into four sections. Part I includes a series of chapters focused on imaging of early disease stages. Part II discusses lesion detection and network analysis methods. Part III focuses on imaging methods used to predict response to antiepileptic drugs and surgery. Finally, Part IV presents imaging techniques used to evaluate disease consequences. The hope of the editors and contributors, who are worldwide recognized experts in the field, is that this book will provide a unique tool to guide the selection of imaging platforms in everyday clinical practice, to improve understanding of the pathophysiology of epilepsy, and to set the stage for future research in neuroimaging of epilepsy.

Contributors

Dace Almane
Department of Neurology, University of
Wisconsin School of Medicine and Public
Health
Madison, WI, USA

Marina K. Alvim
Department of Neurology, FCM, University of
Campinas–UNICAMP
Campinas, SP, Brazil

Jens P. Bankstahl
Preclinical Molecular Imaging, Department of
Nuclear Medicine, Hannover Medical School
Hannover, Germany

Marion Bankstahl
Department of Pharmacology, Toxicology and
Pharmacy, University of Veterinary Medicine
Hannover
Hannover, Germany

Madison M. Berl
Center for Neuroscience, Children's National Medical
Centre, George Washington University School of
Medicine
Washington, DC, USA

Andrea Bernasconi
Neuroimaging of Epilepsy Laboratory, McConnell
Brain Imaging Centre, Montreal Neurological
Institute and Hospital, McGill University
Montreal, Quebec, Canada

Neda Bernasconi
Neuroimaging of Epilepsy Laboratory, McConnell
Brain Imaging Centre, Montreal Neurological
Institute and Hospital, McGill University
Montreal, Quebec, Canada

Boris C. Bernhardt
Multimodal Imaging and Connectome Analysis,
Department of Neurology and McConnell Brain
Imaging Centre, Montreal Neurological Institute,
McGill University
Montreal, Quebec, Canada

Giuseppe Bertini
Department of Neurological and Movement Sciences,
School of Medicine, University of Verona
Verona, Italy

Hal Blumenfeld
Departments of Neurology, Neurobiology and
Neurosurgery, Yale University School of Medicine
New Haven, CT, USA

Sam Bobholz
Department of Neurology, University of Wisconsin
School of Medicine and Public Health
Madison, WI, USA

Silvia B. Bonelli
Department of Neurology, Medical University of Vienna
Vienna, Austria

Benjamin H. Brinkmann
Mayo Systems Electrophysiology Lab, Neurology,
Mayo Clinic
Rochester, MN, USA

Lorenzo Caciagli
Department of Clinical and Experimental Epilepsy,
Queen Square Institute of Neurology, University College
London, UK
MRI Unit, Epileptic Society, Chalfont St Peter, UK

Fernando Cendes
Department of Neurology, FCM, University of
Campinas–UNICAMP
Campinas, SP, Brazil

Ana Carolina Coan
Department of Neurology, FCM, University of
Campinas–UNICAMP
Campinas, SP, Brazil

Luis Concha
Institute of Neurobiology, Universidad Nacional
Autónoma de México
Querétaro, México

Kevin Dabbs
Department of Neurology, University of Wisconsin
School of Medicine and Public Health
Madison, WI, USA

John S. Duncan
Department of Clinical and Experimental Epilepsy,
Queen Square Institute of Neurology, University College
London, UK

Paolo Francesco Fabene
Department of Neurological and Movement
Sciences, School of Medicine, University of Verona
Verona, Italy

Maria Feldmann
Department of Clinical and Experimental Epilepsy,
Queen Square Institute of Neurology, University
College London, UK

William Davis Gaillard
Center for Neuroscience, George Washington
University
Children's National Medical Centre
Washington, DC, USA

Marian Galovic
Department of Clinical and Experimental Epilepsy,
Queen Square Institute of Neurology, University of
London, UK
MRI Unit, Epilepsy Society,
Chalfont St Peter, UK
Department of Neurology, Kantonsspital
St. Gallen, Switzerland

Antonio Gambardella
Institute of Neurology, Department of Medical
and Surgical Sciences, University Magna Graecia,
Catanzaro, Italy

Camille Garcia-Ramos
Medical Physics, University of Wisconsin School of
Medicine and Public Health
Madison, WI, USA

Ravnoor Gill
Neuroimaging of Epilepsy Laboratory, McConnell
Brain Imaging Centre, Montreal Neurological
Institute and Hospital, McGill University
Montreal, Quebec, Canada

Olli Gröhn
A.I. Virtanen Institute for Molecular Sciences,
Biomedical Imaging Unit, University of Eastern
Finland
Kuopio, Finland

Bruce Hermann
Department of Neurology, University of Wisconsin
School of Medicine and Public Health
Madison, WI, USA

Edward Hogan
Washington University School of Medicine in St. Louis
St Louis, MO, USA

Seok-Jun Hong
Neuroimaging of Epilepsy Laboratory, McConnell
Brain Imaging Centre, Montreal Neurological
Institute and Hospital, McGill University
Montreal, Quebec, Canada

Daren Jackson
Department of Neurology, University of Wisconsin
School of Medicine and Public Health
Madison, WI, USA

Graeme D. Jackson
Florey Institute of Neuroscience and Mental
Health, University of Melbourne, Austin Campus
Heidelberg, Victoria, Australia

Jana Jones
Department of Neurology, University of Wisconsin
School of Medicine and Public Health
Madison, WI, USA

Csaba Juhász
Departments of Pediatrics, Neurosurgery and
Neurology, Wayne State University

PET Center and Translational Imaging Laboratory, Children's Hospital of Michigan
Detroit, MI, USA

Matthias Koepp
Department of Clinical and Experimental Epilepsy, Queen Square Institute of Neurology, University College London, UK
MRI Unit, Epilepsy Society, Chalfont St Peter, UK

Angelo Labate
Institute of Neurology, Department of Medical and Surgical Scienes, University of Magna Graecia, Catanzaro, Italy

Jack Lin
Department of Neurology, University of California, Irvine, CA, USA

Min Liu
Neuroimaging of Epilepsy Laboratory, McConnell Brain Imaging Centre, Montreal Neurological Institute and Hospital, McGill University
Montreal, Quebec, Canada

Gilles van Luijtelaar
Donders Centre for Cognition, Radboud University, Nijmegen, The Netherlands

Pasquina Marzola
Department of Computer Science, University of Verona
Verona, Italy

Asht Mangal Mishra
Department of Neurology, Yale University School of Medicine
New Haven, CT, USA

Sandeep Mittal
Department of Neurosurgery, Wayne State University
Detroit, MI, USA

Elena Nicolato
Department of Neurological and Movement Sciences, School of Medicine, University of Verona
Verona, Italy

Amir Omidvarnia
Florey Institute of Neuroscience and Mental Health, University of Melbourne, Austin Campus
Heidelberg, Victoria, Australia

Mangor Pedersen
Florey Institute of Neuroscience and Mental Health, University of Melbourne,
Austin Campus
Heidelberg, Victoria, Australia

Michele Pellitteri
GlaxoSmithKline, Quality Assurance Service, Verona, Italy

Rodney C. Scott
Department of Neurological Sciences, University of Vermont, Burlington, VT, USA

Mike Seidenberg
Department of Psychology, Rosalind Franklin University
North Chicago, IL, USA

Alejandra Sierra
A.I.Virtanen Institute for Molecular Sciences, Biomedical Imaging Unit, University of Eastern Finland
Kuopio, Finland

Elson L. So
Mayo Clinic Division of Epilepsy & Section of Electroencephalography, Neurology, Mayo Clinic
Rochester, MN, USA

Vlastimil Sulc
Department of Neurology, Charles University in Prague, Motol University Hospital
Prague, Czech Republic

William H. Theodore
Clinical Epilepsy Section, National Institute of Neurological Disorders and Stroke, National Institutes of Health
Bethesda, MD, USA

Grygoriy Tsenov
Department of Neurological and Movement Sciences, School of Medicine, University of Verona
Verona, Italy

Britta Wandschneider
Department of Clinical and Experimental Epilepsy,
Queen Square Institute of Neurology, University
College London, UK
MRI Unit, Epilepsy Society, Chalfont St Peter, UK

Gregory Worrell
Mayo Clinic Division of Epilepsy & Section of
Electroencephalography, Neurology,
Mayo Clinic
Rochester, MN, USA

Fenglai Xiao
Department of Clinical and Experimental
Epilepsy, Queen Square University College
London, UK
MRI Unit, Epilepsy Society, Chalfont St Peter, UK

Clarissa Lin Yasuda
Department of Neurology, FCM,
University of Campinas–
UNICAMP
Campinas, SP, Brazil

Imaging Biomarkers for Febrile Status Epilepticus and Other Forms of Convulsive Status Epilepticus

Rodney C. Scott

1.1 Introduction

Status epilepticus was originally conceived as the "maximal expression" of epilepsy, being used to describe seizures that "persist for a sufficient length of time or repeated frequently enough to produce a fixed or enduring condition."[1] Although definitions of status epilepticus have evolved with our understanding of the pathophysiology of seizures and epilepsy, the core idea of a self-sustaining epileptic state remains an important part of most definitions of status epilepticus.

Early experiments in primates found that prolonged seizure activity can lead to permanent neuronal damage,[2] and subsequent studies have shown that the risk of this rises significantly after 30 minutes of seizure activity.[3] Thus, although many clinical studies have moved toward a shorter, operational definition of status epilepticus as seizures lasting longer than 5–10 minutes,[4] in order to prioritize the timely administration of antiseizure medication, when considering the risk of longer term brain injury a longer cutoff of 30 minutes remains important.

The importance of status epilepticus is twofold: first, acute episodes of status epilepticus are associated with significant mortality and morbidity in adults (though less so in children); and second, it is hypothesized that episodes of status epilepticus, in particular prolonged febrile seizures (PFS), cause damage to the hippocampus, leading to the later development of mesial temporal sclerosis and temporal lobe epilepsy.

1.1.1 Mortality

Increased short- and long-term mortality have been reported following status epilepticus, however the extent and significance of this risk vary hugely depending on the age and nature of the population studied. Overall it appears that the most significant determinant of short-term mortality is the underlying etiology, with prolonged seizures due to anoxia and cerebrovascular disease having particularly poor prognoses[5] (Table 1.1). In both adult and pediatric studies, the majority of cases of fatal status epilepticus are associated with acute symptomatic causes, with much lower mortality from other types of status epilepticus, suggesting that the primary cause of mortality is from the acute neurological insult, rather than the prolonged seizure itself. A number of scoring systems have been developed to predict the outcome of acute status epilepticus, although they have mostly focused on clinical and electroencephalogram (EEG) parameters rather than neuroimaging.[6] Magnetic resonance imaging (MRI) is recognized to have an important role in helping to determine the etiology of an episode of status epilepticus,[7] but its independent use as a biomarker to determine prognosis has been relatively unexplored to date.

1.1.2 Epileptogenesis

Febrile seizures, that is, epileptic seizures associated with a febrile illness (temperature >38.0C) without evidence of central nervous system (CNS) infection, are a relatively common phenomenon occurring in childhood that will affect around 2–5% of the general population, with a peak incidence at 18 months age.[9]

Table 1.1 Mortality of Status Epilepticus (Adapted from Neligan and Shorvon[5] and Maytal et al.[8])

Etiology	Estimated mortality
Anoxia	60–100%
Cerebrovascular disease	20–60%
Drug overdose/toxicity	10–20%
Acute CNS infection	0–30%
Idiopathic/cryptogenic	5–20%
Alcohol related	<10%
Prolonged febrile seizures	<2%

1

Up to 5% of these will be prolonged,[10] with durations of >30 minutes.

Mesial temporal sclerosis (MTS) is a structural abnormality that involves not only the hippocampus, but also the entorhinal cortex and the amygdala. Hippocampal sclerosis presents with characteristic radiological (Figure 1.1) and histological (Figure 1.2) appearances. On histological examination there is a loss of neurons in CA1 and CA4 subfields of the hippocampus with atrophy and gliosis. This is reflected in the MRI changes, which show a loss of hippocampal volume on the affected side with an increase in signal intensity on T2-weighted imaging.

MTS is one of the most common lesions found in patients with temporal lobe epilepsy (TLE) and has been described in up to 65% of cases.[11] TLE associated with MTS (TLE-MTS) is frequently refractory to medical treatment with antiepileptic medication. It is one of the most common indications for epilepsy surgery. As 30–50% of patients with TLE-MTS have a history of childhood PFS,[11,12] this strong association has led to the long-standing hypothesis that the two are casually related.

According to this hypothesis, subtle hippocampal injury is caused by the initial PFS that subsequently evolves into TLE-MTS. As the mean age of onset of TLE-MTS is 4–16 years,[13] this presupposes a prolonged silent period of "epileptogenesis" lasting several years, during which the epilepsy develops, thus potentially providing a window of opportunity for preventative intervention.

As the incidence of PFS is orders of magnitude greater than that of TLE-MTS, some further means of identifying those children at risk of developing TLE-MTS is necessary. The development of appropriate biomarkers to reliably identify children who will develop TLE-MTS following PFS would allow trials of targeted antiepileptogenic treatment to take place using surrogate endpoints without the need for a prolonged 8- to 12-year observation period before judging the outcome.

1.2 The Neuroimaging of Status Epilepticus

In order to develop suitable noninvasive biomarkers for use in status epilepticus, it is necessary to understand the physiological changes occurring during and following status epilepticus and how these are reflected by changes to the neuroimaging appearances, MRI being the general technique of choice.

The uncontrolled, synchronous neuronal discharge occurring during an epileptic seizure leads to significant increases in neuronal energy consumption and metabolism. As a result, blood flow increases, driven by this increasing demand. At some point, as a seizure continues, increases in blood supply become unable to keep up with the increased metabolic demands and the cerebral environment becomes increasingly hypoxic and acidotic with the buildup of lactate as a by-product of anaerobic respiration. At the same time, the recurrent neuronal depolarizations cause increases in the concentration of extracellular glutamate due to saturation of reuptake mechanisms at excitatory synapses and massive influx of calcium into neurons, significantly raising the intracellular concentration. Neuronal injury can therefore occur from both hypoxic and excitotoxic mechanisms, with the risks increasing with seizure duration.

Intractable seizures lasting hours to days have been shown in numerous case studies to lead to

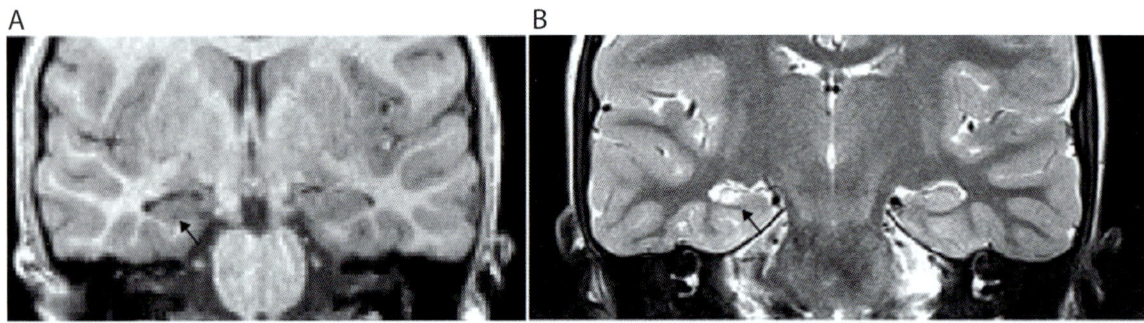

Figure 1.1. MRI images of mesial temporal sclerosis With 3D-FLASH MRI sequence (A) showing volume loss of the right hippocampus (arrow), and T2-weighted coronal MRI sequence (B) showing increased intensity of the right hippocampus (arrow).

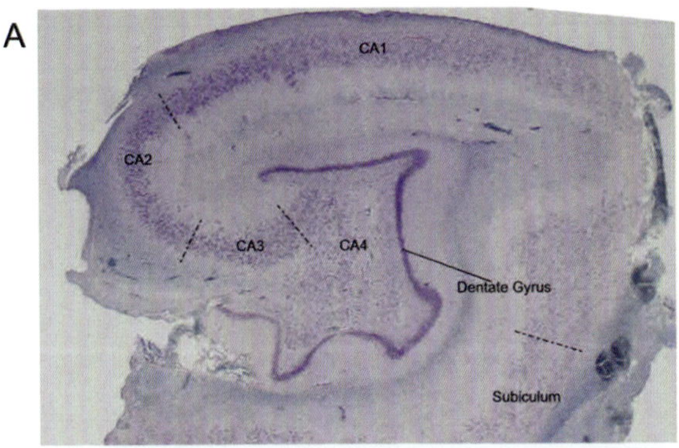

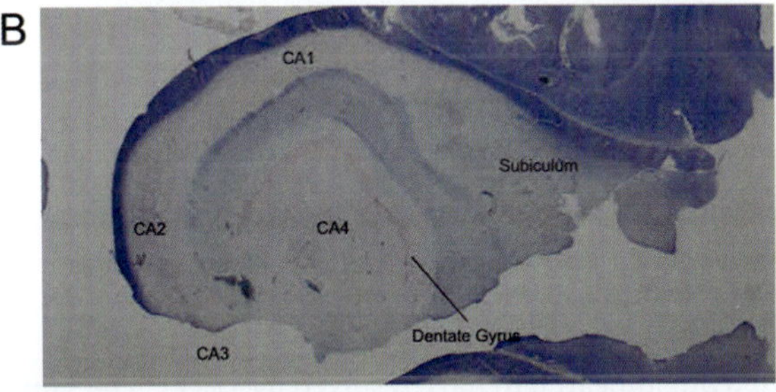

Figure 1.2. Histological appearance of (A) normal hippocampus and (B) MTS (LFB staining).

progressive cerebral edema, reduced conscious level, and death, with brain atrophy and neuronal cell loss visible at postmortem. The hippocampus is posited to be a particularly vulnerable structure due to its rich connectivity and propensity for burst firing and the vulnerability of its blood supply.[14]

1.2.1 Human Studies

1.2.1.1 Hippocampal Changes Following Status Epilepticus

There are now a number of neuroimaging studies that have looked systematically at cohorts of children following an episode of convulsive status epilepticus (CSE) and/or PFS (Table 1.2). There is broad agreement from these studies and a number of smaller case series/reports that in the immediate period following status epilepticus there are a number of acute changes visible on MRI suggestive of vasogenic edema within the hippocampus: an increase in hippocampal volume associated with increased signal on T2-weighted imaging and reductions in the apparent diffusion coefficient (ADC). These changes appear to resolve over time, with the majority of children showing apparent normalization of T2 signal and ADC by 48–72 hours post-CSE. Although most of the attention has been on children following PFS, studies that included children with other forms of status epilepticus did not find significant differences in the hippocampal changes reported, suggesting that they may be due to generic effects of a prolonged seizure rather than specific to PFS.[15,16]

Although hippocampal volumes and T2 changes initially appear to resolve at 48–72 hours, subsequent longitudinal imaging has shown that a proportion of children show ongoing hippocampal volume loss.[15,17,18]

In the largest study, the FEBSTAT group[17] found that 11/130 (8.5%) children showed increased signal on T2-weighted imaging at 1-year follow-up; they felt that 10 of these also showed signs of hippocampal atrophy on visual inspection, meeting criteria for MTS. All of the children with MTS had shown hippocampal T2 changes on their initial MRI.

In a separate study, Yoong et al.[15] looked at 80 children following status epilepticus; no children met clinical criteria for MTS by the end of follow-up at 1 year, but 15 children showed unilateral or bilateral hippocampal volume reduction over this period. The proportions of children showing volume loss were comparable between children with PFS and those with status epilepticus from other etiologies, suggesting that this phenomenon is not restricted to PFS. Previous work by Scott et al.[16] in a smaller cohort of 14 children with PFS had found increased hippocampal asymmetry in 5/14 children at 6-month follow-up, possibly an early sign of developing hippocampal sclerosis.

Taken together, these studies show that there are clearly children who show progressive hippocampal volume loss and develop MTS following PFS and that preexisting MTS as a cause of PFS is very rare. However, this progression appears to occur only in a minority of children (<10%), and at least to the end of existing follow-up periods, relatively few of these children have developed TLE. The long latent period between initial PFS and the later development of TLE-MTS means that it is possible that this remains an underestimate, especially as detecting reduced hippocampal growth rather than an absolute reduction in hippocampal volume is difficult; however, very long-term follow-up of children with status epilepticus[19] and epidemiological studies of children with PFS[20] are broadly supportive of the hypothesis that only a small

Table 1.2 Studies of Convulsive Status Epilepticus

	Study population	Time of initial MRI	Time of follow-up MRI	Findings
FEBSTAT study group 2014[17]	226 children with PFS	<3 days	1 year	22 children showed T2 hyperintensity and increased hippocampal volume on initial MRI, 10 showed MTS at follow-up
Yoong et al. 2013[15]	80 children with SE (33 PFS)	1–8 weeks	6 months and 1 year	15 children showed hippocampal volume loss at 1 year (5 PFS)
Provenzale et al. 2008[18]	11 children with PFS—included within the FEBSTAT population	<3 days	3–23 months	7 children showed T2 hyperintensity at initial scan, 5 children showed hippocampal volume loss at follow-up
Scott et al. 2006[21]	23 children with PFS	<5 days	5.5 months	Increased ADC in children imaged <2 days of CSE
Scott et al. 2002/2003[16,22]	35 children with CSE (21 PFS)	<5 days	4–8 months	Increased hippocampal volumes and increased T2 relaxation time in children imaged <2 days of CSE; 5/14 children had significant hippocampal asymmetry at follow-up

proportion of children develop TLE-MTS following PFS and do not provide any evidence of a "hidden epidemic" developing in later adolescence/adulthood.

1.2.1.2 Extrahippocampal Changes Following Status Epilepticus

Studies of patients with established TLE-MTS have shown that the pathology is not limited to the sclerotic hippocampus, but rather that abnormalities frequently extend throughout the temporal lobe and may affect the contralateral hippocampus and other extrahippocampal structures. These are often not detectable from inspection, but become apparent with quantitative MRI analytic techniques. In particular, studies looking at neuronal microstructure using diffusion tensor imaging (DTI) have found significant reductions in white matter integrity in several different white matter tracts bilaterally[23] and within bilateral thalami.[24]

Interestingly, while these appear to correlate with the degree of cognitive deficit and memory impairment seen in TLE-MTS, they do not appear to influence the response to surgical treatment, suggesting an indirect relationship to seizure generation and control.

Relatively few studies have looked at extrahippocampal pathology following status epilepticus. As part of the STEP IN study, Yoong et al. found extensive white matter changes on DTI following PFS.[25] These changes were apparent at up to 6 months following

PFS but appeared to have resolved by 1 year. The same group of children were also found to show subtle memory impairments,[26] although it has not been shown if these are related. This provides further evidence that there may be a longer term impact of PFS other than MTS and that epilepsy may not be the only outcome of interest.

1.2.2 Animal Studies

There are now several animal models of status epilepticus, including models of PFS.[27] These models are useful for identifying the pathophysiological mechanisms and potential time course of hippocampal injury and for evaluating the relationship between MRI measures and cognitive outcomes.

There are three main pathophysiological processes associated with hippocampal injury following status epilepticus. The first is related to blood flow. MRI perfusion was carried out during pilocarpine-induced status epilepticus in rats and revealed that blood flow in the hippocampus did not change despite the well-described increase in metabolic demand.[28] On the other hand blood flow did increase in the cortex, suggesting that hippocampal homeostatic mechanisms fail during status epilepticus and the ensuing relative hypoperfusion is part of the mechanism that causes hippocampal injury.

The second process is excitotoxicity. Manganese-enhanced MRI (MEMRI) is a tool for exploring whether there has been excess calcium uptake into cells. Manganese competes with calcium, and therefore areas with excessive activation of calcium permeable channels will also allow excessive manganese uptake. The paramagnetic properties of manganese mean that areas with increased manganese will be hyperintense on T1-weighted imaging. Persistent changes in MEMRI imaging are observed from 2 days to 6 weeks after kainic-acid-induced status epilepticus, and the magnitude of these changes inversely correlates with later seizure frequency.[29]

The third potential mechanism is inflammation. Duffy et al. showed that 24 hours after pilocarpine-induced status epilepticus it is possible to identify areas of inflammation with MRI using antibodies to vascular cell adhesion molecule attached to an iron particle.[30] There is evidence for widespread inflammatory changes, although it is not yet known if these changes are predictive of seizure or cognitive outcomes.

Other studies have attempted to identify the time course of MRI changes in the hippocampus using T2-weighted imaging approaches. In the pilocarpine model increases in T2 relaxation time 48 hours after pilocarpine-induced status epilepticus predicts the severity of hippocampal volume reduction. In addition, early studies showed that hyperthermia-induced PFS in rats is associated with the subsequent development of epilepsy in 25–35% of animals.[31] MRI at 1 month post-PFS using a 7.0T field strength showed similar changes to those reported after childhood PFS, with increases in T2 relaxation time.[32] These were shown to be associated with the degree of neuronal cell loss and interictal EEG activity, but only poorly predictive of epilepsy outcome.

By contrast, a more recent study using the same seizure model scanned rats during the acute phase, immediately following PFS using high-magnetic-field MRI (11.7 T). Significant decreases in T2 relaxation time in the amygdala were found at 2 and 18 hours post-PFS, and these were strongly predictive of the later development of spontaneous seizures. Similar reductions in T2 were obtained using a T2* sequence on a 4.7 T scanner, which may be more replicable in a clinical setting. The T2* change is across much of the brain, including the limbic areas. Interestingly, the reduction in T2* signal was inversely related to behavioral performance on a hippocampal dependent place avoidance task. This suggests that there is

a protective mechanism manifest as reductions in T2*, and it is hypothesized that the mechanism is related to efficiency of oxygen extraction in the recovery period following PFS.

1.3 The Use of Biomarkers

As discussed previously, although acute mortality and morbidity are significantly raised following status epilepticus, the role of neuroimaging in predicting these adverse outcomes is limited in the acute situation. At a pragmatic level, tools based on history and examination findings are likely to be more timely and useful in the hospital; although neuroimaging could potentially be used to further refine risk estimates by confirming etiological diagnoses, this remains unexplored and, given the breadth of causes of status epilepticus, is unlikely to result in the identification of a single biomarker.

Similarly, while other aspects of status epilepticus, such as long-term cognitive outcomes, are increasingly being recognized as important, there have been few systematic studies linking these with neuroimaging changes. Data from studies of cognitive impairment in childhood epilepsy,[33] as well as from animal studies,[34] would suggest that it is the global, extrahippocampal changes across distributed brain networks that are affected and drive these changes. While this provides important insight into the neural substrate of cognitive impairment and its relationship with seizure disorders, there are as yet no obvious biomarkers and the development of predictive models is likely to depend on the combination of a number of different factors.

The main use of biomarkers in status epilepticus at present thus remains the identification of TLE-MTS. It has now clearly been demonstrated that a proportion of children sustain progressive hippocampal damage following a PFS, leading to eventual TLE-MTS. While this is an important finding, it appears to occur in only a small minority of children with PFS; therefore it is difficult to recommend routine treatment with an antiepileptogenic agent, even if such were available, at present. Further investigation as to the efficacy and side-effect profile of any such agent would be necessary, and for this to occur in a realistic time frame will necessarily involve the use of biomarkers. Biomarkers for the development of TLE-MTS might also allow better targeting of any putative antiepileptogenic agent to maximize benefit and minimize any potential harm.

Neuroimaging is noninvasive and has the potential to provide direct information on the pathological changes occurring in the brain following PFS. It is therefore one of the most likely investigations to provide useful clinical biomarkers. Neuroimaging is however not without its costs, especially as most children within this age group will require either a general anesthetic or medical sedation to obtain a good quality MRI scan. It is therefore necessary wherever possible to minimize the imaging burden on individual children and maximize the information gained from the investigation; this may mean that further studies are needed before it can be more widely recommended.

The ideal imaging biomarker for use in children following status epilepticus would reliably identify the children who will go on to develop TLE-MTS. Currently the most promising candidates for this are hippocampal T2 relaxometry and serial hippocampal volumetry.

T2 relaxation time has been shown in both animal models and human studies to be a risk factor for the subsequent development of hippocampal damage and MTS. There remain some uncertainties about the optimal timing of MRI and threshold values for increased risk; the FEBSTAT study stratified using visual assessment and did not quantitate T2 relaxation times. Data from other studies are difficult to interpret: long-term T2 changes were demonstrated by Provenzale et al.,[18] whereas Scott et al.[16] reported that the initial T2 changes resolved at the 6-month follow-up. Similarly, earlier animal studies showed increased T2 relaxation times following PFS,[32] whereas more recent studies using the same model found decreases in T2 relaxation time associated with subsequent epilepsy.[35] Much of this could be explained by the use of different time points, magnet strength, and imaging protocols; this serves to underline the fact that T2 changes following status epilepticus are a dynamic and evolving phenomenon and hence the choice of timing and methodology will be critical to determining their predictive value. If the optimal use of T2 imaging can be identified, it has the advantage of requiring only a single time point, and likely one very close to the seizure event to predict outcome.

It makes theoretical sense that early hippocampal volume loss may also be a suitable marker for early injury and the subsequent development of MTS. Progressive loss of hippocampal volume would seem to be a prerequisite for the development of MTS, and longitudinal studies from the FEBSTAT and STEP IN cohorts has now delineated an approximate time scale for this to occur. There are however a number of issues that may make this less suitable as a biomarker in practice: the measurement of hippocampal volume at this age currently relies mainly on manual segmentation, as automated methods have generally been optimized in adult populations and transfer poorly to this age group. Although the development of neonate and infant-specific atlases[36] as well as novel machine-learning techniques[37] means that progress in this area is likely to be rapid, some caution will need to be exercised in applying novel techniques to potentially abnormal hippocampi as sources of error may not be obvious. An additional problem arises due to hippocampal growth rate during this period of development. Hippocampal growth is expected to be highest within the first 2 years of life,[38] and this overlaps with the time of greatest risk for febrile seizures. Detecting an effect on hippocampal volume, whether this be volume loss or a reduction in growth, requires longitudinal imaging at two time points at least. Nevertheless, the time delay required needs to be considered, and whether it would allow the identification of children in a timely enough manner to allow intervention remains unclear.

1.4 Conclusion

The development of a noninvasive biomarker of status epilepticus offers the opportunity to finally attempt to address the neuronal damage and subsequent epileptogenesis that can occur following status epilepticus. We now have plausible candidates for biomarkers for TLE-MTS following PFS, although further development and validation will be required before they are ready for use in a clinical setting or trial.

References

1. Berg AT, Berkovic SF, Brodie MJ, et al. Revised terminology and concepts for organization of seizures and epilepsies: report of the ILAE Commission on Classification and Terminology, 2005–2009. *Epilepsia.* 2010;**51**(4):676–85.

2. Meldrum B. Physiological changes during prolonged seizures and epileptic brain damage. *Neuropadiatrie.* 1978;**9**(3):203–12.

3. Fujikawa DG. The temporal evolution of neuronal damage from pilocarpine-induced status epilepticus. *Brain Res.* 1996;**725**(1):11–22.

4. Lowenstein DH, Bleck T, Macdonald RL. It's time to revise the definition of status epilepticus. *Epilepsia*. 1999;**40**(1):120–2.

5. Neligan A, Shorvon SD. Frequency and prognosis of convulsive status epilepticus of different causes: a systematic review. *Arch Neurol*. 2010;**67**(8):931–40.

6. Sutter R, Kaplan PW, Ruegg S. Outcome predictors for status epilepticus—what really counts. *Nat Rev Neurol*. 2013;**9**(9):525–34.

7. Goyal MK, Sinha S, Ravishankar S, Shivshankar JJ. Role of MR imaging in the evaluation of etiology of status epilepticus. *J Neurol Sci*. 2008;**272**(1–2):143–50.

8. Maytal J, Shinnar S, Moshé SL, Alvarez LA. Low morbidity and mortality of status epilepticus in children. *Pediatrics*. 1989;**83**(3):323.

9. Waruiru C, Appleton R. Febrile seizures: an update. *Arch Dis Child*. 2004;**89**(8):751–6.

10. Nelson K, Ellenberg JH. Prognosis in children with febrile seizures. *Pediatrics*. 1978;**61**:720–7.

11. Valenti A, Alarco G. Mesial temporal lobe epilepsy with hippocampal sclerosis. In: Panayiotopoulos CP, ed. *Atlas of Epilepsies*. London: Springer; 2010:1171–5.

12. Harvey AS, Grattan-Smith JD, Desmond PM, Chow CW, Berkovic SF. Febrile seizures and hippocampal sclerosis: frequent and related findings in intractable temporal lobe epilepsy of childhood. *Pediatr Neurol*. 1995;**12**(3):201–6.

13. Janszky J, Janszky I, Ebner A. Age at onset in mesial temporal lobe epilepsy with a history of febrile seizures. *Neurology*. 2004;**63**(7):1296–8.

14. Arbelaez A, Castillo M, Mukherji SK. Diffusion-weighted MR imaging of global cerebral anoxia. *AJNR Am J Neuroradiol*. 1999;**20**(6):999–1007.

15. Yoong M, Martinos MMM, Chin RRF, Clark CA, Scott RC. Hippocampal volume loss following childhood convulsive status epilepticus is not limited to prolonged febrile seizures. *Epilepsia*. 2013;**54**(12): 2108–15.

16. Scott RC, King MD, Gadian DG, Neville BGR, Connelly A. Hippocampal abnormalities after prolonged febrile convulsion: a longitudinal MRI study. *Brain*. 2003;**126**(pt 11):2551–7.

17. Lewis DV, Shinnar S, Hesdorffer DC, et al. Hippocampal sclerosis after febrile status epilepticus: the FEBSTAT study. *Ann Neurol*. 2014;**75**(2):178–85.

18. Provenzale JM, Barboriak DP, VanLandingham K, MacFall J, Delong D, Lewis DV. Hippocampal MRI signal hyperintensity after febrile status epilepticus is predictive of subsequent mesial temporal sclerosis. *AJR Am J Roentgenol*. 2008;**190**(4):976–83.

19. Pujar SS, Neville BGR, Scott RC, Chin RFM. Death within 8 years after childhood convulsive status epilepticus: a population-based study. *Brain*. 2011;**134** (pt 10):2819–27.

20. Camfield PR, Camfield CS, Gordon K, Dooley J. What types of epilepsy are preceded by febrile seizures? A population-based study of children. *Dev Med Child Neurol*. 1994;**36**:887–92.

21. Scott RC, King MD, Gadian DG, Neville BGR, Connelly A. Prolonged febrile seizures are associated with hippocampal vasogenic edema and developmental changes. *Epilepsia*. 2006;**47**(9):1493–8.

22. Scott RC, Gadian DG, King MD, et al. Magnetic resonance imaging findings within 5 days of status epilepticus in childhood. *Brain*. 2002;**125**(pt 9):1951–9.

23. Gross DW, Concha L, Beaulieu C. Extratemporal white matter abnormalities in mesial temporal lobe epilepsy demonstrated with diffusion tensor imaging. *Epilepsia*. 2006;**47**(8):1360–3.

24. Kimiwada T, Juhász C, Makki M, et al. Hippocampal and thalamic diffusion abnormalities in children with temporal lobe epilepsy. *Epilepsia*. 2006;**47**(1):167–75.

25. Yoong M, Seunarine K, Martinos M, Chin RF, Clark CA, Scott RC. Prolonged febrile seizures cause reversible reductions in white matter integrity. *NeuroImage Clin*. 2013;**24**(3):515–21.

26. Martinos M, Yoong M, Patil S, et al. Recognition memory is impaired in children following prolonged febrile seizures. *Brain*. 2012;**135**(10):3153–64.

27. Dube CM, Ravizza T, Hamamura M, et al. Epileptogenesis provoked by prolonged experimental febrile seizures: mechanisms and biomarkers. *J Neurosci*. 2010;**30**(22):7484–94.

28. Choy M, Wells JA, Thomas DL, Gadian DG, Scott RC, Lythgoe MF. Cerebral blood flow changes during pilocarpine-induced status epilepticus activity in the rat hippocampus. *Exp Neurol*. 2010;**225**(1):196–201.

29. Dedeurwaerdere S, Fang K, Chow M, et al. Manganese-enhanced MRI reflects seizure outcome in a model for mesial temporal lobe epilepsy. *NeuroImage*. 2013;**68**:30–8.

30. Duffy BA, Choy M, Riegler J, et al. Imaging seizure-induced inflammation using an antibody targeted iron oxide contrast agent. *NeuroImage*. 2012;**60**(2):1149–55.

31. Dubé C, Richichi C, Bender RA, Chung G, Litt B, Baram TZ. Temporal lobe epilepsy after experimental prolonged febrile seizures: prospective analysis. *Brain*. 2006;**129**(pt 4):911–22.

32. Dubé C, Yu H, Nalcioglu O, Baram TZ. Serial MRI after experimental febrile seizures: altered T2 signal without neuronal death. *Ann Neurol*. 2004;**56**(5):709–14.

33. Yoong M. Quantifying the deficit—imaging neurobehavioural impairment in childhood epilepsy. *Quant Imaging Med Surg*. 2015;**5**(2):225–37.

34. Barry JM, Choy M, Dube C, et al. T2 relaxation time post febrile status epilepticus predicts cognitive outcome. *Exp Neurol*. 2015;**269**:242–52.

35. Choy M, Dubé CM, Patterson K, et al. A novel, noninvasive, predictive epilepsy biomarker with clinical potential. *J Neurosci*. 2014;**34**(26):8672–84.

36. Gousias IS, Edwards AD, Rutherford MA, et al. Magnetic resonance imaging of the newborn brain: manual segmentation of labelled atlases in term-born and preterm infants. *NeuroImage*. 2012;**62**(3): 1499–509.

37. Guo Y, Wu G, Commander LA, et al. Segmenting hippocampus from infant brains by sparse patch matching with deep-learned features. *Med Image Comput Comput Assist Interv*. 2014;**17**(2):308–15.

38. Evans AC. The NIH MRI study of normal brain development. *NeuroImage*. 2006;**30**(1): 184–202.

Chapter 2

Experimental MRI Approaches to Study Posttraumatic Epilepsy

Olli Gröhn and Alejandra Sierra

2.1 Introduction

The long-term consequences of a brain trauma and the magnitude of this problem are often not recognized. Epilepsy after traumatic brain injury (TBI) develops only in a subpopulation of patients over several years: after severe injuries, epilepsy develops in 17% of patients.[1] There is remarkable variability in terms of severity, brain trauma location, genetic background, and medical history. Therefore, it is currently impossible to predict the outcome of patients after TBI. Imaging biomarker studies of posttraumatic epileptogenesis are very difficult and expensive to perform in patients, as they require long follow-up times in large populations. In practice, identification studies of imaging biomarker candidates in animal models are needed before proceeding to large-scale human studies. This is possible due to dedicated small animal imaging systems and recent development of adequate models for posttraumatic epilepsy. For example, it has recently been shown that only about 30–50% of animals develop epilepsy after experimental fluid-percussion-induced TBI.[2,3]

2.2 MRI Applied to Animal Models of Posttraumatic Epilepsy

Magnetic resonance imaging (MRI) is the most versatile in vivo imaging approach available, capable of providing information about anatomical, microstructural, functional, and metabolic changes in the brain. It is based on inherent properties of certain nuclei that are important in biological systems such as ^{1}H, ^{13}C, or ^{31}P. Most often the signal coming from ^{1}H in water molecules is measured. The signal is based on the nuclear magnetic resonance (NMR) phenomenon, and interaction with nuclei takes place in the radio frequency (RF) area of the electromagnetic spectrum. Therefore, MRI does not involve ionizing radiation, has great tissue penetration, and is almost completely noninvasive. Due to abundance of water in tissue, excellent signal-to-noise ratio and high spatial resolution can be obtained. As signal behavior is modulated by complex interaction of water molecules with tissue structures, MRI has an excellent soft tissue contrast. Considering these properties, it is not surprising that MRI has been used in both animal models and clinical settings to assess multiple facets of posttraumatic alterations in the brain. In the following pages, we introduce some conventional and advanced MRI approaches that have been applied to animal models of TBI and posttraumatic epilepsy to identify potential MRI biomarkers for epileptogenesis. Most of these approaches can be translated to human studies with relatively small modifications.

2.2.1 Specific Features of Small Animal MRI

The human brain weighs ~700 times more than the rat brain. As the water content in both human and rat brains is approximately the same, the amount of MRI signal coming from water is ~700 times larger in the human brain. It is evident that more sensitive MRI scanners are needed for rodent MRI studies to compensate for this difference. Animal MRI scanners typically have a higher magnetic field (4.7 T–16 T) than human scanners, which leads to higher initial polarization and, together with small, dedicated RF coils, helps to compensate for smaller signal. As a result, similar or even better anatomical resolution can be achieved in rodents than in humans. The higher magnetic field also brings in some technical challenges. Artifacts coming from different magnetic susceptibilities in the tissue interfaces are more severe and relaxation times are different from those in human MRI. These technical challenges can be addressed with higher magnetic field gradients, strong shims, and pulse-sequence designs, but MRI techniques used in animal studies often require more tailoring by experienced MRI physicists than similar approaches in human studies. Nevertheless, translation of techniques from clinical settings to experimental ones can be done in most cases when the field-strength-dependent factors are taken into account appropriately.

Another specific feature of animal MRI studies is the use of anesthesia needed to immobilize and comfort the animals during the scanning. While this is not commonly an issue for structural imaging, anesthesia can have a profound effect on brain activity, basal blood flow, and neurovascular coupling, hampering functional and hemodynamic imaging studies. Also, there is increasing awareness that anesthesia may influence the outcome of the animals after brain injuries.[4]

2.2.2 Anatomical MRI and Quantitative Relaxation Time Mapping for Detection of Progressive Tissue Damage

Anatomical imaging is the cornerstone of all MRI protocols. Typical resolution achievable routinely is in order of a few hundred micrometers in plane with 0.5–1.0 mm slice thickness; however, current trend is toward more isotropic voxels that allow better coregistration and group-level analysis. Good contrast between white matter and gray matter and between normal tissue and typical lesion can be obtained with T2 relaxation weighting in high magnetic field used in small animal MRI, while T1 weighting is most often used in clinical MRI. T2 relaxation is an NMR property of nucleus, which for ^{1}H in tissue is predominantly dictated by amount and type of interaction of water molecules with macromolecules in tissue and the presence of strongly paramagnetic or diamagnetic compounds. Therefore, T2-weighted imaging can visualize for example, white matter (high myelin content), edema (high water content), and hemorrhage (high iron content) after TBI. It should be noted that relaxation-based contrast is not very specific and several tissue alterations can lead to apparently similar changes in images. Therefore, interpretation of the results has to be done with caution and may require histological validation.

Volumetric approaches are most commonly exploited in assessment of posttraumatic alterations. Primary lesion area has been found to increase several months after the impact in lateral fluid percussion model in rat.[5] Progressive tissue damage can be seen in atrophy of the hippocampus and thalamus and thinning of the cortical structures, which leads to increased volume of ventricles.[5,6] Conventionally, volumetric analysis is based on manual segmentation of the areas of interests, which is time-consuming and

susceptible for personal bias. The modern analysis approaches rely on atlas-based coregistration and automated algorithms that allow determination of volumes, thicknesses, and shapes of anatomical structures. Large lesion size and extensive atrophy in many TBI models make automated approaches very challenging, and therefore they have been so far seldom used in experimental studies. However, a recent study showed that hippocampal surface shape could differentiate rats that developed epilepsy after fluid percussion TBI.[3]

While relaxation time changes after TBI are most often exploited as a source of contrast in T2- or T1-weighted imaging, T2 and T1 are physical constants that can be quantified. Results can be displayed in the form of quantitative relaxation maps. This is especially useful in estimating the progression of the primary lesion. T2 relaxation has two-phasic behavior in most injury models. T2 peaks typically 1–3 days after impact due to vasogenic edema. During the following 1–2 weeks initial edema is reabsorbed and T2 returns toward the baseline value.[7] In the chronic phase, after 1 month tissue structures in primary lesion area gradually degrade and eventually lesion in severe cases forms a cavity filled with cerebral spinal fluid (CSF) with very high T2.[5,6]

Quantitation of the relaxation times is extremely useful also in biomarker studies. Even though relaxation times are dependent from magnetic field strength, and to a certain extent pulse sequence, with proper harmonization results from different MRI laboratories can be pooled together to reach better statistical power for biomarker analysis. Absolute numerical value of relaxation time can be easily tested, for example, for its predictive value for outcome and included in multiparametric biomarker analysis. This was demonstrated by a recent study showing excellent predictive value of T1ρ relaxation time (rotating frame variant of T1 relaxation) measured in the perilesional cortex after TBI in rats for increased seizure susceptibility (Figure 2.1).[6]

2.2.3 Detection of Hemorrhage and Calcification by MRI Sensitive to Magnetic Susceptibility

Hemorrhage is one of the key features associated with TBI also in animal models. Detection of the hemorrhage is based on local magnetic field distortion caused by paramagnetic iron from blood and its degradation products, which have much higher magnetic susceptibility than surrounding tissue. As small

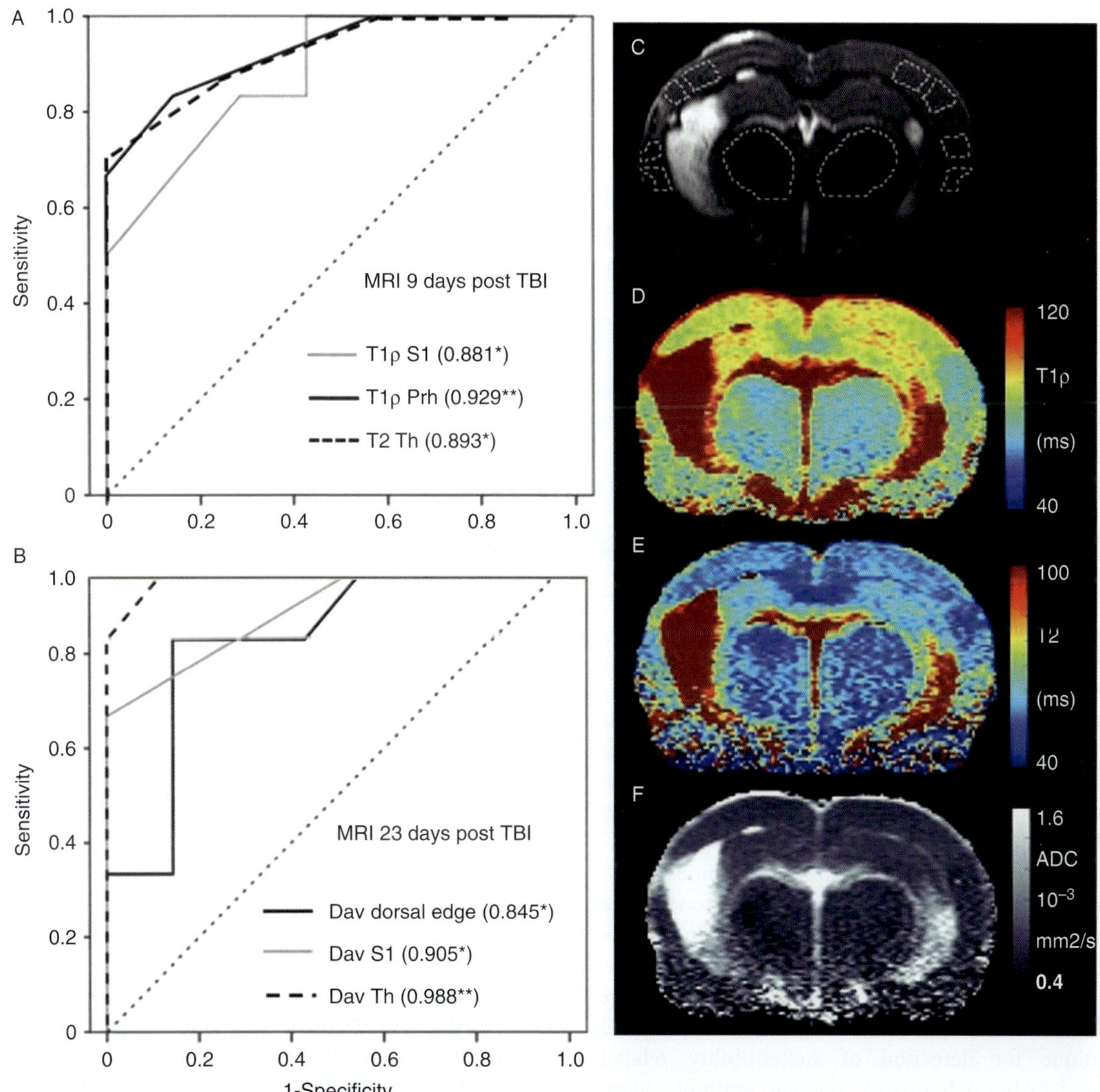

Figure 2.1. Summary of receiver operating characteristic (ROC) for quantitative relaxation and diffusion parameters assessed at 9 days (A) and 23 days post-injury (B), which predicted increased seizure susceptibility at 12 months post-TBI (traumatic brain injury). A diagonal line represents 50% probability for correct prognosis. T2-weighted image (C), T1ρ map (D), T2 map (E), and ADC map from a representative animal 23 days post-TBI. Modified from Immonen et al.[8] with permission of Mary Ann Inc.

animal MRI is generally performed in a high magnetic field which amplifies the effect, MRI has excellent sensitivity to detect hemorrhage in animal models. Large hemorrhages are typically visible already in conventional T2-weighted anatomical images as dark signal void areas. For detection of microhemorrhages gradient echo imaging with T2* weighting with better sensitivity is generally used (Figures 2.2A and

2.2C). The multiecho gradient-echo approach allows quantification of T2* relaxation times and mapping of the magnetic field variation caused by iron. More advanced field-mapping-type approaches (frequency maps, phase maps) allow distinction of paramagnetic substances from diamagnetic ones. For example, diamagnetic calcifications are a common finding in the thalamus after lateral fluid percussion injury, and are

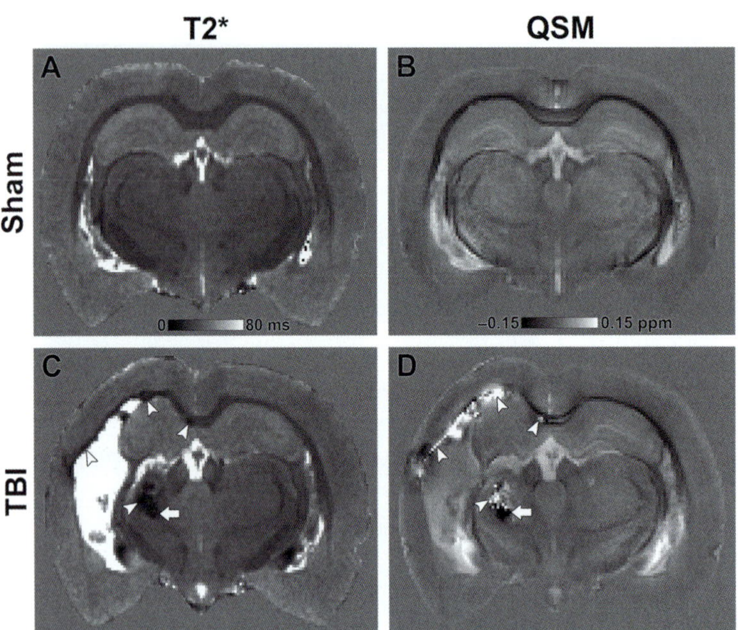

Figure 2.2. Representative T2* (A, C) and QSM maps (B, D) from a sham-operated rat and a rat six months after severe lateral fluid-percussion-induced TBI. White arrowheads point at alterations in susceptibility or relaxation times due to microbleeds, and the white arrow indicates the calcification. Note that in T2* maps both iron and calcium deposits decrease the relaxation time. In QSM maps, areas with high iron content show higher susceptibility values (paramagnetic) and areas with high calcium content have lower susceptibility values (diamagnetic).

often found in the vicinity of hemorrhages. In T2(*)-weighted images, both calcifications and blood appear black, while in field or phase maps opposite dipolar pattern coming from diamagnetic versus paramagnetic compounds can be detected, as recently shown in the fluid percussion TBI model by Lehto et al.[9] Susceptibility-weighted MRI (SWI) combines phase image with magnitude image and can help to visualize hemorrhages and has high translational value as it is widely used in clinical studies.[10] However, quantification is challenging as the contrast presents a mixture of phase and magnitude contrast. The most advanced technique for detection of susceptibility related changes in brain is quantitative susceptibility mapping (QSM). By solving the magnetic field to source the inverse problem, it is possible to produce maps of magnetic susceptibility of the tissue. QSM allows sensitive visualization and more quantitative assessment of the calcification and hemorrhage (Figures 2.2B and 2.2D). Reconstruction of QSM maps is a multistep pipeline involving background field removal and ill posed field to source inversion with no gold standard approach currently available, especially for animal work. The hemorrhages can also be problematic as incomplete inversion can lead to streaking artifacts spreading signal from the site of hemorrhage to surrounding tissue. Despite these challenges, QSM provides a promising yet rapidly evolving tool for assessment of posttraumatic changes in the rodent brain.

2.2.4 Diffusion MRI—Cytotoxic Edema, White Matter Damage, and Beyond

MRI can be sensitized to magnitude and direction of water diffusion by adding a pair of strong magnetic field gradients with opposite polarities into the imaging pulse sequence. The signal coming from water molecules that stay stationary is not changed by bipolar field gradients, while net displacement during the time between gradients leads to signal attenuation. This allows quantification of apparent diffusion coefficient (ADC) of water in tissue. The diffusion coefficient is called "apparent" as diffusion is restricted by surrounding cell and tissue structures. Indeed, this is what makes diffusion a powerful in vivo imaging contrast, as it relays information about water distribution and tissue microstructure rather than just absolute water diffusion constant in given temperature.

The most simple diffusion measurement determines only orientation insensitive ADC in tissue. ADC drops 20–40% within minutes to hours after impact due to cytotoxic edema, when most conventional MRI techniques cannot yet detect any changes in the brain.[5,7] Cytotoxic edema is a result of energy failure in tissue leading to influx of ions followed by water into intracellular space, without net increase of water content in tissue. Diffusion stays decreased from a few hours to several days depending on the animal model and the brain area. Normalization (often called pseudo-normalization as it does not indicate that tissue has recovered) of diffusion is typically followed by increased diffusion weeks—months after the impact when degradation of tissue takes place.[5] This characteristic time course, especially when combined with time course of relaxation changes and relaxation measurements, allows monitoring of progression of lesion and severity of the damage. ADC has potential also to serve as biomarker for outcome after TBI.[11,12] It has been shown that ADC measured in hippocampus as early as 3 hours after fluid percussion injury in rat correlated with increased seizure susceptibility in chronic phase 1 year later.[12]

Nowadays, diffusion is typically measured using at least six different orthogonal orientations of diffusion sensitizing gradient pairs, which allows determination of diffusion tensor and forms the basis for diffusion tensor imaging (DTI). From diffusion tensor, several parameters allowing characterization of TBI-induced tissue changes can be calculated. Most common scalar parameters are fractional anisotropy (FA, which indicates how much diffusion deviates from isotropic diffusion), and axial and radial diffusivity (D_{II}, $D\perp$, the average amount of diffusion in the main diffusion direction and perpendicular to that, respectively). Decreased FA is used as a nonspecific marker related to the degradation of oriented gray and white matter structures, while D_{II} and $D\perp$ are most often used to characterize white matter changes in TBI models. Changes in D_{II} are suggested to be more specifically associated with axonal damage and an increase in $D\perp$ with demyelination.[13] There is growing awareness that this general interpretation is probably oversimplified, however, and multiple factors may affect all scalar DTI metrics after TBI. In a recent study, widespread DTI changes in chronic phase after fluid percussion DTI were detected both in gray matter areas and major white matter tracks outside the primary lesion areas colocalizing with neuronal loss, decrease in myelinated fiber density, and iron accumulation.[14] Very high resolution ex vivo DTI has been used to detected changes in hippocampal network plasticity after TBI.[15] The most robust changes in the CA3 were increased FA and D_{II} and

a change in the orientation of the water diffusion, and in the dentate gyrus a decrease in FA and D_{II}. The alterations in DTI parameters were associated with mossy fiber and perforant pathway axonal reorganization. While this study was performed ex vivo, similar changes were recently detected in longitudinal study after status epilepticus,[16] showing feasibility to detect damage induced axonal reorganization in vivo (Figures 2.3A and 2.3B).

Changes in the orientation of the structures measured by DTI can be used to pinpoint post-TBI structural changes,[16,17] but most often orientation information is used for tractography. Modern approaches go beyond classical DTI approach, and use more complex data acquisition and reconstruction strategies allowing separation of crossing fibers. While tractography provides visualization of white matter tracts, some of the related approaches provide whole-brain tractography using super-resolution techniques, such as track-density imaging (TDI) (Figure 2.3C) and quantitative metrics that allow the characterization of microstructural damage generated by TBI. In a recent work, Wright et al. utilized apparent fiber density (AFD) and track-weighted imaging (TWI) metrics such as track density, average path length, and curvature to detect changes in white matter 12 weeks after fluid percussion TBI in rats.[18] They showed that AFD and TWI metrics were more sensitive than conventional DTI metrics to posttraumatic changes and could find changes also in corticospinal tract that appeared normal in DTI.

2.2.5 Functional MRI—Detection of Functional Defects and Network Reorganization

Functional MRI (fMRI) detects brain activity indirectly through hemodynamic changes that are tightly coupled to neuronal activity. In the close vicinity of activated brain area, cerebral blood flow (CBF) and volume (CBV) are increased in a way that overcompensates elevated oxygen consumption and leads to an increased oxyhemoglobin to deoxyhemoglobin ratio in the area. As oxyhemoglobin is diamagnetic and deoxyhemoglobin paramagnetic, a shift in the balance can be detected through signal changes in T2- or T2*-weighted images. This forms the basis to so-called blood-oxygenation-level-dependent (BOLD) contrast, which is most often used in both human and animal fMRI. In animal fMRI, CBV imaging exploiting intravascular contrast agents is also commonly used due to its greater contrast-to-

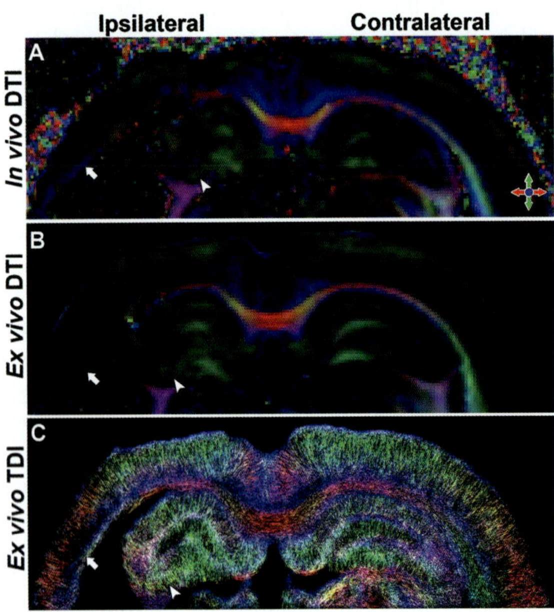

Figure 2.3. Representative directionally encoded color (DEC) fractional anisotropy maps in vivo (A) and ex vivo (B) diffusion tensor imaging (DTI), and a track-density imaging map (TDI) (C) from the same rat five months after severe fluid-percussion-induced TBI. The white arrowheads indicate increase in FA and changes in water diffusion orientation in CA3 of the hippocampus. The white arrow points to a decrease in FA and track density in the perilesional cortex in the ipsilateral side as compared to the contralateral side. Color coding for directions: green: dorsal-ventral, red: lateral-medial, and blue: rostral-caudal.

noise ratio than BOLD. Nevertheless, CBV fMRI relies on the same neurovascular coupling phenomenon as BOLD imaging.

In classical fMRI, rapid imaging pulse sequences with temporal resolution of 1–2 seconds are used to continuously acquire MRI signal during repeated stimulus and resting periods. Measured MRI signal time courses are statistically compared with expected signal time course obtained by convoluting stimulation paradigm with hemodynamic response function. The fact that fMRI is not a direct measure of neuronal activity, but detects activation through neurovascular coupling, should be taken into account when interpreting fMRI results from TBI models. fMRI changes may not always present changes in brain activity, as TBI-induced changes in basal blood flow, vasculature, and complex cascade behind neurovascular coupling may contribute to detected changes or mask the altered brain activity. The situation is further complicated by possible influence of anesthesia on fMRI response in animal models.

In spite of these complications, a recent study showed impaired BOLD fMRI response to fore paw stimulation, in the primary somatosensory area in the rat cortex that appeared intact in structural MRI, 2 days after fluid percussion TBI.[19] Interestingly, functional response partially recovered two weeks later. Simultaneous local field potential and fMRI measurement were able to confirm that changes in BOLD fMRI responses were indeed associated with alterations in electrical activity in brain rather than changes in hemodynamics.

The fMRI signal measured without any external stimulation also has time-dependent signal changes arising from multiple different factors including rhythmic changes in underlying brain activity. This so-called resting state fMRI (rs-fMRI) signal can be used to determine functional connectivity between brain regions. The approach is based on the idea that if two brain areas have synchronous changes in resting state brain activity, i.e., their fMRI time series are correlated, the brain areas are functionally connected. The rs-fMRI signal has been extensively used in human TBI studies, and it is increasingly also used in TBI animal models, as it allows large-scale network-level assessment of posttraumatic alterations in brain. This is a highly interesting approach as many long-term outcomes of TBI, including epilepsy, are increasingly recognized to be network-level disorders. In a recent study, rats with increased seizure susceptibility in pentylenetetrazol test, in chronic phase after fluid percussion injury, were studied using rs-fMRI.[20] The group statistics revealed decreased connectivity between the ipsilateral and contralateral parietal cortex and between the parietal cortex and hippocampus on the side of injury as compared to sham-operated animals. Injured animals also had abnormal negative connectivity between the ipsilateral and contralateral parietal cortex and other regions. Novel rs-fMRI analyses, such as graph theory analysis, can also quantify topological features of functional networks. Recently, this approach was applied to controlled cortical impact rat TBI model in days 7–28 postinjury.[21] Exploratory global network analysis showed changes in network parameters indicative of possible acutely increased random connectivity and temporary reductions in modularity that were matched by local increases in connectedness and increased efficiency among more weakly connected regions. The global network parameters of shortest path length, clustering coefficient, and modularity that were most affected by trauma also scaled with the severity of injury.

As demonstrated also by the examples above, resting state analysis almost always relies on group-level analysis, in order to compensate for inherently weak contrast-to-noise ratio of the approach. Group-level analysis has produced highly interesting and relevant results, shedding light on the network-level mechanism in posttraumatic cascade. However, a major challenge for rs-fMRI is to be able to detect reliably and reproducibly functional connectivity changes in individual subjects, which is a basic requirement for rs-fMRI to be used in the search for a biomarker for posttraumatic alterations.

2.2.6 Contrast Agents and Arterial Spin Labeling—Measuring Blood Flow and Blood-Brain Barrier Leakage

Gadolinum (Gd) chelate T1-contrast agents are widely used to measure hemodynamic parameters such as relative CBF, CBV, and mean transit time (MTT) in both experimental and clinical settings. Gd contrast agents are typically used in the dynamic contrast enhanced (DCE, also called bolus tracking) approach. In this approach, signal intensity changes during first pass bolus of contrast agent are measured using rapid imaging with temporal resolution of 0.5–1 seconds and relative CBV, CBF, and MTT are calculated from signal time courses presenting. In addition, superparamagnetic iron oxide particles are often used in small animal MRI, as they provide very strong T2 and T2* contrast. Due to the large size of these contrast agents, they stay in the intravascular pool for an extended period of time, which makes possible high-resolution mapping of relative CBV. This type of CBV imaging was recently used to characterize temporal profile of CBV in the perilesional area, hippocampus, and primary lesion after weight drop induced TBI in the acute-subacute phase.[22] While the first response to injury was a decrease of CBV in acute phase all brain areas, the recovery profile varied in subacute phase including also overshoot of CBV in the hippocampus at 2 weeks. The CBV changes in the subacute phase were associated with motor recovery emphasizing the value of hemodynamic changes as a potential biomarker of functional outcome after TBI.

CBF can also be measured without an external contrast agent using arterial spin labeling (ASL). As this approach provides quantitative CBF values in absolute units and can be used in both animals and humans, it has great potential in the search for a biomarker.

The technique is based on noninvasive labeling of inflowing blood with RF pulses, typically in the level of the neck. The small difference in signal intensity between images with labeled and images with non-labeled inflowing blood is used to calculate CBF maps. Due to small differences in the signal intensity between labeled and nonlabeled images, ASL has inherently lower contrast-to-noise ratio and thus requires a longer acquisition time than the DCE approach with contrast agent. Nevertheless, ASL has been used to characterize longitudinal changes in CBF in the acute-subacute[23] and chronic phases[24] in posttraumatic epilepsy model.

Impaired integrity of blood-brain barrier (BBB) is an important hallmark of TBI, and it has recently been described as one of the potential mechanisms causing posttraumatic epilepsy.[25] BBB integrity can be measured using MRI contrast agents both in animals and in humans, therefore potentially providing a promising translatable imaging biomarker for epileptogenesis. In the conventional approach, T1-weighted images are measured before and after gadolinium-bolus injection. As an intact BBB keeps contrast agent in the vasculature, leakage of Gd to the extraluminal space can be used as a marker for damaged BBB. Using this approach, Frey et al. showed that BBB damage in injured cortex at 72 hours after fluid percussion injury was significantly correlated with EEG-based measures of seizure susceptibility to chemo convulsant challenge 3 months later.[11] A recent improvement to the technique that incorporates step-down infusion of gadolinium instead of bolus injection and quantitative relaxation mapping can detect markedly more subtle BBB damage.[26]

2.3 Conclusions and Future Perspectives

In spite of being actively used for decades, MRI is still a rapidly evolving technique. Higher magnetic field strength, together with advanced parallel imaging technology and other hardware development produces higher spatial resolution and faster imaging sequences. Together with advanced MRI contrasts, this allows finding more abnormalities in the brain structures also outside the primary lesion areas that look normal in conventional MRI after TBI.[9] The major challenge in interpreting MRI results is the lack of specificity, as a number of different pathological events can cause similar MRI changes. Novel microstructural imaging techniques that go beyond conventional DTI in diffusion imaging or exploit more advanced relaxation based contrast, such as relaxation along fictitious field (RAFF),[27] can greatly increase the specificity of the MRI. However, to understand the complex interrelationship between MRI contrast and tissue pathology, extensive validation in animal models with histological and electron microscopy characterization, together with computer simulations of MRI contrast behavior, has to be performed. This is also an important step in the biomarker search. Linking the imaging result to specific cellular-level change or pathological process, rather than using imaging results blindly as a surrogate marker, will greatly increase the value and translatability of the results.

Being able to find focal abnormalities in the brain, post-TBI, is the cornerstone for the biomarker search; however, what eventually counts is how the brain is functioning. Heterogeneity of the brain traumas and complexity of the launched cascades leads to an almost infinite number of combinations of focal changes; still the brain can have only a limited number of functional states. In the future, these states may be measured more accurately by advanced rs-fMRI approaches. It is now understood that functional connectivity varies in time and dynamics of the connectivity may change in epilepsy[28] and provide potential biomarkers for posttraumatic epilepsy. The major challenge is to find changes that would be robust enough to be determined reliably in individuals rather than in group-level analysis.

Major limitations in many current biomarker studies are the lack of proper outcome measures and compromised statistical power due to the small number of animals. The majority of preclinical MRI studies describe changes after TBI but do not have long enough follow-up times and/or lack video EEG monitoring to detect epilepsy.[29] Currently, it is difficult to evaluate which of the many detected MRI changes after TBI would indeed have value as a biomarker for posttraumatic epilepsy. To address these issues, there are currently several large-scale posttraumatic epileptogenesis biomarker studies ongoing including also MRI. With a wealth of structural and functional information available from the entire brain, contemporary MRI is likely to play, together with EEG and molecular approaches, an important role in the search for a biomarker for posttraumatic epileptogenesis and epilepsy, in the future.

References

1. Annegers JF, Hauser WA, Coan SP, Rocca WA. A population-based study of seizures after traumatic brain injuries. *N Engl J Med*. 1998;**338**(1):20–4.

2. Kharatishvili I, Nissinen JP, McIntosh TK, Pitkanen A. A model of posttraumatic epilepsy induced by lateral fluid-percussion brain injury in rats. *Neuroscience*. 2006;**140**(2):685–97.

3. Shultz SR, Cardamone L, Liu YR, et al. Can structural or functional changes following traumatic brain injury in the rat predict epileptic outcome? *Epilepsia*. 2013;**54**(7):1240–50.

4. Bar-Klein G, Klee R, Brandt C, et al. Isoflurane prevents acquired epilepsy in rat models of temporal lobe epilepsy. *Ann Neurol*. 2016;**80**(6):896–908.

5. Immonen RJ, Kharatishvili I, Niskanen JP, Grohn H, Pitkanen A, Grohn OH. Distinct MRI pattern in lesional and perilesional area after traumatic brain injury in rat—11 months follow-up. *Exper Neurol*. 2009;**215**(1):29–40.

6. Immonen RJ, Kharatishvili I, Grohn H, Pitkanen A, Grohn OH. Quantitative MRI predicts long-term structural and functional outcome after experimental traumatic brain injury. *NeuroImage*. 2009;**45**(1):1–9.

7. Long JA, Watts LT, Chemello J, Huang S, Shen Q, Duong TQ. Multiparametric and longitudinal MRI characterization of mild traumatic brain injury in rats. *J Neurotrauma*. 2015;**32**(8):598–607.

8. Immonen R, Kharatishvili I, Grohn O, Pitkanen A. MRI biomarkers for post-traumatic epileptogenesis. *J Neurotrauma*. 2013;**30**(14):1305–9.

9. Lehto LJ, Sierra A, Corum CA, et al. Detection of calcifications in vivo and ex vivo after brain injury in rat using SWIFT. *NeuroImage*. 2012;**61**(4):761–72.

10. Liu S, Buch S, Chen Y, et al. Susceptibility-weighted imaging: Current status and future directions. *NMR Biomed*. 2017;**30**(4). doi:10.1002/nbm.3552.

11. Frey L, Lepkin A, Schickedanz A, Huber K, Brown MS, Serkova N. ADC mapping and T1-weighted signal changes on post-injury MRI predict seizure susceptibility after experimental traumatic brain injury. *Neurol Res*. 2014;**36**(1):26–37.

12. Kharatishvili I, Immonen R, Grohn O, Pitkanen A. Quantitative diffusion MRI of hippocampus as a surrogate marker for post-traumatic epileptogenesis. *Brain*. 2007;**130**(pt 12):3155–68.

13. Song SK, Sun SW, Ramsbottom MJ, Chang C, Russell J, Cross AH. Dysmyelination revealed through MRI as increased radial (but unchanged axial) diffusion of water. *NeuroImage*. 2002;**17**(3):1429–36.

14. Laitinen T, Sierra A, Bolkvadze T, Pitkanen A, Grohn O. Diffusion tensor imaging detects chronic

microstructural changes in white and gray matter after traumatic brain injury in rat. *Front Neurosci.* 2015; **9**:128.

15. Sierra A, Laitinen T, Grohn O, Pitkanen A. Diffusion tensor imaging of hippocampal network plasticity. *Brain Struct Funct.* 2015;**220**(2):781–801.

16. Salo RA, Miettinen T, Laitinen T, Grohn O, Sierra A. Diffusion tensor MRI shows progressive changes in the hippocampus and dentate gyrus after status epilepticus in rat—histological validation with Fourier-based analysis. *NeuroImage.* 2017;**152**:221–36.

17. Hutchinson EB, Rutecki PA, Alexander AL, Sutula TP. Fisher statistics for analysis of diffusion tensor directional information. *J Neurosci Methods.* 2012;**206** (1):40–5.

18. Wright R, Raimondo JV, Akerman CJ. Spatial and temporal dynamics in the ionic driving force for GABA(A) receptors. *Neural Plast.* 2011;2011:728395.

19. Niskanen JP, Airaksinen AM, Sierra A, et al. Monitoring functional impairment and recovery after traumatic brain injury in rats by FMRI. *J Neurotrauma.* 2013;**30**(7):546–56.

20. Mishra AM, Bai X, Sanganahalli BG, et al. Decreased resting functional connectivity after traumatic brain injury in the rat. *PLOS ONE.* 2014;**9**(4):e95280.

21. Harris NG, Verley DR, Gutman BA, Thompson PM, Yeh HJ, Brown JA. Disconnection and hyper-connectivity underlie reorganization after TBI: A rodent functional connectomic analysis. *Exper Neurol.* 2016;**277**:124–38.

22. Immonen R, Heikkinen T, Tahtivaara L, et al. Cerebral blood volume alterations in the perilesional areas in the rat brain after traumatic brain injury—comparison

with behavioral outcome. *J Cerebral Blood Flow Metab.* 2010;**30**(7):1318–28.

23. Hayward NM, Tuunanen PI, Immonen R, Ndode-Ekane XE, Pitkanen A, Grohn O. Magnetic resonance imaging of regional hemodynamic and cerebrovascular recovery after lateral fluid-percussion brain injury in rats. *J Cerebral Blood Flow Metab.* 2011;**31**(1):166–77.

24. Hayward NM, Immonen R, Tuunanen PI, Ndode-Ekane XE, Grohn O, Pitkanen A. Association of chronic vascular changes with functional outcome after traumatic brain injury in rats. *J Neurotrauma.* 2010;**27**(12):2203–19.

25. Weissberg I, Reichert A, Heinemann U, Friedman A. Blood-brain barrier dysfunction in epileptogenesis of the temporal lobe. *Epilepsy Res Treat.* 2011;**2011**:143908.

26. van Vliet EA, Otte WM, Gorter JA, Dijkhuizen RM, Wadman WJ. Longitudinal assessment of blood-brain barrier leakage during epileptogenesis in rats. A quantitative MRI study. *Neurobiol Dis.* 2014;**63**:74–84.

27. Hakkarainen H, Sierra A, Mangia S, et al. MRI relaxation in the presence of fictitious fields correlates with myelin content in normal rat brain. *Magn Reson Med.* 2016;**75**(1):161–8.

28. Robinson LF, He X, Barnett P, et al. The temporal instability of resting state network connectivity in intractable epilepsy. *Hum Brain Mapp.* 2017;**38**(1): 528–40.

29. Pitkanen A, Loscher W, Vezzani A, et al. Advances in the development of biomarkers for epilepsy. *Lancet Neurol.* 2016;**15**(8):843–56.

Chapter 3

Imaging Biomarkers of Acquired Epilepsies

Marian Galovic and Matthias Koepp

3.1 Introduction

Between 30% and 49% of all epilepsies are secondary to acute or chronic brain diseases (Figures 3.1 and 3.2).[1,2] This number is even higher in elderly patients, who most commonly develop seizures after cerebrovascular or neurodegenerative disorders. In contrast, young adults are more likely to acquire epilepsy due to traumatic brain injury or tumors or multiple sclerosis (Figure 3.2). In this chapter we focus on imaging biomarkers of acquired epilepsies in humans as a consequence of stroke, trauma, brain tumors, multiple sclerosis, and Alzheimer's disease (Table 3.1).

3.1.1 Acute and Remote Symptomatic Seizures

Two types of seizures after acute brain insults need to be distinguished.[3] Acute symptomatic seizures, also termed early-onset seizures, occur in close temporal relationship to an acute brain disease and are directly provoked by neuronal, metabolic, toxic, or other conditions. Typically, seizures are considered acute symptomatic if they occur within 7 days of cerebrovascular disease or trauma; however, longer intervals might be appropriate for tumors or hemorrhages.

In contrast, remote symptomatic seizures, commonly named late-onset seizures, occur as a consequence of a distant brain insult and are considered to be unprovoked. By definition, the latent period between an insult and the seizure needs to be at least 7 days. The discrimination of early- and late-onset seizures is paramount because both entities differ in pathophysiological mechanisms, risk factors, recurrence rates, and outcome. There are also significant distinctions in biomarkers between acute and remote symptomatic seizures.

The recurrence rate significantly differs between acute (19%) and remote (65%) symptomatic seizures.[4] Seizure recurrence is particularly common in patients with a first late-onset seizure after stroke or traumatic brain injury with rates ranging from 70% to 90%.[4–8]

These frequencies are similar to those observed in non-lesional patients after two unprovoked seizures. However, according to the new practical definition of epilepsy, the clinician does not have to wait for a second seizure to occur to diagnose epilepsy and can do this after a first seizure if the recurrence risk is higher than 60% within 10 years.[9] Hence, a patient with a single late-onset seizure after stroke can be defined as having epilepsy and antiepileptic treatment might be initiated.

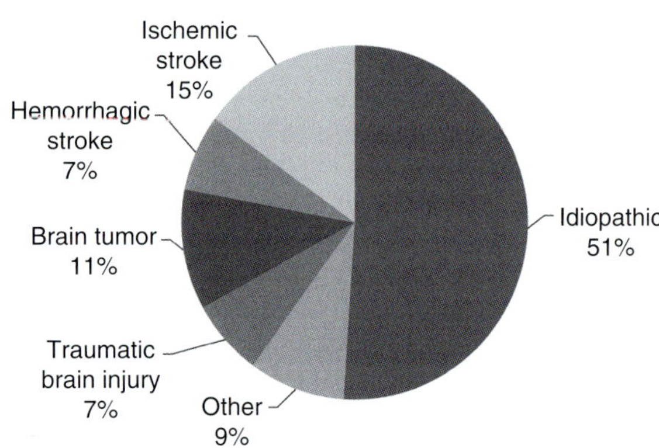

Figure 3.1. Causes of unprovoked seizures in persons aged ≥17 years. Data based on Forsgren.[1]

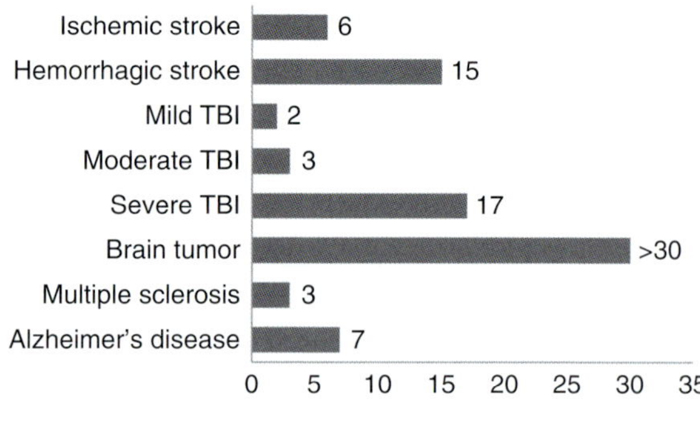

Figure 3.2. Relative risk of unprovoked seizures after brain insults.

Table 3.1 Evidence Levels of Imaging Biomarkers in Acquired Epilepsies

Imaging biomarker	Evidence level
Stroke	
Cortical location	++
Location in anterior circulation/MCA territory	++
Parietotemporal lesions	(+)
Lesion size	++
Primary or secondary hemorrhage	++
Lobar brain microbleeds	(+)
Increased ^{55}Co uptake	(+)
Decreased regional cerebral blood flow and oxygen consumption	(+)
Traumatic brain injury	
Focal lesions on initial brain scan	++
Intracerebral hemorrhage	++
Hemosiderin deposits surrounded by gliotic wall	+
Lesions requiring surgical treatment	+
Depressed skull fractures	+
Hydrocephalus	(+)
Open skull injury with brain prolapse	(+)
MT > T2 mismatch	(+)
Blood-brain barrier dysfunction	+
Brain tumors	
Visual tumor grading	++
Frontal, temporal, or parietal tumor location	+
Cortical involvement	+
Tumor size	++
Hemosiderin deposits	+
Multiple sclerosis	
Intracortical lesions	++
Alzheimer's disease	
Hippocampal atrophy	(+)

++ = robust evidence from several studies; + = evidence from a few studies, (+) = single small study or conflicting evidence. MCA = medial cerebral artery; MT = magnetization transfer MR sequence.

3.1.2 Biomarkers of Epileptogenesis

Several months or years may pass between a brain insult and the occurrence of seizures. This latent period might extend to 20 years in gradually developing neurodegenerative disorders. Epileptogenesis being a slow process, there is plenty of time for intervention. However, most medical trials to prevent epileptogenesis were unsuccessful.[10]

One of the results of these failures is the lack of a suitable biomarker to select patients at risk of developing epilepsy, to specifically target treatments for this enhanced population, and to use this biomarker to monitor treatment outcomes. The specific clinical question to be asked is, "Which patients are likely to develop recurrent seizures after a brain insult, and which of them might benefit from medical treatment?"

Imaging will play an important role to provide answers to this problem. However, there is still a long way to go. Currently, imaging can identify risk factors for acquired epilepsies, but these would not yet qualify as valid biomarkers. In the future, pathophysiological knowledge on epileptogenesis will need to be translated from experimental and animal studies into clinical imaging research to provide more insights into the development of symptomatic epilepsy.

3.2 Stroke

3.2.1 Epidemiology

Stroke is the most common cause of seizures and epilepsy in the elderly,[11] mainly due to the high prevalence of stroke in this population. Nevertheless, seizures are not a very common consequence of stroke. The incidence of both acute (≤7 days) and remote (>7 days) symptomatic seizures ranges between 2% and 6%.[12] The frequency of recurrent seizures is approximately 2% to 4%.[12] The incidence of acute and remote symptomatic seizures after childhood stroke ranges from 20% to 25% and is markedly higher compared to adults.[13]

The frequency of seizures and epilepsy after stroke varies from one subtype to another.[14–16] The highest incidence is observed in venous infarction, followed by intracerebral hemorrhage and cerebral infarction with hemorrhagic transformation. Primary ischemic stroke without hemorrhage has the lowest incidence of seizures.

The majority of acute symptomatic seizures occur within the first 24 hours after stroke.[17] They are an important risk factor for delayed seizures,[6,18] and these are most likely to occur within the first year after stroke.[17]

3.2.2 Imaging Biomarkers

Imaging plays an important role in determining the risk of poststroke seizures and is needed to establish the location and size of the lesion, stroke subtype, and presence of hemorrhage.

3.2.2.1 Stroke Location

Cortical lesion location is the best characterized risk factor for seizures and epilepsy after stroke. This result is supported by a variety of studies with different designs and applies to both ischemic and hemorrhagic stroke.[15,18–21] However, it remains controversial whether seizures also occur after subcortical stroke. Some studies reported infrequent seizures after subcortical stroke,[22–25] whereas others did not find an association.[26,27] Possibly, some of the patients with seizures after lacunar infarcts also had cortical involvement not evident on CT scans. On the other hand, subcortical stroke could also provoke seizures through the neurotoxic effects of acute glutamate release, reduction of cortical inhibition due to diaschisis, or disruption of thalamocortical loops.

Poststroke seizures are also associated with infarcts of the anterior circulation[6,23,28] and, more specifically, of the medial cerebral artery territory.[16,29] Conversely, seizures are unlikely in posterior circulation stroke, especially if only infratentorial structures are affected. Only few studies analyzed the association of stroke location with seizures in more detail. Some found an association on a lobar level with parietotemporal cortical lesions.[30–32] Others did not find a correlation at all.[33] Although a detailed voxel-based lesion symptom mapping study was not yet completed, these results argue against a specific cerebral region susceptible to generating seizures after stroke. The findings rather suggest that seizures can occur after disruption of any part of the cerebral cortex without a preference to a specific location.

3.2.2.2 Lesion Size

Several studies found an association of poststroke seizures with size of the infarct.[17,18,20,29] A volumetric analysis showed a mean lesion volume in patients with and without seizures of 4 ml and 16 ml, respectively[34] (Figure 3.3). A lesion size $\geq$ 70 ml increased the odds ratio of seizures fourfold. Conversely, the authors also demonstrate that although seizures are more likely in large lesions, also small infarcts can cause seizures if the cerebral cortex is strategically affected.

3.2.2.3 Intracerebral Hemorrhage

Remote symptomatic seizures in patients with intracerebral hemorrhage were recently shown to be independently associated with lobar brain microbleeds.[21] Lobar microbleeds are commonly a manifestation of an underlying vasculopathy, i.e., cerebral amyloid angiopathy. Hence, these results suggest an increased risk of late-onset seizures in patients with intracerebral hemorrhage due to cerebral amyloid angiopathy. However, further studies will be needed to demonstrate a direct correlation of seizures with amyloid angiopathy.

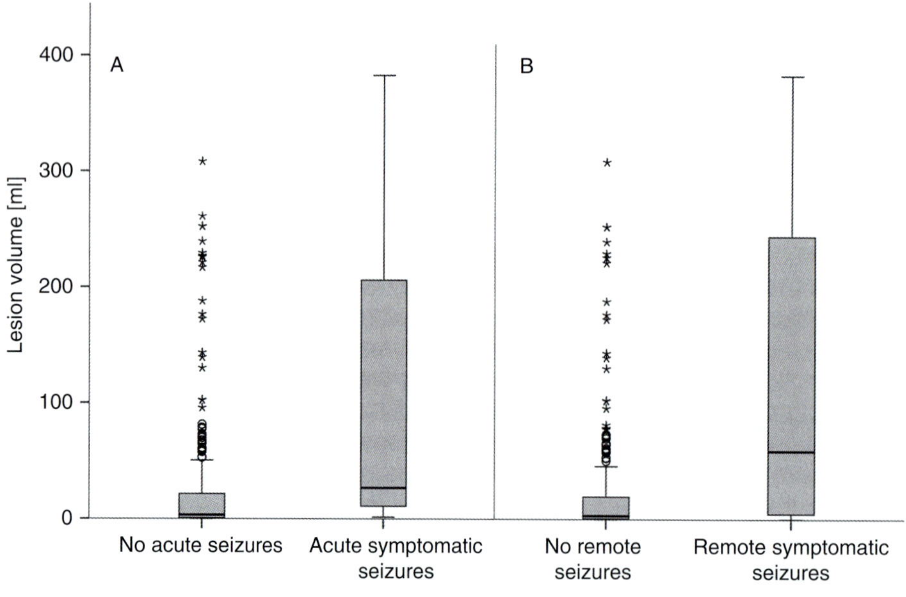

Figure 3.3. Lesion volume in patients with and without (A) acute symptomatic or (B) remote symptomatic seizures after stroke. Based on personal unpublished data and Wagner et al.[34]

3.2.2.4 Functional Imaging Studies

Few functional imaging studies have compared stroke patients with and without poststroke seizures. Two $^{15}O_2$ PET studies showed a decreased regional cerebral blood flow (rCBF) and oxygen consumption (rCMRO$_2$) in stroke patients with seizures.[35,36] The authors concluded that a more severe brain ischemia was associated with poststroke seizures. However, one cannot differentiate whether these functional changes are a cause rather than consequence of the seizures.

In another study, increased binding of cobalt-55 (^{55}Co) was demonstrated in 60% of patients with remote symptomatic seizures after stroke, whereas no increases were seen in stroke patients without seizures.[37] The degree of ^{55}Co uptake is known to correlate with inflammatory response and the severity of ischemic damage within the prior 2 months. Hence, the authors interpret these findings as evidence for recurrent stroke as a cause for poststroke seizures. However, a more likely explanation is that increased ^{55}Co uptake corresponds to excitotoxic damage after seizures, and is not a cause but a consequence of epilepsy.

3.2.2.5 Other Biomarkers

Clinical biomarkers most commonly shown to be associated with seizures after cerebral infarcts are stroke severity and etiology. Some authors have demonstrated that severe strokes correlate with seizures.[17,19,33] Others argue that increased stroke severity is only a manifestation of bigger lesion size and cortical infarct location in patients with seizures.[38]

Also, cardioembolic infarcts are likely to cause large lesions involving the cortex, and thus an association with seizures seems reasonable. Whereas cardioembolic stroke etiology was shown to increase the risk of seizures in some studies,[22,39,40] others did not find a correlation.[15,17]

3.2.2.6 Differentiation of Acute Seizures from Stroke

Seizures are a common mimic of stroke and can complicate rapid decision making at the emergency department.[41] Several imaging techniques were devised to enable a differentiation between postictal deficits and stroke. Perfusion CT commonly shows an ictal hyperperfusion, whereas stroke is associated with reduced regional blood flow.[42] This pattern enables the differentiation between ongoing seizures or status epilepticus and stroke. However, shortly after termination of seizures the perfusion is rapidly reduced and a regional hypoperfusion can be observed.[43] Therefore, perfusion CT might be unreliable to distinguish between postictal deficits and stroke.

Certain MRI sequences might prove more valuable than perfusion CT in the differentiation of late seizures from stroke recurrence. Diffusion-weighted imaging (DWI) predominantly shows increased signal intensity after late-onset seizures around the previously infarcted cortex.[44] On arterial spin labeling (ASL), hyperperfusion can be observed after seizures,[45] whereas hypoperfusion is seen in acute stroke patients.[46] In a small study, changes on ASL were more sensitive for seizures than abnormalities found on DWI.[45] Although these results need to be confirmed in larger studies, the high sensitivity of ASL and its easy implementation without the need of a contrast agent might make this technique a valuable tool in differentiating poststroke seizures from acute stroke recurrence in the emergency room.

3.3 Traumatic Brain Injury

3.3.1 Epidemiology

Traumatic brain injury (TBI) accounts for 3% to 7% of all epilepsies[1,47,48] (Figure 3.1). Similarly, only 2% to 3% of patients presenting with their first unexplained seizure showed traumatic injuries on brain scans.[49] These numbers may seem low at first sight, however, TBI is an important cause of epilepsy from a clinical perspective. Seizures are a common problem in TBI patients and for no other cause of epilepsy are the risk factors so well described and easily predictable.[50]

The relative risk of seizures after severe TBI is markedly elevated by a factor of >90 during the first year and of 17 during the subsequent five years, compared to the general population.[51] Especially vulnerable are veterans with penetrating war injuries, causing a relative risk of 580 during the first year and 25 during the subsequent 10 years.[52] The risk of seizures clearly correlates with the severity of the injury. The standardized incidence ratios of moderate (7 during the first year and 3 during the subsequent five years) and mild (3 during the first year and 2 during the subsequent five years) TBI are distinctly lower.[51] Even for mild injuries the elevated risk of epilepsy persists for at least 10 years.[53]

Interestingly, nonconvulsive seizures after TBI are common in the intensive care unit and they can be easily missed due to the bad general condition and impaired consciousness of patients. Albeit, if continuous EEG monitoring is used, seizures can be detected in approximately a quarter of monitored patients.[54]

3.3.2 Imaging Biomarkers

3.3.2.1 Type of Injury

One of the first reports on the value of early imaging in posttraumatic epilepsy showed that focal parenchymal damage or intracerebral hemorrhage on CT scans was associated with posttraumatic seizures.[55] This and subsequent studies found that focal lesions on initial CT have a prognostic role for epilepsy and seizures after TBI.[51,52,56–59] These authors also recognized, that the type of injury markedly influences the likelihood of seizures.

The best evidence exists for the presence of intracerebral hemorrhage and the accumulation of iron in hemosiderin-complexes.[51,52,56–62] Cortical application of iron is known to elicit seizures and has been used in animal models of posttraumatic epilepsy.[63] Hemosiderin can be easily spotted on MRI images with gradient echo (GRE) T2* or susceptibility-weighted imaging (SWI) sequences. One would expect that signs of hemosiderin deposits on brain imaging were associated with poststroke epilepsy, however this seems to be only part of the story. Several investigations showed that the mere presence of hemosiderin does not increase the odds of seizures. However, there was a significant association with hemosiderin deposits surrounded by a hyperintense gliotic wall.[61,62,64] This observation suggests that not hemosiderin itself but the formation of a gliotic scar after hemorrhage promotes epileptogenesis.

Other well described biomarkers of posttraumatic epilepsy are lesions on CT scans requiring surgical treatment[56,58–61] and depressed skull fractures.[51,57,58,65] Some authors also mentioned hydrocephalus[60] and open skull injury with brain prolapse as relevant factors.[65]

The use of specialized MRI sequences might further localize focal changes prone to generating seizures. Magnetization transfer (MT) MR Imaging might be more sensitive to axonal damage than T2 imaging. In a study of MT MRI in TBI patients, the authors demonstrated that patients with MT abnormalities exceeding those seen on T2 imaging were of increased risk of epilepsy after TBI.[61] One might speculate, whether this "mismatch" represents particularly vulnerable tissue capable of generating seizures.

3.3.2.2 Hippocampal Sclerosis

A complex association exists between posttraumatic epilepsy and changes of the temporal lobe. Patients with TBI in childhood or young adulthood are at higher risk to develop temporal lobe epilepsy including mesial temporal lobe sclerosis.[66] In addition, early posttraumatic seizures can cause structural damage to the temporal lobes. Acute nonconvulsive seizures after TBI were associated with ipsilateral hippocampal sclerosis on follow-up MRI imaging after 6 months.[54] Also, posttraumatic epilepsy correlated with the degree of hypoperfusion in the temporal lobes on SPECT one year after the trauma.[60] These results suggest that structural changes in the temporal lobes, i.e., hippocampal sclerosis, after brain injury might lead to posttraumatic temporal lobe epilepsy that would, in turn, generate further seizures.

3.3.2.3 Blood-Brain Barrier Dysfunction

Increased scientific interest has recently focused on the role of blood-brain barrier (BBB) disruption in several neurological disorders, including epilepsy and TBI.[67] Animal experiments showed that opening the BBB leads to network changes, long-lasting epileptiform activity and eventual neurodegeneration.[68,69] In humans, an MRI based imaging method was devised to semiquantitatively analyze BBB function in vivo.[70] Clinical studies show that altered BBB permeability is commonly observed in neurological patients, including those with TBI.[71]

Based on these findings in TBI and epilepsy patients, two studies analyzed BBB disturbances in posttraumatic epilepsy.[72,73] Both showed a significantly increased BBB permeability in patients with seizures after TBI. The size of BBB dysfunction was also associated with posttraumatic epilepsy and the location of BBB disturbances correlated with focal EEG changes (Figure 3.4). These findings support the notion that lasting BBB disruption may be causally related to the emergence of cortical dysfunction. It is less likely that BBB opening was a consequence of seizures, as increased BBB permeability was observed in patients despite apparent complete medical control of seizures, and no seizures were reported at least 24 hours before the brain scans.

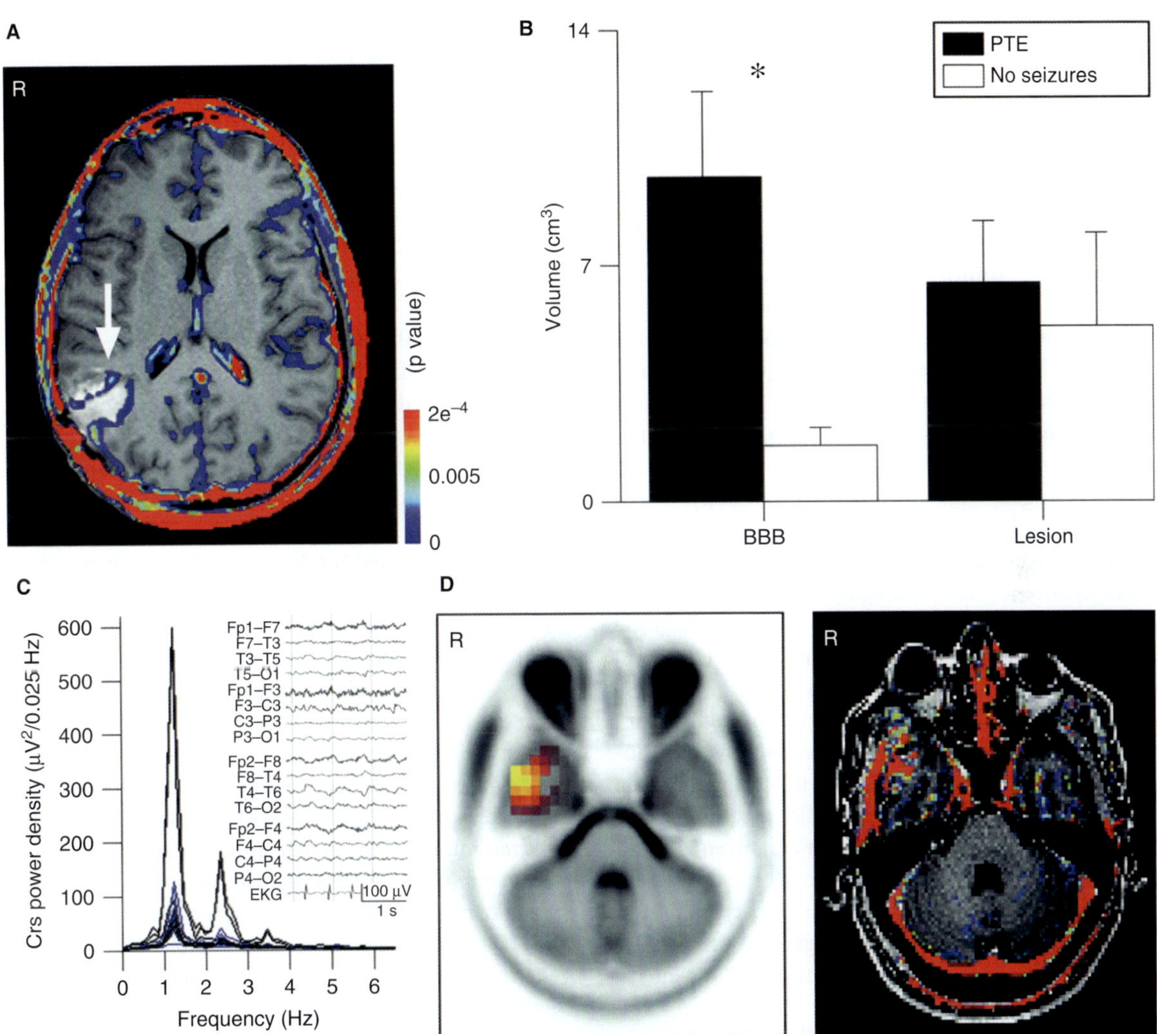

Figure 3.4. Relationship of blood-brain barrier (BBB) dysfunction and seizures after traumatic brain injury (TBI). (A) Statistically significant enhancement of T1 MRI scan in the region surrounding the cortical lesion in patient 21, 10 days following the trauma. (B) Among patients with posttraumatic epilepsy (PTE), the volume of BBB disruption was significantly larger than that of nonepileptic patients. (C) A 34-year-old patient with PTE one month following mild TBI. Power spectrum showing a marked increase in power at 1.125 Hz, taken from an electroencephalography recording, demonstrating abnormal slowing maximal at right frontal and temporal electrodes (inset). (D) sLORETA localizing the pathological signal to the anterior parts of the right middle temporal gyrus (Brodmann area 21, left), and MRI signal enhancement indicating increased BBB permeability localized to the same region (right). *$p < .05$. Reproduced with permission from Tomkins et al.[72]

3.4 Brain Tumors

3.4.1 Epidemiology

More than 30% of patients with brain tumors will suffer from seizures during the course of their illness[47] (Figure 3.2). Seizures are the presenting symptom of brain tumors in 20% to 40% of patients.[74] This makes brain tumors a cerebral disease with the highest likelihood to develop acquired epilepsy, surpassing TBI or cerebrovascular disease.[75] Seizures are likely to aggravate neuropsychological problems of tumor patients. Lack of complete seizure control reduces health-related quality of life whereas the use of antiepileptic drugs leads to significant cognitive reductions.[76]

3.4.2 Imaging Biomarkers

The role of imaging in tumor-associated epilepsy involves the definition of tumor type, tumor location, and changes in peritumoral environment.

3.4.2.1 Tumor Type

There is a clear inverse relationship between the tumor growth rate and the incidence of epilepsy.[77,78] Thus, the primary role of imaging involves visually grading the tumor before histopathological diagnosis can be done. Epileptogenesis is a slow process, requiring months to years to establish focal network changes in the development of epilepsy.[78] Low-grade slow-growing tumors have the highest seizure frequency. These include dysembryoblastic neuro-epithelial tumors (up to 100%), oligodendroglioma (60–85%), low-grade astrocytomas (60–85%), and anaplastic astrocytoma (45–70%).[74,78–81] MRI features of these tumors include a homogenous structure with no or minimal contrast enhancement. Oligodendroglial tumors characteristically show calcifications, best spotted on CT or susceptibility-weighted MR images.

Nevertheless, a nonnegligible proportion of patients with faster growing tumors like glioblastoma multiforme (15–30%) or cerebral metastasis (20–35%) are also likely to develop seizures.[74,75,77–79,81] These tumors are thought to induce epilepsy through abrupt tissue damage due to necrosis or deposition of hemosiderin caused by their rapid growth.[78] Glioblastoma multiforme typically appears on MRI as a T2 hyperintense mass with irregular contrast enhancement and vasogenic edema. MR features of cerebral metastases involve single or multiple nodular T2 hyperintense structures surrounded with vasogenic edema and, commonly, annular or uniform contrast enhancement.

3.4.2.2 Tumor Location, Size, and Hemosiderin Deposits

Location of the brain tumor significantly influences the incidence of seizures.[80] Tumors located in the frontal, temporal, and parietal lobes are most likely to be associated with epilepsy, whereas occipital or infratentorial tumors have a low propensity for seizures.[74,80] Seizures also occur infrequently, if only white matter structures are affected.[77]

Other factors that influence the occurrence of seizures are the size of the tumor and the presence of hemorrhage. Tumor size and the number of lesions are positively associated with epilepsy.[82–84] The presence of hemosiderin deposits on susceptibility-weighted MR images (SWI) predicted seizures in patients with brain metastasis and glioblastoma. An even stronger association was seen if SWI lesions involved the cerebral cortex.[84]

3.4.2.3 Tuberous Sclerosis Complex

Tuberous sclerosis complex (TSC) is an autosomal dominant disorder with a high spontaneous mutation rate, leading to tumorous growth in multiple organs including the brain. More than 80% of children with TSC have epilepsy due to multiple brain tubers and some can be treated surgically. To ensure the success of surgery, the clinician needs to determine which of these tubers are epileptogenic and will need to be resected.

A functional imaging method was devised to noninvasively localize the epileptogenic tubers. PET with the α-[^{11}C]methyl-L-tryptophan ([^{11}C]AMT) tracer showed decreased binding in those tubers, which were not expected to generate seizures. Conversely, epileptogenic tubers had an equal or increased uptake ratio compared to the normal cortex[85] (Figure 3.5). For tubers with an uptake ratio of ≥ 1 compared to normal tissue, [^{11}C]AMT-PET shows a specificity of 80–90% for the detection of epileptogenicity.[86,87] Although this concept has also been utilized in children with intractable epilepsy, it has not yet been successfully adapted to patients with seizures due to other types of brain tumors.[88]

Another noninvasive method to identify epileptogenic areas in TSC is EEG-fMRI. A small study in five children found interictal activations in and around epileptogenic tubers.[89] These activations were more widespread compared to the abnormalities found on PET or SPECT studies, suggesting that a multifocal network generates interictal discharges in TSC. Although this method might be more sensitive for hyperexcitable networks distant to the lesions, it might be less specific for the evaluation of critical tubers that should be surgically resected.

3.5 Other Diseases

3.5.1 Multiple Sclerosis

Traditionally, multiple sclerosis (MS) was regarded as a disease selectively affecting the white matter. As such, a subcortical disorder would be unlikely to have a relevant propensity for epileptic seizures. However, the increased frequency of epilepsy in MS has been well documented within the last decades. Population-based studies show a prevalence of around 3% (Figure 3.2), corresponding to a relative risk of 3 to 4 compared to the general population.[90,91] Occurrence of seizures is associated with early MS onset and increased disease severity.[92,93]

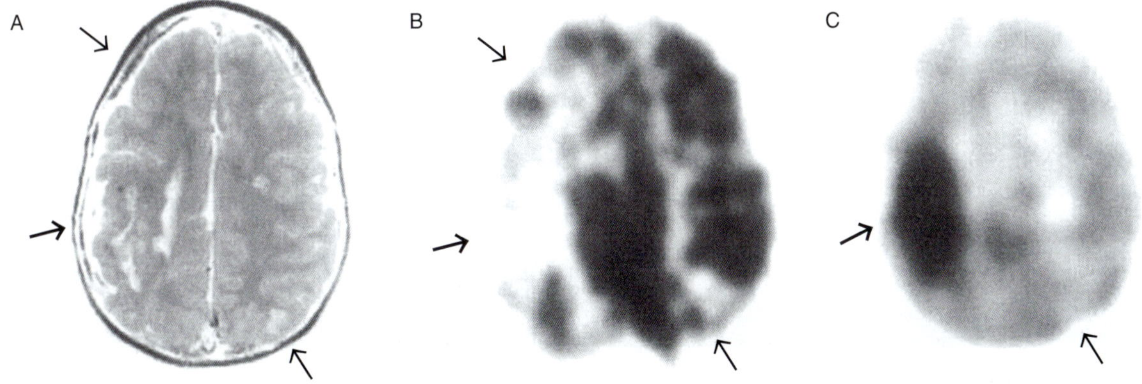

Figure 3.5. Comparison of MRI with fluorodeoxyglucose (FDG) PET and [^{11}C]AMT-PET in an 8-year-old girl with tuberous sclerosis complex and intractable epilepsy. (A) MRI scan showing large tuber in the right central/parietal region (bold arrow) as well as other small lesions (thin arrows). (B) The FDG PET shows cortical hypometabolism in the same regions as the lesions on the MRI (arrows). (C) The [^{11}C]AMT-PET displays decreased tracer uptake in the location of small tubers (thin arrow). The large right central tuber (bold arrow) shows a 110% increase of [^{11}C] AMT uptake. Reproduced with permission from Chugani et al.[85]

However, as evidence for cortical involvement in MS started to accumulate during the last years, there was a major paradigm shift.[94,95] Nowadays, there is no doubt that MS also affects cortical structures. Although older findings inconsistently showed an association of seizures with corticosubcortical lesions,[96–101] these studies have likely underestimated the prevalence of isolated cortical lesions. Conventional MR sequences used in these studies, including fluid-attenuated inversion recovery (FLAIR) and T2-weighted spin echo (T2SE), cannot detect isolated intracortical inflammation, as was demonstrated with histopathological evidence.

The advent of modern double inversion recovery (DIR) sequences markedly improved the detection of cortical plaques (Figure 3.6). Studies performed with this technique consistently find a correlation of intracortical inflammation with seizures. Intracortical lesions were observed in 90% of MS patients with seizures, compared to 48% of those without ($p = .001$).[102] MS patients with epilepsy showed five times more intracortical lesions, and their volume was six times larger. A "worm-like" shape particularly correlated with seizures. On the other hand, no significant differences were observed with regard to the number and volume of juxtacortical lesions and T2 white-matter lesion volume.

These results underline the importance of cortical inflammation for epileptogenesis in MS. Additionally, MS patients with seizures have a more rapidly and severely evolving cortical pathology. During a 3-year study, these patients had a higher accumulation of intracortical lesions and a faster reduction of gray matter fraction compared to those without seizures.[103] Conversely, the two groups did not differ in the number of new white matter lesions. Further evidence comes from a histopathological study that found cortical plaques and atrophy of the temporal lobe to be associated with seizures in MS.[104]

3.5.2 Alzheimer's Disease

Seizures as a comorbidity of neurodegenerative disorders are particularly well described for Alzheimer's disease (AD). Patients with Alzheimer's dementia are at sevenfold risk of seizures compared to age-matched controls[105] and account for around 2% of all epilepsy.[106] Due to the high prevalence of AD in the aging population, these numbers are of increasing relevance in elderly patients.

Several clinical variables have been associated with seizures in AD. The most robust correlation is described for young age at dementia onset.[107–109] This is particularly apparent in genetic forms of dementia as many human AD gene mutations leading to increased β-amyloid also cause epilepsy.[110] Other clinical variables correlating with seizures were antipsychotic drug use, African American race, epileptic discharges on EEG, and greater cognitive impairment at baseline.[107,111]

Imaging predictors of epilepsy in AD have not been described yet. One short report found hippocampal atrophy in all AD patients with seizures.[112]

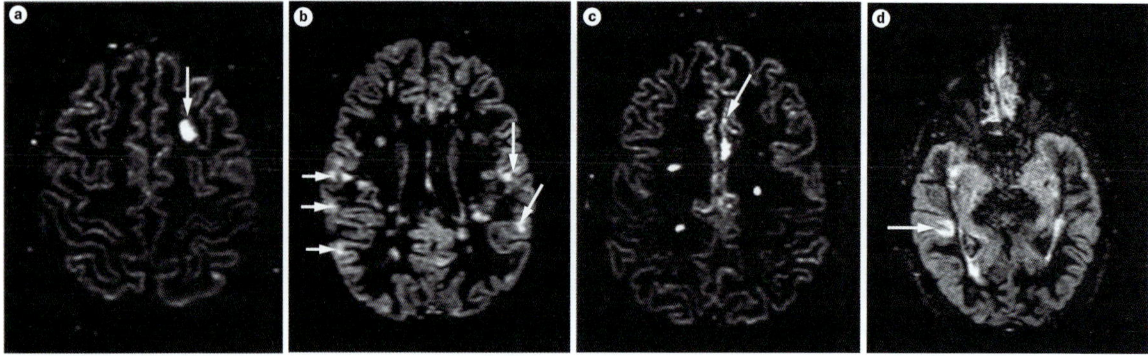

Figure 3.6. Four types of cortical lesion identified in multiple sclerosis patients by double inversion recovery (DIR) imaging. (A) Mixed white matter and gray matter lesion (type I). (B) Intracortical lesions (type II). (C) Subpial lesion (type III). (D) "Worm-like" lesion (type IV). Reproduced with permission from Calabrese et al.[95]

However, this could also be attributed to increased disease severity in these patients. This significant gap of knowledge will need to be addressed by future studies. These will concentrate on the complex bidirectional interplay between dementia and epilepsy. On one hand AD can cause seizures; on the other hand epilepsy patients are likely to develop progressive cognitive deficits and, ultimately, dementia. Future neuroimaging studies will focus on the common pathomechanisms of these disorders including hyperexcitable temporal networks, amyloid and tau deposition, BBB disruption, and changes of NMDA receptor transmission.[110,113,114]

References

1. Forsgren L. Prospective incidence study and clinical characterization of seizures in newly referred adults. *Epilepsia.* 1990;**31**:292–301.

2. Hauser WA, Annegers JF, Rocca WA. Descriptive epidemiology of epilepsy: contributions of population-based studies from Rochester, Minnesota. *Mayo Clin Proc.* 1996;**71**:576–86. doi:10.1016/S0025-6196(11)64115-3.

3. Beghi E, Carpio A, Forsgren L, et al. Recommendation for a definition of acute symptomatic seizure. *Epilepsia.* 2010;**51**:671–5. doi:10.1111/j.1528–1167.2009.02285.x.

4. Hesdorffer DC, Benn EKT, Cascino GD, Hauser WA. Is a first acute symptomatic seizure epilepsy? Mortality and risk for recurrent seizure. *Epilepsia.* 2009;**50**:1102–8. doi:10.1111/j.1528–1167.2008.01945.x.

5. Haltiner AM, Temkin NR, Dikmen SS. Risk of seizure recurrence after the first late posttraumatic seizure. *Arch Phys Med Rehabil.* 1997;**78**:835–40.

6. So EL, Annegers JF, Hauser WA, O'Brien PC, Whisnant JP. Population-based study of seizure disorders after cerebral infarction. *Neurology.* 1996;**46**:350–5.

7. Hart YM, Sander JW, Johnson AL, Shorvon SD. National General Practice Study of Epilepsy: recurrence after a first seizure. *Lancet.* 1990;**336**:1271–4.

8. Annegers JF, Shirts SB, Hauser WA, Kurland LT. Risk of recurrence after an initial unprovoked seizure. *Epilepsia.* 1986;**27**:43–50.

9. Fisher RS, Acevedo C, Arzimanoglou A, et al. ILAE official report: a practical clinical definition of epilepsy. *Epilepsia.* 2014;**55**:475–82. doi:10.1111/epi.12550.

10. Temkin NR. Antiepileptogenesis and seizure prevention trials with antiepileptic drugs: meta-analysis of controlled trials. *Epilepsia.* 2001;**42**:515–24.

11. Lühdorf K, Jensen LK, Plesner AM. Etiology of seizures in the elderly. *Epilepsia.* 1986;**27**:458–63.

12. Camilo O, Goldstein LB. Seizures and epilepsy after ischemic stroke. *Stroke.* 2004;**35**:1769–75. doi:10.1161/01.STR.0000130989.17100.96.

13. Hsu C-J, Weng W-C, Peng SS-F, Lee W-T. Early-onset seizures are correlated with late-onset seizures in children with arterial ischemic stroke. *Stroke.* 2014;**45**:1161–3. doi:10.1161/STROKEAHA.113.004015.

14. Chen T-C, Chen Y-Y, Cheng P-Y, Lai C-H. The incidence rate of post-stroke epilepsy: a 5-year follow-up study in Taiwan. *Epilepsy Res.* 2012;**102**:188–94. doi:10.1016/j.eplepsyres.2012.06.003.

15. Beghi E, D'Alessandro R, Beretta S, et al. Incidence and predictors of acute symptomatic seizures after stroke. *Neurology.* 2011;**77**:1785–93. doi:10.1212/WNL.0b013e3182364878.

16. Benbir G, Ince B, Bozluolcay M. The epidemiology of post-stroke epilepsy according to stroke subtypes. *Acta Neurol Scand*. 2006;**114**:8–12. doi:10.1111/j.1600–0404.2006.00642.x.

17. Bladin CF, Alexandrov AV, Bellavance A, et al. Seizures after stroke: a prospective multicenter study. *Arch Neurol*. 2000;**57**:1617–22. doi:10–1001/pubs.Arch Neurol.-ISSN-0003–9942-57–11-noc90169.

18. Lamy C, Domigo V, Semah F, et al. Early and late seizures after cryptogenic ischemic stroke in young adults. *Neurology*. 2003;**60**:400–4. doi:10.1212/WNL.60.3.400.

19. Zhang C, Wang X, Wang Y, et al. Risk factors for post-stroke seizures: a systematic review and meta-analysis. *Epilepsy Res*. 2014;**108**:1806–16. doi:10.1016/j.eplepsyres.2014.09.030.

20. Leone MA, Tonini MC, Bogliun G, et al. Risk factors for a first epileptic seizure after stroke: a case control study. *J Neurol Sci*. 2009;**277**:138–42. doi:10.1016/j.jns.2008.11.004.

21. Rossi C, De Herdt V, Dequatre-Ponchelle N, Hénon H, Leys D, Cordonnier C. Incidence and predictors of late seizures in intracerebral hemorrhages. *Stroke*. 2013;**44**:1723–5. doi:10.1161/STROKEAHA.111.000232.

22. Giroud M, Gras P, Fayolle H, André N, Soichot P, Dumas R. Early seizures after acute stroke: a study of 1,640 cases. *Epilepsia*. 1994;**35**:959–64.

23. Burn J, Dennis M, Bamford J, Sandercock P, Wade D, Warlow C. Epileptic seizures after a first stroke: the Oxfordshire Community Stroke Project. *BMJ*. 1997;**315**:1582–7.

24. Avrahami E, Drory VE, Rabey MJ, Cohn DF. Generalized epileptic seizures as the presenting symptom of lacunar infarction in the brain. *J Neurol*. 1988;**235**:472–4.

25. Bentes C, Pimentel J, Ferro JM. Epileptic seizures following subcortical infarcts. *Cerebrovasc Dis*. 2001;**12**:331–4.

26. Kilpatrick CJ, Davis SM, Tress BM, Rossiter SC, Hopper JL, Vandendriesen ML. Epileptic seizures in acute stroke. *Arch Neurol*. 1990;**47**:157–60.

27. De Reuck J, Nagy E, Van Maele G. Seizures and epilepsy in patients with lacunar strokes. *J Neurol Sci*. 2007;**263**:75–8. doi:10.1016/j.jns.2007.06.004.

28. Lo YK, Yiu CH, Hu HH, Su MS, Laeuchli SC. Frequency and characteristics of early seizures in Chinese acute stroke. *Acta Neurol Scand*. 1994;**90**:83–5.

29. Okuda S, Takano S, Ueno M, Hamaguchi H, Kanda F. Clinical features of late-onset poststroke seizures. *J Stroke Cerebrovasc Dis*. 2012;**21**:583–6. doi:10.1016/j.jstrokecerebrovasdis.2011.01.006.

30. De Reuck J, Claeys I, Martens S, et al. Computed tomographic changes of the brain and clinical outcome of patients with seizures and epilepsy after an ischaemic hemispheric stroke. *Eur J Neurol*. 2006;**13**:402–7. doi:10.1111/j.1468–1331.2006.01253.x.

31. Arboix A, Comes E, Massons J, García L, Oliveres M. Relevance of early seizures for in-hospital mortality in acute cerebrovascular disease. *Neurology*. 1996;**47**:1429–35.

32. Heuts-van Raak L, Lodder J, Kessels F. Late seizures following a first symptomatic brain infarct are related to large infarcts involving the posterior area around the lateral sulcus. *Seizure*. 1996;**5**:185–94.

33. Szaflarski JP, Rackley AY, Kleindorfer DO, et al. Incidence of seizures in the acute phase of stroke: a population-based study. *Epilepsia*. 2008;**49**:974–81. doi:10.1111/j.1528–1167.2007.01513.x.

34. Wagner F, Erdelyi B, Siebel P, Weber J, Tettenborn B, Felbecker A. Post-stroke epilepsy: Does stroke volume matter? *Eur J Neurol*. 2011;**18**(suppl 2):20–65.

35. De Reuck J, Decoo D, Algoed L, et al. Epileptic Seizures after Thromboembolic Cerebral Infarcts: A Positron Emission Tomographic Study. *Cerebrovasc Dis*. 1995;**5**:328–33. doi:10.1159/000107877.

36. De Reuck J, Algoed L, Decoo D, et al. Positron emission tomographic study of postinfarction seizures. *J Stroke Cerebrovasc Dis*. 1994;**4**:262–6. doi:10.1016/S1052-3057(10)80104–1.

37. De Reuck J, Vonck K, Santens P, et al. Cobalt-55 positron emission tomography in late-onset epileptic seizures after thrombo-embolic middle cerebral artery infarction. *J Neurol Sci*. 2000;**181**:13–8.

38. Labovitz DL, Hauser WA, Sacco RL. Prevalence and predictors of early seizure and status epilepticus after first stroke. *Neurology*. 2001;**57**:200–6. doi:10.1212/WNL.57.2.200.

39. So EL, Annegers JF, Hauser WA, O'Brien PC, Whisnant JP. Population-based study of seizure disorders after cerebral infarction. *Neurology*. 1996;**46**:350–5. doi:10.1212/WNL.46.2.350.

40. Misirli H, Özge A, Somay G, Erdogan N, Erkal H, Erenoglu NY. Seizure development after stroke. *Int J Clin Pract*. 2006;**60**:1536–41. doi:10.1111/j.1742–1241.2005.00782.x.

41. Winkler DT, Fluri F, Fuhr P, et al. Thrombolysis in stroke mimics: frequency, clinical characteristics, and outcome. *Stroke*. 2009;**40**:1522–5. doi:10.1161/STROKEAHA.108.530352.

42. Masterson K, Vargas MI, Delavelle J. Postictal deficit mimicking stroke: role of perfusion CT. *J Neuroradiol*. 2009;**36**:48–51. doi:10.1016/j.neurad.2008.08.006.

43. Gelfand JM, Wintermark M, Josephson SA. Cerebral perfusion-CT patterns following seizure. *Eur J Neurol*.

2010;17:594–601. doi:10.1111/j.1468–1331.2009.02869.x.

44. Salmenpera TM, Symms MR, Boulby PA, Barker GJ, Duncan JS. Postictal diffusion weighted imaging. *Epilepsy Res.* 2006;70:133–43. doi:10.1016/j.eplepsyres.2006.03.010.

45. Miyaji Y, Yokoyama M, Kawabata Y, et al. Arterial spin-labeling magnetic resonance imaging for diagnosis of late seizure after stroke. *J Neurol Sci.* 2014;339:87–90. doi:10.1016/j.jns.2014.01.026.

46. Chalela JA, Alsop DC, Gonzalez-Atavales JB, Maldjian JA, Kasner SE, Detre JA. Magnetic resonance perfusion imaging in acute ischemic stroke using continuous arterial spin labeling. *Stroke.* 2000;31:680–7.

47. Hauser WA, Annegers JF, Kurland LT. Incidence of epilepsy and unprovoked seizures in Rochester, Minnesota: 1935–1984. *Epilepsia.* 1993;34:453–8. doi:10.1111/j.1528–1157.1993.tb02586.x.

48. Sander JW, Hart YM, Johnson AL, Shorvon SD. National General Practice Study of Epilepsy: newly diagnosed epileptic seizures in a general population. *Lancet.* 1990;336:1267–71.

49. King MA, Newton MR, Jackson GD, et al. Epileptology of the first-seizure presentation: a clinical, electroencephalographic, and magnetic resonance imaging study of 300 consecutive patients. *Lancet.* 1998;352:1007–11. doi:10.1016/S0140-6736(98)03543–0.

50. Chadwick D. Seizures and epilepsy after traumatic brain injury. *Lancet.* 2000;355:334–6. doi:10.1016/S0140-6736(99)00452–3.

51. Annegers JF, Hauser WA, Coan SP, Rocca WA. A population-based study of seizures after traumatic brain injuries. *N Engl J Med.* 1998;338:20–4. doi:10.1056/NEJM199801013380104.

52. Salazar AM, Jabbari B, Vance SC, Grafman J, Amin D, Dillon JD. Epilepsy after penetrating head injury. I. Clinical correlates: a report of the Vietnam Head Injury Study. *Neurology.* 1985;35:1406–14.

53. Christensen J, Pedersen MG, Pedersen CB, Sidenius P, Olsen J, Vestergaard M. Long-term risk of epilepsy after traumatic brain injury in children and young adults: a population-based cohort study. *Lancet.* 2009;373:1105–10. doi:10.1016/S0140-6736(09)60214–2.

54. Vespa PM, McArthur DL, Xu Y, et al. Nonconvulsive seizures after traumatic brain injury are associated with hippocampal atrophy. *Neurology.* 2010;75:792–8. doi:10.1212/WNL.0b013e3181f07334.

55. D'Alessandro R, Tinuper P, Ferrara R, et al. CT scan prediction of late post-traumatic epilepsy. *J Neurol Neurosurg Psychiatr.* 1982;45:1153–5.

56. De Santis A, Sganzerla E, Spagnoli D, Bello L, Tiberio F. Risk factors for late posttraumatic epilepsy. *Acta Neurochir Suppl (Wien).* 1992;55:64–7.

57. Pohlmann-Eden B, Bruckmeir J. Predictors and dynamics of posttraumatic epilepsy. *Acta Neurol Scand.* 1997;95:257–62.

58. Englander J, Bushnik T, Duong TT, et al. Analyzing risk factors for late posttraumatic seizures: a prospective, multicenter investigation. *Arch Phys Med Rehabil.* 2003;84:365–73. doi:10.1053/apmr.2003.50022.

59. Temkin NR. Risk factors for posttraumatic seizures in adults. *Epilepsia.* 2003;44(suppl 10):18–20.

60. Mazzini L, Cossa FM, Angelino E, Campini R, Pastore I, Monaco F. Posttraumatic epilepsy: neuroradiologic and neuropsychological assessment of long-term outcome. *Epilepsia.* 2003;44:569–74.

61. Kumar R, Gupta RK, Husain M, et al. Magnetization transfer MR imaging in patients with posttraumatic epilepsy. *AJNR Am J Neuroradiol.* 2003;24:218–24.

62. Messori A, Polonara G, Carle F, Gesuita R, Salvolini U. Predicting posttraumatic epilepsy with MRI: prospective longitudinal morphologic study in adults. *Epilepsia.* 2005;46:1472–81. doi:10.1111/j.1528–1167.2005.34004.x.

63. Willmore LJ, Sypert GW, Munson JB. Recurrent seizures induced by cortical iron injection: a model of posttraumatic epilepsy. *Ann Neurol.* 1978;4:329–36. doi:10.1002/ana.410040408.

64. Angeleri F, Majkowski J, Cacchiò G, et al. Posttraumatic epilepsy risk factors: one-year prospective study after head injury. *Epilepsia.* 1999;40:1222–30.

65. Asikainen I, Kaste M, Sarna S. Early and late posttraumatic seizures in traumatic brain injury rehabilitation patients: brain injury factors causing late seizures and influence of seizures on long-term outcome. *Epilepsia.* 1999;40:584–9.

66. Diaz-Arrastia R, Agostini MA, Frol AB, et al. Neurophysiologic and neuroradiologic features of intractable epilepsy after traumatic brain injury in adults. *Arch Neurol.* 2000;57:1611–6.

67. Shlosberg D, Benifla M, Kaufer D, Friedman A. Blood-brain barrier breakdown as a therapeutic target in traumatic brain injury. *Nat Rev Neurol.* 2010;6:393–403. doi:10.1038/nrneurol.2010.74.

68. Ivens S, Kaufer D, Flores LP, et al. TGF-beta receptor-mediated albumin uptake into astrocytes is involved in neocortical epileptogenesis. *Brain.* 2007;130:535–47. doi:10.1093/brain/awl317.

69. Seiffert E, Dreier JP, Ivens S, et al. Lasting blood-brain barrier disruption induces epileptic focus in the rat somatosensory cortex. *J Neurosci.* 2004;24:7829–36. doi:10.1523/JNEUROSCI.1751–04.2004.

70. Prager O, Chassidim Y, Klein C, Levi H, Shelef I, Friedman A. Dynamic in vivo imaging of cerebral

blood flow and blood-brain barrier permeability. *NeuroImage*. 2010;**49**:337–44. doi:10.1016/j. neuroimage.2009.08.009.

71. Tomkins O, Kaufer D, Korn A, et al. Frequent blood-brain barrier disruption in the human cerebral cortex. *Cell Mol Neurobiol*. 2001;**21**:675–91. doi:10.1023/A:1015147920283.

72. Tomkins O, Shelef I, Kaizerman I, et al. Blood-brain barrier disruption in post-traumatic epilepsy. *J Neurol Neurosurg Psychiatr*. 2008;**79**:774–7. doi:10.1136/jnnp.2007.126425.

73. Tomkins O, Feintuch A, Benifla M, Cohen A, Friedman A, Shelef I. Blood-brain barrier breakdown following traumatic brain injury: a possible role in posttraumatic epilepsy. *Cardiovasc Psychiatry Neurol*. 2011;**2011**:765923. doi:10.1155/2011/765923.

74. Lynam LM, Lyons MK, Drazkowski JF, et al. Frequency of seizures in patients with newly diagnosed brain tumors: a retrospective review. *Clin Neurol Neurosurg*. 2007;**109**:634–8. doi:10.1016/j.clineuro.2007.05.017.

75. Herman ST. Epilepsy after brain insult: targeting epileptogenesis. *Neurology*. 2002;**59**:S21–6. doi:10.1212/WNL.59.9_suppl_5.S21.

76. Klein M, Engelberts NHJ, van der Ploeg HM, et al. Epilepsy in low-grade gliomas: the impact on cognitive function and quality of life. *Ann Neurol*. 2003;**54**:514–20. doi:10.1002/ana.10712.

77. Sperling MR, Ko J. Seizures and brain tumors. *Semin Oncol*. 2006;**33**:333–41. doi:10.1053/j.seminoncol.2006.03.009.

78. van Breemen MSM, Wilms EB, Vecht CJ. Epilepsy in patients with brain tumours: epidemiology, mechanisms, and management. *Lancet Neurol*. 2007;**6**:421–30. doi:10.1016/S1474-4422(07)70103-5.

79. Lote K, Stenwig AE, Skullerud K, Hirschberg H. Prevalence and prognostic significance of epilepsy in patients with gliomas. *Eur J Cancer*. 1998;**34**:98–102.

80. Liigant A, Haldre S, Oun A, et al. Seizure disorders in patients with brain tumors. *Eur Neurol*. 2001;**45**:46–51.

81. Hildebrand J, Lecaille C, Perennes J, Delattre J-Y. Epileptic seizures during follow-up of patients treated for primary brain tumors. *Neurology*. 2005;**65**:212–5. doi:10.1212/01.wnl.0000168903.09277.8 f.

82. Whittle IR, Beaumont A. Seizures in patients with supratentorial oligodendroglial tumours. Clinicopathological features and management considerations. *Acta Neurochir (Wien)*. 1995;**135**:19–24.

83. Moots PL, Maciunas RJ, Eisert DR, Parker RA, Laporte K, Abou-Khalil B. The course of seizure disorders in patients with malignant gliomas. *Arch Neurol*. 1995;**52**:717–24.

84. Roelcke U, Boxheimer L, Fathi AR, et al. Cortical hemosiderin is associated with seizures in patients with newly diagnosed malignant brain tumors. *J Neurooncol*. 2013;**115**:463–8. doi:10.1007/s11060-013-1247-7.

85. Chugani DC, Chugani HT, Muzik O, et al. Imaging epileptogenic tubers in children with tuberous sclerosis complex using alpha-[^{11}C]methyl-L-tryptophan positron emission tomography. *Ann Neurol*. 1998;**44**:858–66. doi:10.1002/ana.410440603.

86. Asano E, Chugani DC, Muzik O, et al. Multimodality imaging for improved detection of epileptogenic foci in tuberous sclerosis complex. *Neurology*. 2000;**54**:1976–84.

87. Kagawa K, Chugani DC, Asano E, et al. Epilepsy surgery outcome in children with tuberous sclerosis complex evaluated with alpha-[^{11}C]methyl-L-tryptophan positron emission tomography (PET). *J Child Neurol*. 2005;**20**:429–38.

88. Juhász C, Chugani DC, Muzik O, et al. Alpha-methyl-L-tryptophan PET detects epileptogenic cortex in children with intractable epilepsy. *Neurology*. 2003;**60**:960–8.

89. Jacobs J, Rohr A, Moeller F, et al. Evaluation of epileptogenic networks in children with tuberous sclerosis complex using EEG-fMRI. *Epilepsia*. 2008;**49**:816–25. doi:10.1111/j.1528–1167.2007.01486.x.

90. Koch M, Uyttenboogaart M, Polman S, De Keyser J. Seizures in multiple sclerosis. *Epilepsia*. 2008;**49**:948–53. doi:10.1111/j.1528–1167.2008.01565.x.

91. Marrie RA, Reider N, Cohen J, et al. A systematic review of the incidence and prevalence of sleep disorders and seizure disorders in multiple sclerosis. *Mult Scler*. 2015;**21**:342–9. doi:10.1177/1352458514564486.

92. Durmus H, Kurtuncu M, Tuzun E, et al. Comparative clinical characteristics of early- and adult-onset multiple sclerosis patients with seizures. *Acta Neurol Belg*. 2013;**113**:421–6. doi:10.1007/s13760-013–0210-x.

93. Catenoix H, Marignier R, Ritleng C, et al. Multiple sclerosis and epileptic seizures. *Mult Scler*. 2011;**17**:96–102. doi:10.1177/1352458510382246.

94. van Munster CEP. Gray matter damage in multiple sclerosis: Impact on clinical symptoms. *Neuroscience*. 2015;**303**:446–61. doi:10.1016/j.neuroscience.2015.07.006.

95. Calabrese M, Filippi M, Gallo P. Cortical lesions in multiple sclerosis. *Nat Rev Neurol*. 2010;**6**:438–44. doi:10.1038/nrneurol.2010.93.

96. Gambardella A, Valentino P, Labate A, et al. Temporal lobe epilepsy as a unique manifestation of multiple sclerosis. *Can J Neurol Sci*. 2003;**30**:228–32.

97. Sokic DV, Stojsavljevic N, Drulovic J, et al. Seizures in multiple sclerosis. *Epilepsia*. 2001;**42**:72–9.

98. Moreau T, Sochurkova D, Lemesle M, et al. Epilepsy in patients with multiple sclerosis: radiological-clinical correlations. *Epilepsia*. 1998;**39**:893–6.

99. Thompson AJ, Kermode AG, Moseley IF, MacManus DG, McDonald WI. Seizures due to multiple sclerosis: seven patients with MRI correlations. *J Neurol Neurosurg Psychiatr*. 1993;**56**:1317–20.

100. Ghezzi A, Montanini R, Basso PF, Zaffaroni M, Massimo E, Cazzullo CL. Epilepsy in multiple sclerosis. *Eur Neurol*. 1990;**30**:218–23.

101. Büttner T, Hornig CR, Dorndorf W. [Multiple sclerosis and epilepsy. An analysis of 14 case histories]. *Der Nervenarzt*. 1989;**60**:262–7.

102. Calabrese M, De Stefano N, Atzori M, et al. Extensive cortical inflammation is associated with epilepsy in multiple sclerosis. *J Neurol*. 2008;**255**:581–6. doi:10.1007/s00415-008-0752-7.

103. Calabrese M, Grossi P, Favaretto A, et al. Cortical pathology in multiple sclerosis patients with epilepsy: a 3 year longitudinal study. *J Neurol Neurosurg Psychiatr*. 2012;**83**:49–54. doi:10.1136/jnnp-2011-300414.

104. Nicholas R, Magliozzi R, Campbell G, Mahad D, Reynolds R. Temporal lobe cortical pathology and inhibitory GABA interneuron cell loss are associated with seizures in multiple sclerosis. *Mult Scler*. 2016;**22**:25–35. doi:10.1177/1352458515579445.

105. Imfeld P, Bodmer M, Schuerch M, Jick SS, Meier CR. Seizures in patients with Alzheimer's disease or vascular dementia: a population-based nested case-control analysis. *Epilepsia*. 2013;**54**:700–7. doi:10.1111/epi.12045.

106. Annegers JF, Rocca WA, Hauser WA. Causes of epilepsy: contributions of the Rochester epidemiology project. *Mayo Clin Proc*. 1996;**71**:570–5. doi:10.1016/S0025-6196(11)64114-1.

107. Amatniek JC, Hauser WA, DelCastillo-Castaneda C, et al. Incidence and predictors of seizures in patients with Alzheimer's disease. *Epilepsia*. 2006;**47**:867–72. doi:10.1111/j.1528-1167.2006.00554.x.

108. Scarmeas N, Honig LS, Choi H, et al. Seizures in Alzheimer disease: who, when, and how common? *Arch Neurol*. 2009;**66**:992–7. doi:10.1001/archneurol.2009.130.

109. Vossel KA, Beagle AJ, Rabinovici GD, et al. Seizures and epileptiform activity in the early stages of Alzheimer disease. *JAMA Neurol*. 2013;**70**:1158–66. doi:10.1001/jamaneurol.2013.136.

110. Noebels J. A perfect storm: Converging paths of epilepsy and Alzheimer's dementia intersect in the hippocampal formation. *Epilepsia*. 2011;**52** (suppl 1):39–46. doi:10.1111/j.1528-1167.2010.02909.x.

111. Irizarry MC, Jin S, He F, et al. Incidence of new-onset seizures in mild to moderate Alzheimer disease. *Arch Neurol*. 2012;**69**:368–72. doi:10.1001/archneurol.2011.830.

112. Dhikav V, Anand K. Hippocampal atrophy may be a predictor of seizures in Alzheimer's disease. *Med Hypotheses*. 2007;**69**:234–5. doi:10.1016/j.mehy.2006.11.031.

113. Chin J, Scharfman HE. Shared cognitive and behavioral impairments in epilepsy and Alzheimer's disease and potential underlying mechanisms. *Epilepsy Behav*. 2013;**26**:343–51. doi:10.1016/j.yebeh.2012.11.040.

114. Palop JJ, Mucke L. Epilepsy and cognitive impairments in Alzheimer disease. *Arch Neurol*. 2009;**66**:435–40. doi:10.1001/archneurol.2009.15.

Chapter

4

Imaging and Cognition in Children with New-Onset Epilepsies

Kevin Dabbs, Camille Garcia-Ramos, Daren Jackson, Jack Lin, Sam Bobholz, Dace Almane, Jana Jones, Mike Seidenberg, and Bruce Hermann

4.1 Introduction

While defined by the presence of recurrent seizures, epilepsy can be associated with abnormalities in cognition, psychiatric status, and social-adaptive behaviors.[1] These complications of epilepsy, especially alterations in cognition including memory, have been shown to be associated with alterations in brain structure and function.[2] While the comorbidities of epilepsy exert major adverse effects on quality of life of people with epilepsy,[3] it has proven difficult to fully understand the etiology, timing, and course of abnormalities in both brain structure and cognition. A growing opinion has been that to better understand these issues it would be useful to examine patients near the time of the onset of their epilepsy and follow patients prospectively with the hope of identifying biomarkers of problematic courses.[4,5]

In that spirit, the focus here is on children with epilepsy and what has been learned about the origin, development, and course of abnormalities in brain structure and cognition. We will begin by first briefly reviewing several trends that have been identified regarding cognitive and other neurobehavioral comorbidities of epilepsy and their potential neurodevelopmental roots. We then move on to what has been learned about the origin and course of these problems through studying children at or near the time of the onset of their epilepsy and following them prospectively over time using both objective measures of cognition and neuroimaging.

4.2 Childhood Onset Epilepsy

As noted, given the diverse medical, social, psychiatric, cognitive, and quality of life complications of the chronic adult epilepsies[6–11] and their associated direct and indirect costs,[12,13] an increasing focus of interest has been on when, why, and how these problems develop and progress. At least three sets of evidence suggest a significant neurodevelopmental contribution.

First there is the *indirect* evidence linking the risk of comorbidities to the age of onset of recurrent seizures. For example, studying *adults* with chronic epilepsy, many neuropsychological studies have shown that an earlier age of onset of recurrent seizures is associated with poorer cognitive function than is chronic epilepsy with a later onset.[14–19] This same relationship has been reported in some but not all neuropsychological studies of younger patients with focal impaired awareness and other types of seizures.[20–22] Complementing these cognitive studies are neuroimaging investigations of brain structure. Although this is a more limited literature, it has been shown that childhood onset epilepsy is associated with reductions in whole-brain volume among patients with a history of complicated febrile convulsions,[23] altered volumes of the corpus callosum, whole-brain white and/or gray matter tissue, and specific disrupted patterns of gray and white matter connectivity in individuals with childhood compared to adult onset chronic partial epilepsy.[17,24–27]

Second is the compelling evidence that childhood onset epilepsy adversely impacts life course. An important indication of the public health significance of childhood epilepsy is its association with problematic outcomes in later youth and adulthood. This impact is reflected in an international literature published over the course of decades with studies emanating from the United Kingdom, Canada, Finland, Japan, Australia, and the Netherlands. Collectively, these reports include both uncontrolled[28–34] and controlled community- and population-based investigations[8,35–41] as well as long-term (adult) follow-up of national birth cohorts.[42–46] Groups of children with epilepsy have been followed prospectively for up to 30+ years in controlled investigations,[39,40] which has yielded a number of insights: (1) problems in a diversity of major life outcomes have been reported including rates of marriage, employment, income, psychiatric status, independent living status, and other critical aspects of quality of life; (2) while it would not be surprising that children with severe and complicated epilepsies would exhibit poor life span psychosocial outcomes,[30–32] many but not all[34] investigations report adverse life span outcomes in persons with intact intelligence, so-called benign epilepsies, and even persons with remitted childhood epilepsy who are no longer being treated with antiepilepsy medications;[28,36,40,41] (3) the causes underlying these adult outcomes and the factors associated with favorable versus unfavorable outcomes remain to be fully clarified, and these problematic life span outcomes are variably associated with clinical epilepsy features (e.g., seizure control); and (4) adverse life span outcomes have been found to be associated with histories of neurobehavioral cognitive and psychiatric comorbidities,[28,30,37,38] including what is now called attention deficit hyperactivity disorder (ADHD).[30]

Third, there is an extensive and widely appreciated set of background data that involve the direct examination of comorbidities among children with established epilepsy.[47–49] This literature benefits from both population-based and careful tertiary care investigations (see Lin et al.[1] for review).

However, to directly assess neurodevelopmental contributions to specific comorbidities of epilepsy and anomalies in brain structure, connectivity, and function and their course over time, a useful line of research would be to examine children at or near the time of diagnosis and follow them prospectively. There is a developing literature on this topic in regard to both cognition and imaging that will be reviewed below.

4.2.1 Cognition

Prior to 2000 only a small number of studies examined cognition in children with new-onset epilepsy.[50–53] Several of these early papers tested children at or near diagnosis in order to serve as a drug-naïve baseline with which to assess subsequent treatment or comparative drug effects.[52,53] Others focused solely on intelligence and were primarily interested in characterizing clinical seizure factors associated with prospective IQ changes, especially declines.[50] These studies examined children with epilepsy of various types, age ranges, test batteries, and clinical seizure characteristics. Comments that follow focus only on the baseline findings.

Bourgeois et al.[50] examined intelligence in 72 children with mixed epilepsies within two weeks of diagnosis, age 18 months to 16 years (with IQ 40+), compared to 45 siblings. The mean IQ of the children with epilepsy (98.0) was lower than that of matched siblings (104.4), although this difference was not statistically significant. Children with idiopathic epilepsy had a higher IQ than symptomatic epilepsy (102.5 vs. 89.1) at baseline.

Stores[52] examined 63 children (ages 7–12) with new-onset epilepsy both before and after initiation of antiepileptic drugs (AED) treatment and 47 similarly aged healthy controls. Measures of specific cognitive abilities (attention, memory, motor, and psychomotor speed) and general intelligence as well as reading and arithmetic performance were administered. At baseline, prior to AED initiation, the focal epilepsy group exhibited significantly poorer attention (ability to establish and focus attention) as well as poorer motor and psychomotor ability. Children with absence and other genetic generalized epilepsies exhibited fewer significant baseline cognitive abnormalities, with differences limited to motor ability and psychomotor speed.

Williams et al.[53] compared 37 children with new-onset epilepsy (22 focal and 15 generalized epilepsy, ages 6–17) prior to initiation of AEDs and a control group of 26 children with new-onset diabetes on measures of attention and concentration, memory, complex motor processing speed, and behavior. The children with epilepsy performed more poorly than the diabetic children across 9 of 11 measures, but none of the group differences were statistically significant.

Kolk et al.[51] compared 14 healthy controls, 12 children with focal seizures, and 18 with congenital hemiparesis and seizures (mean age = 6 years), the children with epilepsy said to be of normal psychometric intelligence. The NEPSY was used for cognitive assessment, which consisted of 37 tests falling into five cognitive domains (attention and executive function, language, sensorimotor, visuospatial, learning and memory). The children with new-onset partial epilepsy showed significantly poorer performance across four of the five composite cognitive domain scores.

This has become an area of increased interest and several recent investigations have appeared. In the seminal paper by Oostrom et al.,[54] children (n = 51, ages 5–16) with newly diagnosed epilepsy were seen and assessed prior to the administration of antiseizure medications (tested within 48 hours of diagnosis).

The children with epilepsy had at least two unprovoked nonfebrile seizures of idiopathic or unclear cause, attended mainstream schools, and were without any identified neurological disease, a diagnosis of another chronic illness, or previous use of AEDs. Controls were healthy classmates invited by the child with epilepsy and/or their family ($n = 48$). A comprehensive test battery was administered that was reduced through factor analysis to measures of attention, reaction time, intelligence, academic skills, learning, and behavior. Significantly poorer performance on the part of the children with new-onset epilepsy was observed in the areas of attention, reaction time, visual memory, and behavior with a trend ($p = .07$) of poorer academic achievement.[54] From the time of this investigation there has been considerably more interest in children with new-onset epilepsies, and in the material to follow we will consider only those new-onset studies that included a control group for direct comparison.

Fastenau et al.[55] examined a large community-based sample of children with first recognized seizures ($n = 282$ children, ages 6–14 years, IQ $\geq$ 70) and 147 sibling controls, characterizing their neuropsychological and academic status. They found that 27% of children with a single seizure and up to 40% of those with risk factors exhibited neuropsychological deficits at or near onset. Risk factors for neuropsychological deficit included multiple seizures (OR = 1.96), use of AEDs, symptomatic/cryptogenic etiology (OR = 2.15), and epileptiform activity on the initial EEG (OR = 1.9). Absence epilepsy also carried increased odds for neuropsychological impairment (OR = 2.0). Academic achievement appeared unaffected, suggesting an opportunity for early intervention (see Dunn et al.[56] and McNelis et al.[57] for additional information regarding academic achievement in both new-onset and chronic cases from the Indianapolis series).

Vintan et al.[58] compared 18 children with drug-naïve new-onset benign epilepsy with centrotemporal spikes (BECTS) to 18 controls (mean age 8.88 and 8.22, respectively). Overall intellectual ability was assessed using Raven's Progressive Matrices, and specific cognitive abilities were assessed using subtests from the Cambridge Neuropsychological Test Automated Battery (CANTAB), which included (1) induction: Motor Screening Test (MOT); (2) visual memory: Delayed Matching to Sample (DMS), Paired Associates Learning (PAL), Pattern Recognition Memory (PRM), and Spatial Recognition Memory (SRM) tests; and (3) executive functions: Spatial Span (SSP) and Spatial Working Memory (SWM) tests. The BECTS group exhibited normal performances for induction and executive functions, compared with control children. There were significant differences on the visual memory subtests (PRM percentage correct and SRM percentage correct), suggesting that children with BECTS had difficulties determining the accuracy of the pattern and spatial recognition memory.

Lee et al.[59] examined 30 children with drug-naïve new-onset epilepsy (21 focal and 9 generalized epilepsies) and 25 health controls, ages 7–16, and compared them on the Korean version of the Wechsler Intelligence Scale for Children–III (WISC-III), Stroop Interference, and Trail Making Tests (A&B). The children with epilepsy had significantly worse WISC-III freedom from distraction and verbal IQ scores, as well as significantly longer response times on the Stroop Test color condition. Comparing the focal and generalized epilepsy groups, the only difference was lower verbal IQ in the focal epilepsy group.

Filippini et al.[60] compared 15 children with drug-naïve new-onset BECTS and 15 healthy controls (mean ages 8.8 and 9.2 years, respectively) across a comprehensive test battery assessing language, executive function, academic skills, visuomotor and visuospatial skills, and short-term memory. The children with BECTS performed significantly worse on tests of language function (color naming, spoonerism, phonemic synthesis) and academic skills (word reading errors as well as mistakes in nonword writing and reading and nonword reading speed) and tests of executive function (five-point test correct and errors). They concluded that children with new-onset untreated BECTS showed a range of abnormalities in cognition, especially executive attention, despite normal IQ.

Cheng et al.[61] examined 37 children with absence epilepsy (15 of whom were new-onset and drug-naïve) and 37 healthy controls (ages 8.0 and 8.5, respectively). A series of 10 cognitive tests were administered including Raven's Progressive Matrices (fluid intelligence), Wisconsin Card Sorting Test (response inhibition and mental flexibility), visual tracing (visual attention), with control tasks including choice reaction time, number magnitude comparison, mental rotation, simple subtraction, word semantics, paired associate learning, and word rhyming. The drug-naïve group performed worse than controls on Raven's Matrices, visual tracing, and the

In addition, patients with BECTS displayed significantly larger volumes of bilateral putamen and amygdala compared to healthy controls. Finally, BECTS patients with ADHD ($n = 8$) were shown to possess significantly thicker left caudal anterior and posterior cingulate gyri and a significantly larger left pars opercularis gyral volume compared to BECTS patients without ADHD.

Luo et al.[76] examined 21 children with BECTS (average of 9 years old), 12 of whom were new-onset, and 20 healthy controls. Cortical and subcortical volumes were compared using voxel-based morphology (VBM) and resting-state functional connectivity analyses. Relative to controls, patients with BECTS showed significantly increased volume in the bilateral putamen, paracentral lobule, right anterior insula/frontal operculum, right supplementary motor area (SMA), right inferior temporal gyrus, and left cerebellum. Negative correlations were found between bilateral putamen volume and onset age of epilepsy; positive correlations were apparent between the volume of the right anterior insula/frontal operculum with the verbal and full-scale IQ scores.

Lee et al.[59] examined children with new-onset epilepsy, average 3.4 months duration including 21 focal epilepsy (13 BECTS), 9 generalized epilepsy (9 JME), and 25 controls (aged ~10.5 years) and focused on gray and white matter volumes using VBM. They reported GM volume decreases in the left inferior frontal and right middle frontal gyri of patients compared with controls. Correlations with cognition found that lower verbal IQ scores were associated with decreased GM volumes in the left superior temporal and anterior cingulate gyri and decreased WM volumes in the left superior temporal and right parahippocampal gyri in the control group. However, no correlations were found between the verbal IQ score and GM/WM volumes in the epilepsy group. In addition, lower scores in freedom from distractibility were correlated with decreased WM volumes in the left frontal subgyral area, precuneus, and the superior parietal lobule as well as the right parahippocampal and middle temporal gyri in the control group. Lower scores in freedom from distractibility also were correlated with decreased GM volumes in the left postcentral gyrus in the epilepsy group. A longer response time on the Stroop Color Test was correlated with decreased GM volume of the right posterior lobe and decreased WM volumes of the right frontal subgyral and insular areas and part of the left sublobar region

in the control group. In contrast, a correlation between Stroop Color Test response time and GM volume was observed only in the right superior temporal gyrus in the epilepsy group.

Wang et al.[65] examined regional gray matter volume differences (VBM) in 20 drug-naïve children with new-onset CAE and 20 healthy age- and gender-matched controls (mean age 8.65 and 8.83, respectively). Compared to controls, the children with CAE demonstrated reduced gray matter volume in the bilateral thalami.

Ekmekci et al.[66] examined white matter integrity (DTI) in 24 youth with new-onset, but not drug-naïve JME and 28 healthy control children (mean ages = 14.9 and 14.7, respectively). The investigators found significant decreases in FA value in dorsolateral prefrontal cortex, SMA, right thalamus, posterior cingulate, anterior corpus callosum, anterior corona radiata, and middle frontal white matter. Increased ADC was less significant in uncinate fasciculus and anterior. Furthermore, the authors found associations between cognitive measures and white matter integrity, which speaks to the clinical significance of the findings.

Perani et al.[77] examined quantitative measures of gray matter volume in the thalamus, putamen, caudate, pallidum, hippocampus, precuneus, prefrontal cortex, precentral cortex, and cingulate in 29 AED-naïve patients with new-onset genetic generalized epilepsies and 32 age-matched healthy controls (mean ages 15.07 and 16.9, respectively). The thalamus was the only region that demonstrated reduced gray matter volume and shape analysis showed deflation which particularly affected a circumferential strip involving anterior, superior, posterior, and inferior regions with sparing of medial and lateral regions.

In summary, comparable to studies of cognition in children with new-onset epilepsies, a limited number of studies have demonstrated that brain abnormalities are present very early in the course of the disorder.

4.3 Tracking the Course of Cognition and Brain Development

Over the last 15 years we have been investigating a large cohort of children with new and recent onset idiopathic epilepsies using objective neuropsychological assessment, diverse neuroimaging techniques, and psychiatric assessment. These procedures have been repeated over time (2 years, 5 years) with examination of longer term (10 years) behavioral and psychosocial outcomes. By comparing the results to normally developing children, we have attempted to better characterize the presence, nature, and course of abnormalities in cognitive and brain development and their prognostic significance. Here we present direct comparisons of baseline and prospective cognitive and brain development in controls and participants with epilepsy.

4.3.1 Participants

Research participants consisted of 241 youth aged 8–18, including 134 with new and recent-onset epilepsy and 107 healthy first-degree cousin controls (see Table 4.1). The groups did not differ in terms of age, sex, or grade level. Full-scale IQ was lower in children with epilepsy (mean = 101.93, SD = 13.66) than in controls (mean = 109.35, SD = 11.31; $p < .001$), and academic problems were more common (epilepsy group = 48.5%, control group = 18.7%; $p < .001$; see Table 4.1 for demographic and clinical variables broken down by epilepsy subsyndrome). At baseline, all participants attended regular schools. Children with epilepsy were recruited from pediatric neurology clinics at three Midwestern medical centers (University of Wisconsin–Madison, Marshfield Clinic, Dean Clinic) and met the following inclusion criteria: (1) diagnosis of epilepsy within the past 12 months, (2) no other developmental disabilities (e.g., intellectual impairment, autism), (3) no other neurological disorder, and (4) normal clinical MRI. All children entered the study with active epilepsy

diagnosed by their treating pediatric neurologists and confirmed by medical record review of the research study pediatric neurologist. We did not exclude children on the basis of psychiatric comorbidities (including ADHD) or learning disabilities. We did however exclude children with intellectual disorder, autism, and/or other neurological disorders. Details regarding the subject selection process have been described in detail in previous publications.[68] In general, we tried to stay true to the concept of "epilepsy only" as defined broadly in the literature: normal neurological exams, intelligence, and attendance at regular schools. Each child's epilepsy syndrome was defined in a research consensus meeting by the research pediatric neurologist who reviewed all available clinical data (e.g., seizure description and phenomenology, EEG, clinical imaging, neurodevelopmental history) while blinded to all research cognitive, behavioral, and neuroimaging data.

First-degree cousins were used as controls and exclusion criteria were as follows: (1) history of any initial precipitating insult (e.g., simple or complex febrile seizures, cerebral infections, perinatal stroke), (2) any seizure or seizure-like episode, (3) diagnosed neurological disease, (4) loss of consciousness for a period greater than 5 min, (5) history of a first-degree relative with epilepsy or febrile convulsions. We used cousin controls rather than siblings or other potential control groups for the following reasons: (1) first-degree cousins are more genetically distant from the participants with epilepsy and thus less predisposed than siblings to shared genetic factors that

Table 4.1 Participant Characteristics of Controls and Epilepsy Participants by Syndrome

Variable	HC (n = 107)	BECTS (n = 41)	Focal (n = 51)	JME (n = 42)
Age (years)	12.46 (3.00)	10.32 (1.67)	12.36 (3.01)	14.85 (2.75)
Gender (#/% female)	54 (51%)	15 (37%)	24 (47%)	25 (60%)
FSIQ	109.35 (11.31)	104.44 (15.08)	99.55 (12.83)	102.38 (12.99)
Academic problems	19 (18%)	22 (38%)	27 (53%)	19 (45%)
Age of seizure onset (years)	—	9.55 (1.68)	11.34 (2.98)	14.03 (2.88)
Epilepsy duration (months)	—	7.49 (3.33)	8.37 (3.47)	8.36 (3.44)
Antiepileptic drugs (0/1/2+)	—	17/24/0	8/38/5	1/39/2

Values are means/standard deviations or frequencies/%. HC = healthy controls; BECTS = benign epilepsy with centrotemporal spikes; focal = benign occipital epilepsy (n = 3), temporal lobe epilepsy (n = 20), frontal lobe epilepsy (n = 8), focal epilepsy NOS (n = 20); JME = juvenile myoclonic epilepsy.

may contribute to anomalies in brain structure and cognition, (2) a greater number of first-degree cousins are available than siblings in the target age range, and (3) the family link was anticipated to facilitate participant recruitment and especially retention over time (which is our intent) compared to more general control populations (e.g., unrelated school mates).

4.3.2 Neuropsychological Assessment

All participants were administered a comprehensive test battery that included measures of intelligence, academic achievement, language, immediate and delayed verbal memory, executive function, and speeded fine motor dexterity (see Table 4.2). Tests were selected not only for their pertinence to the cognitive domains of interest, but also for their broad applicable age ranges, so that the item pools were identical across the broad age range investigated here (as opposed to administering different versions of a test containing varying item pools to children across age categories), ensuring ability to directly and quantitatively characterize cognitive development. For the purposes of this presentation, all groups were compared across a subset of administered tests that were selected as representative of fundamental cognitive domains. Psychomotor function was assessed using a symbol substitution task (WISC-III Digit Symbol–Coding) and a simple motor task (Grooved Pegboard-dominant hand). The language domain was represented by the Boston Naming Test, a confrontation naming task. Verbal memory was assessed via a list-learning task from the Children's Memory Scale. Finally, executive function was examined by measuring ability to inhibit prepotent responses on a Stroop-like task from the Delis-Kaplan Executive Function System (see Table 4.2 for full citations of all tests). All analyses used raw test scores and covaried age and sex. Patient performance in each group (BECTS, Focal, JME) were compared to controls using a 2 (Group: Patient vs. Control) × 2 (Time: Baseline vs. Follow-Up) mixed design ANCOVA for each of the six selected tests. Differences between

Table 4.2 Neuropsychological Tests by Domain

Domain	Ability	Test
Academic achievement	Letter/word recognition	WRAT-3 (reading)[78]
	Letter and word writing	WRAT-3 (spelling)
	Basic arithmetic	WRAT-3 (arithmetic)
Intelligence	Verbal	WASI (verbal IQ)[79]
	Nonverbal	WASI (performance IQ)
Language	Confrontation naming	Boston Naming Test[80]
	Expressive naming	Expressive Vocabulary Test[81]
	Receptive language	Peabody Picture Vocabulary Test–III[82]
	Generative naming	D-KEFS (letter fluency)[83]
Memory	Verbal memory	CMS (word lists learning)[84]
		CMS (word lists delayed)
Executive function	Problem solving	D-KEFS (confirmed correct sorts)
	Response inhibition	D-KEFS (color-word interference test Inhibition)
	Divided attention	D-KEFS (category switching accuracy)
	Inattentiveness	CCPT-II (omission and commission errors)[85]
Motor function	Speeded fine motor dexterity	Grooved Pegboard[86]
	Psychomotor speed	WISC-III (digit symbol coding)[87]

WRAT-3 = Wide Range Achievement Test–3; WASI = Wechsler Abbreviated Scale of Intelligence; D-KEFS = Delis-Kaplan Executive Function System; CMS = Children's Memory Scale; CCPT-II = Conners' Continuous Performance Test–II; WISC-III = Wechsler Intelligence Scale for Children–III.

patients and controls were examined at each time point. Significance was evaluated at $p < .05$; for all tests reaching significance, p values are given in parentheses. In all cases, patient performance was worse than that of control participants.

4.3.3 Neuroimaging

Images were obtained on a 1.5 T GE Signa MRI scanner (GE Healthcare, Waukesha, WI, USA). Sequences acquired for each participant included (1) T1-weighted, three-dimensional spoiled gradient recall (SPGR) acquired with the following parameters: TE = 5 ms, TR = 24 ms, flip angle = 40 degrees, NEX = 1, slice thickness = 1.5 mm, slices = 124, plane = coronal, FOV = 200 mm, matrix = 256 × 256; (2) proton density (PD); and (3) T2-weighted images acquired with the following parameters: TE = 36 ms (for PD) or 96 ms (for T2), TR = 3,000 ms, NEX = 1, slice thickness = 3.0 mm, slices = 64, slice plane = coronal, FOV = 200 mm, matrix = 256 × 256. Images were transferred to a Mac OS X computer for processing with the FreeSurfer image analysis suite, a set of software tools for the study of cortical and subcortical anatomy that is documented and freely available for download online (http://surfer.nmr.mgh.harvard.edu). A brief technical presentation of these procedures as used in our laboratory is presented in Dabbs et al.[88]

4.4 Cognitive and Imaging Findings

In the material to follow we review the baseline cognitive and imaging findings at baseline and over a prospective 2-year period. This information will be reviewed by epilepsy syndrome, examining BECTS first, then JME, followed by a group of children with focal epilepsies (FE).

4.4.1 BECTS

Figures 4.1 and 4.2 depict the cognitive performances for children with BECTS and controls across all six measures of interest. Significant group main effects were found for tests of verbal learning (.032), coding (.027), and Grooved Pegboard (.006). No Group × Time interactions were found. Children with BECTS showed baseline deficits in confrontation naming (.037) and psychomotor speed/coding (.005), which were no longer significantly different from controls at follow-up. Deficits in verbal learning (.012) and simple motor function (Grooved Pegboard; .015), conversely, were present at both baseline and follow-up.

In terms of imaging results, multiple regions of thinner cortex in BECTS subjects compared to control participants are shown in blue-cyan (Figure 4.7, top row). At baseline, thinner regions include bilateral rostral middle frontal gyrus (.005), right inferior frontal gyrus (.01), left inferior temporal gyrus (.05), left lateral occipital (.001), and right cuneus (.005). Figure 4.8 presents the annualized percent change in cortical thickness over two years for each group. Control subjects (top row) demonstrate widespread cortical thinning (blue-cyan). In contrast, BECTS subjects had only one small region of cortical thinning in the left isthmus cingulate (.020) and one region of cortical thickening (red) in the right precentral gyrus (.025). For the subcortical structures, a main effect of Time

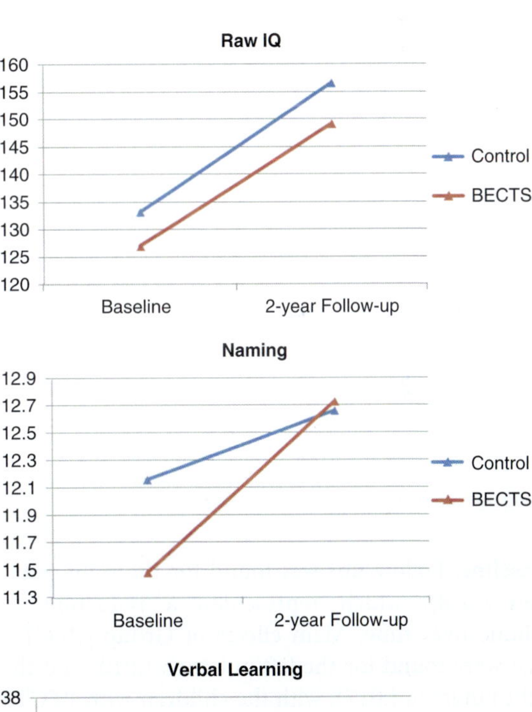

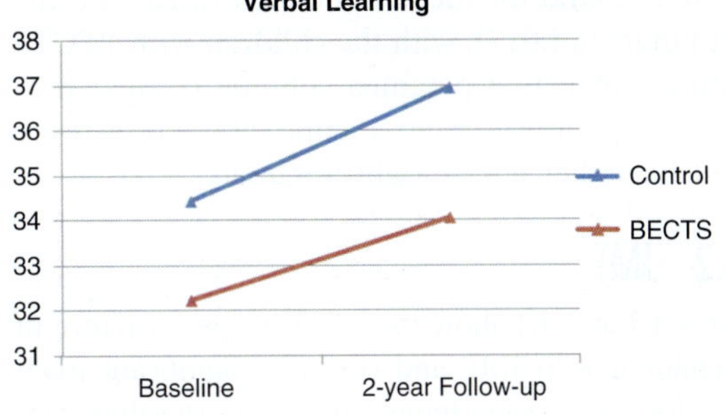

Figure 4.1. Plot of prospective changes in age- and gender-adjusted raw scores in intelligence, confrontation naming, and verbal learning for BECTS patients and healthy controls.

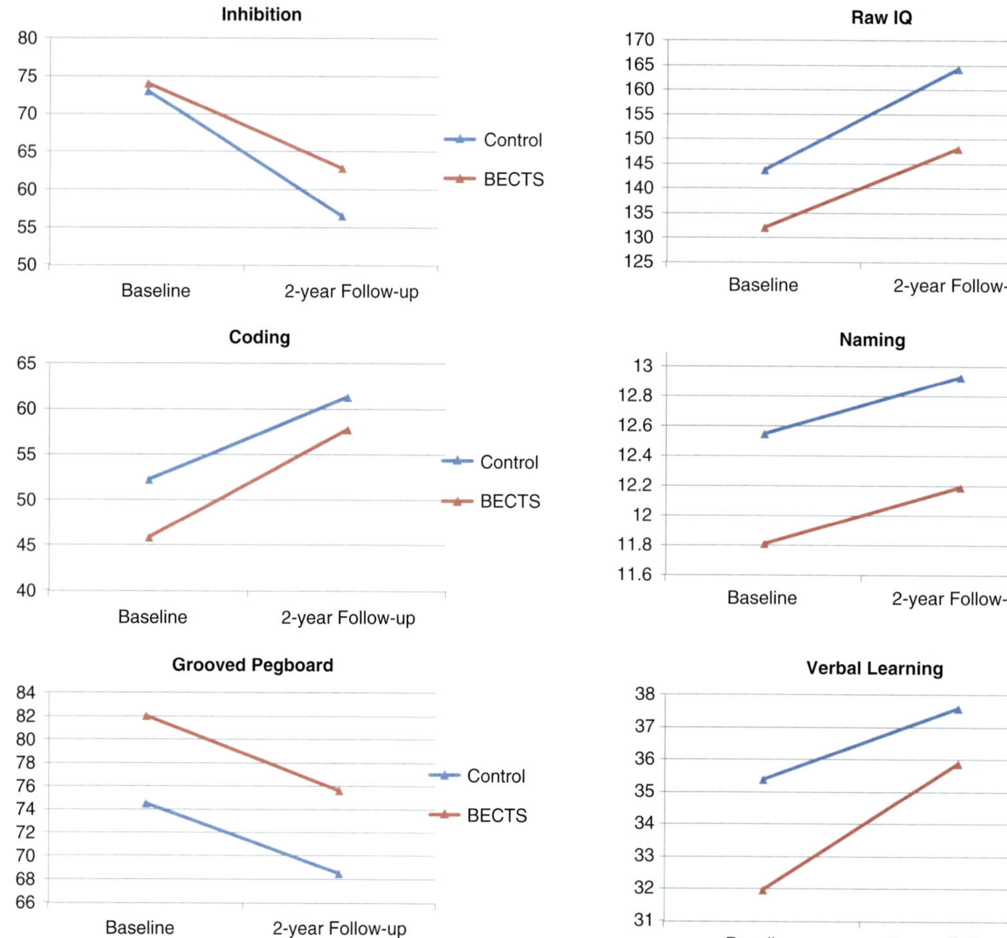

Figure 4.2. Plot of prospective changes in age- and gender-adjusted raw scores in inhibition (time to complete task), coding (number of correct responses), and motor sequencing (time to complete task) for BECTS patients and healthy controls.

Figure 4.3. Plot of prospective changes in age- and gender-adjusted raw scores in intelligence, confrontation naming, and verbal learning for JME patients and healthy controls.

(Baseline, Follow-up) was found for the right putamen (.028), which represented a reduction in volume over time. Main effects of Group (BECTS, HC) were found for the left putamen (.010) and the right putamen (.017), with the children with BECTS exhibiting increased putamen volumes compared to HC. No significant Group × Time interactions were found for either left or right putamen.

4.4.2 JME

Figures 4.3 and 4.4 show the cognitive performance of the children with JME and controls. Significant main effects for Group were found for all tests (*p* values ranging from .001 to .029). There were no Group × Time interactions. JME patients performed significantly worse than controls on tests of inhibitory control (.003) and speeded motor performance (Coding, < .001; Grooved Pegboard, < .001) in addition to general IQ (.015). Performance on these tests continued to differentiate children with JME from healthy controls at follow-up (Inhibition, .046; Coding, .004; Grooved Pegboard, .023; IQ, .004).

Turning to imaging findings, multiple regions of thinner cortex in JME subjects compared to control participants are shown in blue-cyan (Figure 4.7, bottom row). At baseline, thinner regions include left lateral occipital (.0001), left precentral (.015), and right superior parietal (.0008) regions. The cortical thinning over two years so evident in the control population is again absent in the JME subjects (Figure 4.8 top and bottom rows, respectively).

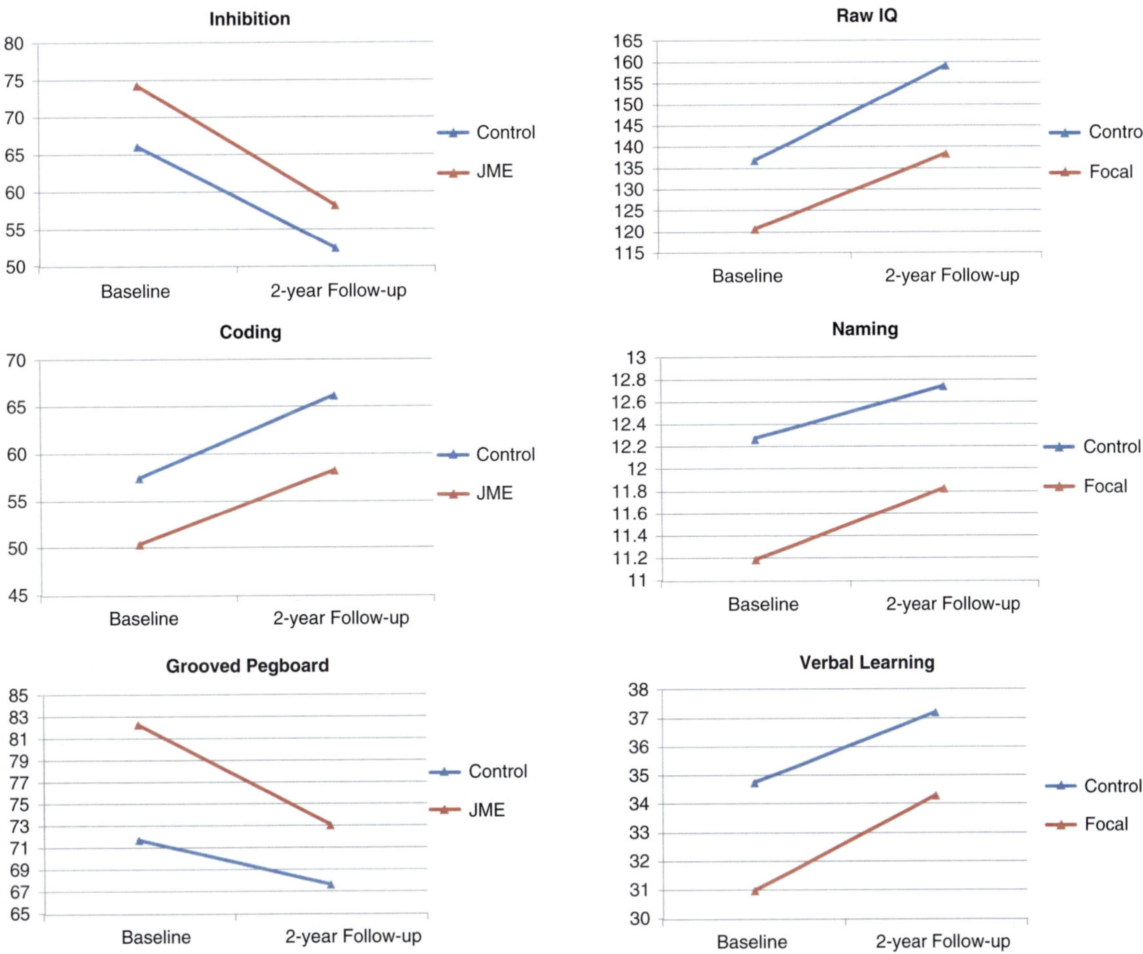

Figure 4.4. Plot of prospective changes in age- and gender-adjusted raw scores in inhibition (time to complete task), coding (number of correct responses) and motor sequencing (time to complete task) for JME patients and healthy controls.

Figure 4.5. Plot of prospective changes in age- and gender-adjusted raw scores in intelligence, confrontation naming, and verbal learning for children with focal epilepsy and healthy controls.

4.4.3 Focal Epilepsies

Figures 4.5 and 4.6 show performance of children with focal (temporal, frontal, benign occipital) epilepsy and controls. Significant main effects for Group were found for all tests (p values < .003) but inhibitory control. There were no Group × Time interactions. Focal patients scored lower than patients on each of these measures at both time points (ps from < .001 to − .049).

In terms of imaging, compared to control subjects, the FE group exhibited thinner cortex in an area extending from inferior temporal to the fusiform gyrus (.009) (Figure 4.7, second row). There was bilateral cortical thinning with FE, albeit more restricted than in the controls. Cortical thinning regions are found in the left lateral occipital (.0001), left middle temporal (.038), right lateral occipital (.0001), right posterior cingulate (.033), and right caudal middle frontal (.0001) regions.

Children with FE also exhibited larger bilateral putamen and right pallidum than controls (p < .05) at baseline. A significant Group × Time interaction was found for the right hippocampus (.011). The control subjects increased right hippocampus volume faster than the FE subjects.

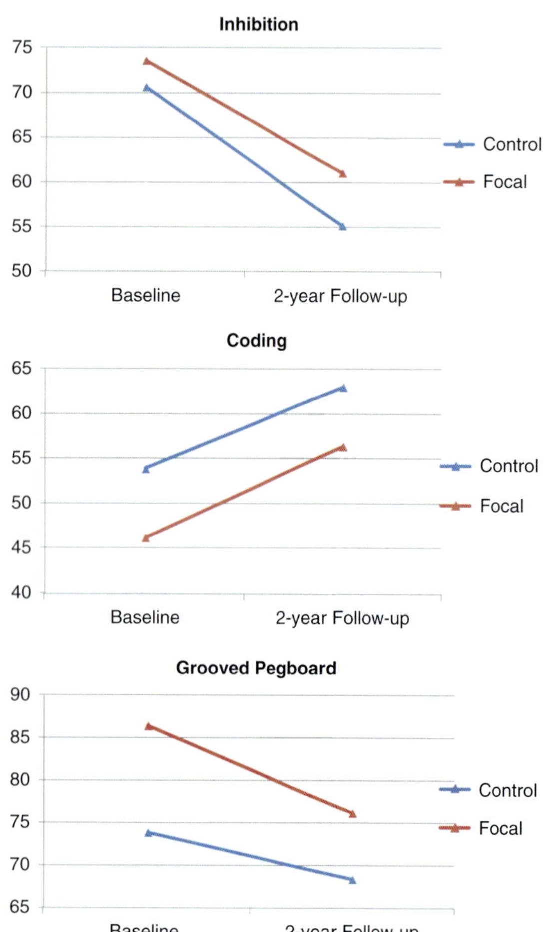

Figure 4.6. Plot of prospective changes in age- and gender-adjusted raw scores in inhibition (time to complete task), coding (number of correct responses), and motor sequencing (time to complete task) for children with focal epilepsy and healthy controls.

4.5 Conclusions

Across these common childhood epilepsy syndromes, cognitive abnormalities are present from the outset/very early in the course of the disorder. While classically there has been considerable interest in the unique cognitive effects of discrete epilepsy syndrome, these data help to appreciate the overlap that may be present and adversely affect cognition across epilepsy syndromes. While there appear to be few syndrome-specific effects, there are differences in the apparent magnitude of impact across syndromes. More importantly, following the children prospectively it is clear that there is little in the way of adverse progressive cognitive effects over time. That is, a common finding is that the differences between healthy controls and children with epilepsy present at baseline, the differences maintained two years later, with both groups often showing normal developmental change over time. Hence the cognitive "offset" noted at baseline is observed two years later to a similar degree, this seen in the context of normal cognitive development (or performance improvement) over the two-year interval. This pattern of results applies of course only to the syndromes examined here. It is surely possible and even likely that children with more severe and intractable epilepsies may show different prospective courses. At a minimum, however, these results imply that cognitive challenges are evident in children with even uncomplicated epilepsies early in the course of the disorder.

The pattern of imaging results is different compared to cognition. Differences in cortical thickness are present to a modest degree in all syndromes presented with an arguably modest degree of syndrome specificity. A primary signal is that there is a difference in the prospective trajectories of brain development in the children with epilepsy compared to normally developing children with the diffuse pattern of "cortical thinning" observed in controls being markedly attenuated in the children with active epilepsy, with some differences in these patterns across syndromes. This was found to be the case across the common childhood epilepsy syndromes examined here, true for JME, BECTS, and FE.

Interesting is the "dissociation" between the trajectories of the imaging and cognitive data. The imaging results in isolation would lead one to suspect that the cognitive trajectories would be quite different in the children with epilepsy, perhaps reflected in relatively arrested cognitive development, or slowed slopes of cognitive development over time, something clearly not the case here.

The long-term implications of these findings remain to be determined. For example, do patterns of cognitive impairment at baseline (or their short-term [2-year] trajectories), or imaging abnormalities, speak to epilepsy course and the potential for epilepsy remission in the future? Do the baseline and/or prospective imaging results predict longer term academic, psychiatric, or social outcomes?

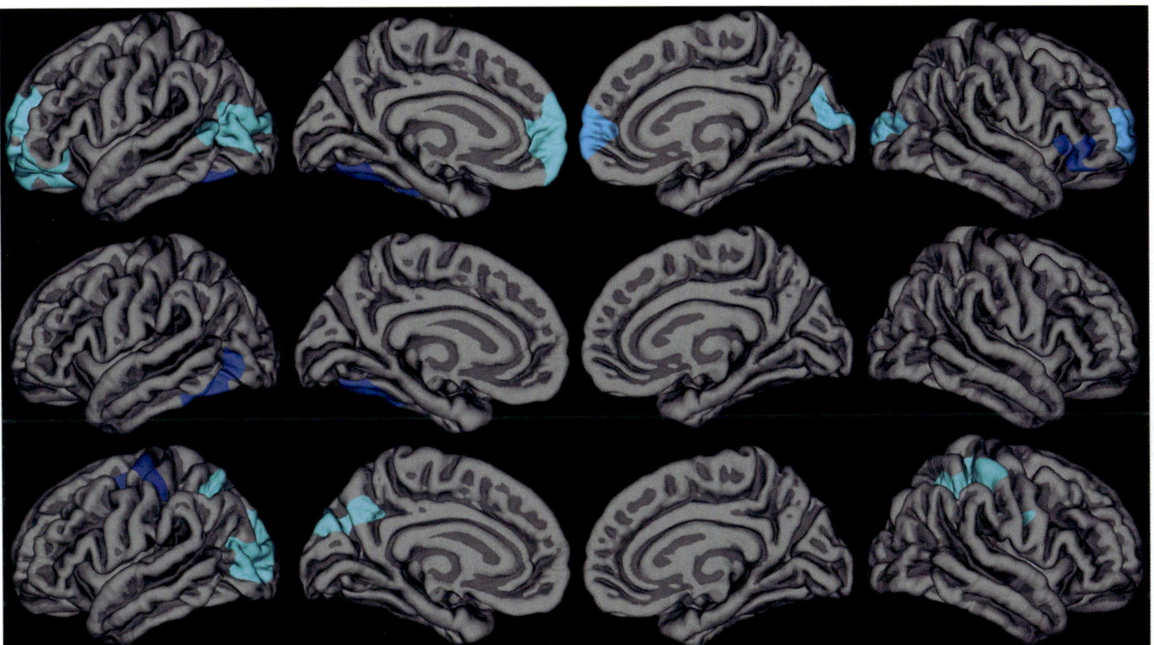

Figure 4.7. Thickness differences at baseline. Areas of thinner cortex compared to controls for the left and right hemispheres are indicated by the blue regions. Regions are found by a whole-brain vertex-wise analysis. Vertex-wise results have been cluster corrected while controlling for age and gender. Top row is BECTS, second row focal, and bottom row for the JME subjects.

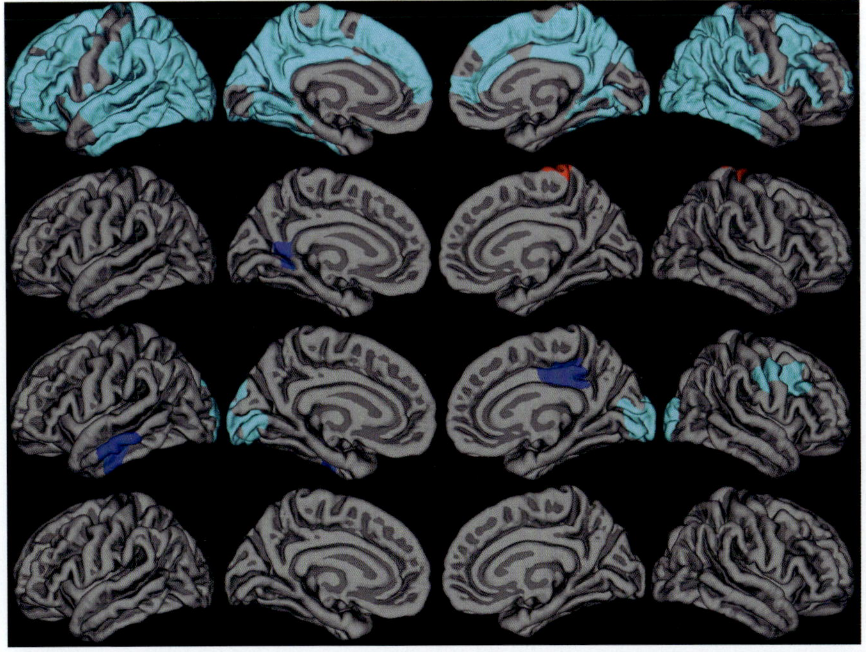

Figure 4.8. Prospective change in cortical thickness for each group. Areas of the percentage change over two years for the left and right hemispheres are indicated by the colored regions. Blue represents cortical thinning; red represents cortical thickening. Regions are found by a whole-brain vertex-wise analysis. Vertex-wise results have been cluster corrected while controlling for age and gender. Top row is healthy controls, second row BECTS, third row focal, and bottom row for the JME subjects.

The mean or average cognitive and imaging findings for the epilepsy syndromes examined here are just that, mean profiles. Within these average profiles is of course considerable individual variability, with some children more adversely affected than others, and some children not adversely affected at all. Of interest is the presence and significance of underlying cognitive or imaging phenotypes that may be present within the epilepsy syndromes and their implications. These and many other issues require careful investigation as we search for biomarkers of the neurobehavioral complications of epilepsy.

References

1. Lin JJ, Mula M, Hermann BP. Uncovering the neurobehavioural comorbidities of epilepsy over the lifespan. *Lancet*. 2012;**380**(9848):1180–92.

2. Hermann BP, Lin JJ, Jones JE, Seidenberg M. The emerging architecture of neuropsychological impairment in epilepsy. *Neurol Clin*. 2009;**27**(4):881–907.

3. Institute of Medicine. *Epilepsy Across the Spectrum: Promoting Health and Understanding*. Washington, DC: Institute of Medicine Committee on the Public Health Dimension of Epilepsies; 2012.

4. Helmstaedter C, Aldenkamp AP, Baker GA, Mazarati A, Ryvlin P, Sankar R. Disentangling the relationship between epilepsy and its behavioral comorbidities—the need for prospective studies in new-onset epilepsies. *Epilepsy Behav*. 2014;**31**:43–7.

5. Pohlmann-Eden B, Aldenkamp A, Baker GA, et al. The relevance of neuropsychiatric symptoms and cognitive problems in new-onset epilepsy—current knowledge and understanding. *Epilepsy Behav*. 2015;**51**:199–209.

6. Hinnell C, Williams J, Metcalfe A, et al. Health status and health-related behaviors in epilepsy compared to other chronic conditions—a national population-based study. *Epilepsia*. 2010;**51**(5):853–61.

7. Jalava M, Sillanpaa M. Concurrent illnesses in adults with childhood-onset epilepsy: a population-based 35-year follow-up study. *Epilepsia*. 1996;**37**(12):1155–63.

8. Kobau R, Zahran H, Thurman DJ, et al. Epilepsy surveillance among adults—19 states. Behavioral Risk Factor Surveillance System 2005. *MMWR Surveill Summ*. 2008;**57**(6):1–20.

9. Ottman R, Lipton RB, Ettinger AB, et al. Comorbidities of epilepsy: results from the Epilepsy Comorbidities and Health (EPIC) survey. *Epilepsia*. 2011;**52**(2):308–15.

10. Tellez-Zenteno JF, Matijevic S, Wiebe S. Somatic comorbidity of epilepsy in the general population in Canada. *Epilepsia*. 2005;**46**(12):1955–62.

11. Tellez-Zenteno JF, Patten SB, Jette N, Williams J, Wiebe S. Psychiatric comorbidity in epilepsy: a population-based analysis. *Epilepsia*. 2007;**48**(12):2336–44.

12. Begley CE, Famulari M, Annegers JF, et al. The cost of epilepsy in the United States: an estimate from population-based clinical and survey data. *Epilepsia*. 2000;**41**(3):342–51.

13. Jennum P, Gyllenborg J, Kjellberg J. The social and economic consequences of epilepsy: a controlled national study. *Epilepsia*. 2011;**52**(5):949–56.

14. Dikmen S, Matthews CG, Harley JP. The effect of early versus late onset of major motor epilepsy upon cognitive-intellectual performance. *Epilepsia*. 1975;**16**(1):73–81.

15. Dikmen S, Matthews CG, Harley JP. Effect of early versus late onset of major motor epilepsy on cognitive-intellectual performance: further considerations. *Epilepsia*. 1977;**18**(1):31–6.

16. Dodrill CB, Matthews CG. The role of neuropsychology in the assessment and treatment of persons with epilepsy. *Am Psychol*. 1992;**47**(9):1139–42.

17. Hermann B, Seidenberg M, Bell B, et al. The neurodevelopmental impact of childhood-onset temporal lobe epilepsy on brain structure and function. *Epilepsia*. 2002;**43**(9):1062–71.

18. Kaaden S, Helmstaedter C. Age at onset of epilepsy as a determinant of intellectual impairment in temporal lobe epilepsy. *Epilepsy Behav*. 2009;**15**(2):213–7.

19. Lennox WG. *Epilepsy and Related Disorders*. Vol. 2. Boston: Little, Brown; 1960.

20. Cormack F, Cross JH, Isaacs E, et al. The development of intellectual abilities in pediatric temporal lobe epilepsy. *Epilepsia*. 2007;**48**(1):201–4.

21. O'Leary DS, Seidenberg M, Berent S, Boll TJ. Effects of age of onset of tonic-clonic seizures on neuropsychological performance in children. *Epilepsia*. 1981;**22**(2):197–204.

22. Schoenfeld J, Seidenberg M, Woodard A, et al. Neuropsychological and behavioral status of children with complex partial seizures. *Dev Med Child Neurol*. 1999;**41**(11):724–31.

23. Theodore WH, DeCarli C, Gaillard WD. Total cerebral volume is reduced in patients with localization-related

epilepsy and a history of complex febrile seizures. *Arch Neurol*. 2003;**60**(2):250–2.

24. Hermann B, Hansen R, Seidenberg M, Magnotta V, O'Leary D. Neurodevelopmental vulnerability of the corpus callosum to childhood onset localization-related epilepsy. *NeuroImage*. 2003;**18**(2): 284–92.

25. Kaaden S, Quesada CM, Urbach H, et al. Neurodevelopmental disruption in early-onset temporal lobe epilepsy: evidence from a voxel-based morphometry study. *Epilepsy Behav*. 2011;**20**(4): 694–9.

26. Riley JD, Franklin DL, Choi V, et al. Altered white matter integrity in temporal lobe epilepsy: association with cognitive and clinical profiles. *Epilepsia*. 2010;**51** (4):536–45.

27. Weber B, Luders E, Faber J, et al. Distinct regional atrophy in the corpus callosum of patients with temporal lobe epilepsy. *Brain*. 2007;**130**:3149–54.

28. Camfield C, Camfield P, Smith B, Gordon K, Dooley J. Biologic factors as predictors of social outcome of epilepsy in intellectually normal children: a population-based study. *J Pediatrics*. 1993;**122**(6): 869–73.

29. Harrison RM, Taylor DC. Childhood seizures: a 25-year follow up. Social and medical prognosis. *Lancet*. 1976;**1**(7966):948–51.

30. Lindsay J, Ounsted C, Richards P. Long-term outcome in children with temporal lobe seizures. I: Social outcome and childhood factors. *Dev Med Child Neurol*. 1979;**21**(3):285–98.

31. Lindsay J, Ounsted C, Richards P. Long-term outcome in children with temporal lobe seizures. II: Marriage, parenthood and sexual indifference. *Dev Med Child Neurol*. 1979;**21**(4):433–40.

32. Lindsay J, Ounsted C, Richards P. Long-term outcome in children with temporal lobe seizures. III: Psychiatric aspects in childhood and adult life. *Dev Med Child Neurol*. 1979;**21**(5):630–6.

33. Micallef S, Spooner CG, Harvey AS, Wrennall JA, Wilson SJ. Psychological outcome profiles in childhood-onset temporal lobe epilepsy. *Epilepsia*. 2010;**51**(10):2066–73.

34. Wakamoto H, Nagao H, Hayashi M, Morimoto T. Long-term medical, educational, and social prognoses of childhood-onset epilepsy: a population-based study in a rural district of Japan. *Brain Dev*. 2000;**22**(4): 246–55.

35. Geerts A, Brouwer O, van Donselaar C, et al. Health perception and socioeconomic status following childhood-onset epilepsy: the Dutch study of epilepsy in childhood. *Epilepsia*. 2011;**52**(12):2192–2202.

36. Jalava M, Sillanpaa M, Camfield C, Camfield P. Social adjustment and competence 35 years after onset of childhood epilepsy: a prospective controlled study. *Epilepsia*. 1997;**38**(6):708–15.

37. Kokkonen J, Kokkonen ER, Saukkonen AL, Pennanen P. Psychosocial outcome of young adults with epilepsy in childhood. *J Neurol Neurosurg Psychiatry*. 1997;**62**(3):265–8.

38. Koponen A, Seppala U, Eriksson K, et al. Social functioning and psychological well-being of 347 young adults with epilepsy only—population-based, controlled study from Finland. *Epilepsia*. 2007;**48**(5): 907–12.

39. Shackleton DP, Kasteleijn-Nolst Trenite DG, de Craen AJ, Vandenbroucke JP, Westendorp RG. Living with epilepsy: long-term prognosis and psychosocial outcomes. *Neurology*. 2003;**61**(1):64–70.

40. Sillanpaa M, Jalava M, Kaleva O, Shinnar S. Long-term prognosis of seizures with onset in childhood. *N Engl J Med*. 1998;**338**(24):1715–22.

41. Wirrell EC, Camfield CS, Camfield PR, Dooley JM, Gordon KE, Smith B. Long-term psychosocial outcome in typical absence epilepsy. Sometimes a wolf in sheeps' clothing. *Arch Pediatr Adolesc Med*. 1997;**151**(2):152–8.

42. Britten N, Morgan K, Fenwick PB, Britten H. Epilepsy and handicap from birth to age 36. *Dev Med Child Neurol*. 1986;**28**(6):719–28.

43. Chin RF, Cumberland PM, Pujar SS, Peckham C, Ross EM, Scott RC. Outcomes of childhood epilepsy at age 33 years: a population-based birth-cohort study. *Epilepsia*. 2011;**52**(8):1513–21.

44. Cooper JE. Epilepsy in a longitudinal survey of 5,000 children. *Br Med J*. 1965;**1**(5441):1020–2.

45. Ross EM, Peckham CS. School children with epilepsy. In: Parsonage M, Grant RHE, Craig AG, eds. *Advances in Epileptology*. New York: Rave Press; 1983: 215–20.

46. Ross EM, Peckham CS, West PB, Butler NR. Epilepsy in childhood: findings from the National Child Development Study. *Br Med J*. 1980;**280**(6209): 207–10.

47. Gleissner U, Sassen R, Lendt M, Clusmann H, Elger CE, Helmstaedter C. Pre- and postoperative verbal memory in pediatric patients with temporal lobe epilepsy. *Epilepsy Res*. 2002;**51**(3):287–96.

48. Lassonde M, Sauerwein HC, Jambaque I, Smith ML, Helmstaedter C. Neuropsychology of childhood

epilepsy: pre- and postsurgical assessment. *Epileptic Disord.* 2000;**2**(1):3–13.

49. MacAllister WS, Schaffer SG. Neuropsychological deficits in childhood epilepsy syndromes. *Neuropsychol Rev.* 2007;**17**(4):427–44.

50. Bourgeois BF, Prensky AL, Palkes HS, Talent BK, Busch SG. Intelligence in epilepsy: a prospective study in children. *Ann Neurol.* 1983;**14**(4):438–44.

51. Kolk A, Beilmann A, Tomberg T, Napa A, Talvik T. Neurocognitive development of children with congenital unilateral brain lesion and epilepsy. *Brain Dev.* 2001;**23**(2):88–96.

52. Stores G. A clinical approach to poorly controlled seizures in children. *Br J Hosp Med.* 1992;**48**(2):93–8.

53. Williams J, Bates S, Griebel ML, et al. Does short-term antiepileptic drug treatment in children result in cognitive or behavioral changes? *Epilepsia.* 1998;**39**(10):1064–9.

54. Oostrom KJ, Smeets-Schouten A, Kruitwagen CL, et al. Not only a matter of epilepsy: early problems of cognition and behavior in children with "epilepsy only"—a prospective, longitudinal, controlled study starting at diagnosis. *Pediatrics.* 2003;**112**(6 pt 1):1338–44.

55. Fastenau PS, Johnson CS, Perkins SM, et al. Neuropsychological status at seizure onset in children: risk factors for early cognitive deficits. *Neurology.* 2009;**73**(7):526–34.

56. Dunn DW, Harezlak J, Ambrosius WT, Austin JK, Hale B. Teacher assessment of behaviour in children with new-onset seizures. *Seizure.* 2002;**11**(3):169–75.

57. McNelis AM, Dunn DW, Johnson CS, Austin JK, Perkins SM. Academic performance in children with new-onset seizures and asthma: a prospective study. *Epilepsy Behav.* 2007;**10**(2):311–8.

58. Vintan M, Palade S, Cristea A, Benga I, Muresanu D. A neuropsychological assessment, using computerized battery tests (CANTAB) in children with benign rolandic epilepsy before AED therapy. *J Med Life.* 2012;**5**(1):114–9.

59. Lee JH, Kim SE, Park CH, Yoo JH, Lee HW. Gray and white matter volumes and cognitive dysfunction in drug-naive newly diagnosed pediatric epilepsy. *Biomed Res Int.* 2015;**2015**:923861.

60. Filippini M, Ardu E, Stefanelli S, Boni A, Gobbi G, Benso F. Neuropsychological profile in new-onset benign epilepsy with centrotemporal spikes (BECTS): focusing on executive functions. *Epilepsy Behav.* 2016;**54**:71–9.

61. Cheng D, Yan X, Gao Z, Xu K, Zhou X, Chen Q. Neurocognitive profiles in childhood absence epilepsy. *J Child Neurol.* 2017;**32**(1):46–52.

62. Matricardi S, Deleo F, Ragona F, et al. Neuropsychological profiles and outcomes in children with new onset frontal lobe epilepsy. *Epilepsy Behav.* 2016;**55**:79–83.

63. Vannest J, Tenney JR, Altaye M, et al. Impact of frequency and lateralization of interictal discharges on neuropsychological and fine motor status in children with benign epilepsy with centrotemporal spikes. *Epilepsia.* 2016;**57**(8):e161–7.

64. Vannest J, Maloney TC, Tenney JR, et al. Changes in functional organization and functional connectivity during story listening in children with benign childhood epilepsy with centro-temporal spikes. *Brain Lang.* 2017. doi:10.1016/j.bandl.2017.01.009.

65. Wang G, Dai ZY, Song W, et al. Grey matter anomalies in drug-naive childhood absence epilepsy: A voxel-based morphometry study with MRI at 3.0 T. *Epilepsy Res.* 2016;**124**:63–6.

66. Ekmekci B, Bulut HT, Gumustas F, Yildirim A, Kustepe A. The relationship between white matter abnormalities and cognitive functions in new-onset juvenile myoclonic epilepsy. *Epilepsy Behav.* 2016;**62**:166–70.

67. Cheng D, Yan X, Gao Z, Xu K, Chen Q. Attention contributes to arithmetic deficits in new-onset childhood absence epilepsy. *Front Psychiatry.* 2017;**8**:166.

68. Hermann B, Jones J, Sheth R, Dow C, Koehn M, Seidenberg M. Children with new-onset epilepsy: neuropsychological status and brain structure. *Brain.* 2006;**129**(pt 10):2609–19.

69. Pulsipher DT, Seidenberg M, Guidotti L, et al. Thalamofrontal circuitry and executive dysfunction in recent-onset juvenile myoclonic epilepsy. *Epilepsia.* 2009;**50**(5):1210–9.

70. Hutchinson E, Pulsipher D, Dabbs K, et al. Children with new-onset epilepsy exhibit diffusion abnormalities in cerebral white matter in the absence of volumetric differences. *Epilepsy Res.* 2010;**88**(2–3),208–14.

71. Jackson DC, Irwin W, Dabbs K, et al. Ventricular enlargement in new-onset pediatric epilepsies. *Epilepsia.* 2011;**52**(12):2225–32.

72. Widjaja E, Zarei Mahmoodabadi S, Go C, et al. Reduced cortical thickness in children with new-onset seizures. *AJNR Am J Neuroradiol.* 2012;**33**(4):673–7.

73. Yang T, Guo Z, Luo C, et al. White matter impairment in the basal ganglia-thalamocortical circuit of drug-naive childhood absence epilepsy. *Epilepsy Res.* 2012;**99**(3):267–73.

74. Widjaja E, Kis A, Go C, Raybaud C, Snead OC, Smith ML. Abnormal white matter on diffusion tensor

imaging in children with new-onset seizures. *Epilepsy Res.* 2013;**104**(1–2):105–11.

75. Kim EH, Yum MS, Shim WH, Yoon HK, Lee YJ, Ko TS. Structural abnormalities in benign childhood epilepsy with centrotemporal spikes (BCECTS). *Seizure.* 2015;**27**:40–6.

76. Luo C, Zhang Y, Cao W, et al. Altered structural and functional feature of striato-cortical circuit in benign epilepsy with centrotemporal spikes. *Int J Neural Syst.* 2015;**25**(6):1550027.

77. Perani S, Tierney TM, Centeno M, et al. Thalamic volume reduction in drug-naive patients with new-onset genetic generalized epilepsy. *Epilepsia.* 2018;**59**(1):226–34.

78. Wilkinson GS. *Wide Range Achievement Test: Manual.* Wilmington, DE: Wide Range, Inc.; 1993.

79. Wechsler D. *Wechsler Abbreviated Scale of Intelligence.* San Antonio, TX: Psychological Corporation; 1999.

80. Kaplan E, Goodglass, H, Weintraub, S. *Boston Naming Test.* Philadelphia: Lea & Febiger; 1983.

81. Williams KT. *Expressive Vocabulary Test.* Circle Pines, MN: American Guidance Service; 1997.

82. Dunn L, Dunn L, Williams KT. *Peabody Picture Vocabulary Test.* Circle Pines, MN: American Guidance Service; 1997.

83. Delis DC, Kaplan E, Kramer JH. *The Delis-Kaplan Executive Function System.* San Antonio, TX: Psychological Corporation; 2001.

84. Cohen MJ. *Children's Memory Scale.* San Antonio, TX: Psychological Corporation; 1997.

85. Conners CK. *The Connors' Continuous Performance Test.* Toronto: Multi-Heath Systems; 1995.

86. Lafayette Instrument Company. *Grooved Peg Board Test.* Lafayette, IN; 2002.

87. Wechsler D. *Wechsler Intelligence Scale for Children.* San Antonio, TX: Psychological Corporation; 1991.

88. Dabbs K, Jones JE, Jackson DC, Seidenberg M, Hermann BP. Patterns of cortical thickness and the Child Behavior Checklist in childhood epilepsy. *Epilepsy Behav.* 2013;**29**(1):198–204.

Chapter 5

Imaging Genetics for Benign Mesial Temporal Lobe Epilepsy

Antonio Gambardella and Angelo Labate

5.1 Introduction

Temporal lobe epilepsy (TLE) represents the most common type of focal epilepsy in adulthood.[1] Based on seizure semiology, TLE is divided into two main broad categories, the more common form with mesial temporal lobe symptoms, and the second with lateral temporal lobe symptoms. Traditionally, mesial TLE with hippocampal sclerosis (MTLE-HS) has been viewed as acquired, usually drug-resistant epilepsy associated with a history of prolonged febrile convulsions and usually drug-resistant epilepsy, which often requires surgical treatment.[2] According to literature, MTLE with MTLE-HS carries a worse prognosis than other forms of epilepsy, reinforced by intensive investigation of refractory epilepsy patients for consideration of surgery.[3] Indeed, long-term outcome studies from tertiary care centers provided evidence that only 10% of MTLE-HS patients were seizure-free, demonstrating the importance of HS as a major prognostic factor.[4] This finding concurs with another study showing that with 10 years or more of follow-up about 50% of children with new-onset TLE became seizure-controlled with medications if the magnetic resonance imaging (MRI) was normal, and of those with lesions on neuroimaging none was likely to be seizure-free.[5]

Over the past two decades, studies from nonsurgical series of MTLE patients confirmed the existence of a benign form of MTLE (*bMTLE*), which is characterized by at least 24 months of seizure freedom with or without antiepileptic medication.[6,7] This entity of *bMTLE* appears to be a common and often unrecognized clinical entity, in which genetic factors play a major etiopathogenetic role.[8] Moreover, the identification of MRI evidence of *Hs* in many cases with *bMTLE* has shown that *Hs* itself does not always mean intractable epilepsy.[9] Very recently, however, a prospective longitudinal cohort study illustrated that *bMTLE* patients with radiologic evidence of HS since recruitment carried a three times higher likelihood of becoming refractory later in life than those without HS.[10]

Additional risk factors that contributed to refractoriness at long-term follow-up were earlier age of onset, longer duration of epilepsy, and antecedents with febrile seizures.[10] These findings indicate that *bMTLE* is a syndrome representing the mildest form of the wide spectrum of MTLE, which includes both drug-responsive and refractory MTLE along a biological continuum.[10] Most importantly, *bMTLE* does not symbolize a prerefractory stage of MTLE.

This group of *bMTLE* patients offers a superb opportunity not only to determine the biologic substrates underlying MTLE, but also to identify biomarkers for this epilepsy syndrome. Predictive biomarkers are becoming increasingly important tools in drug development and clinical research, and represent the new frontier for researchers in epilepsy to definitively improve the management of people with epilepsy.[11] It has been recently claimed that a first step to identify potential biomarkers for pharmacoresistance may be to classify several well-defined epilepsy syndromes that are associated with drug resistance but in which there are also patients who are well controlled, which this cohort of *bMTLE* suitably symbolizes.[12] This would, in turn, allow the development of biomarkers that can determine future drug response in new-onset cases, thus helping specialists to guide more precise decision making.

5.2 Genetics of *bMTLE*

Familial MTLE was first recognized in twin families, where its strong genetic basis was demonstrated by high concordance in monozygotic twins.[13] It is noteworthy that the clinical, EEG, and imaging features, as well as seizure outcome of familial MTLE, where cases were selected because of a family history, are very similar to those described in unrelated cases with *bMTLE*.[6,8,9] Indeed, the clinical features of familial MTLE included onset typically during the teenage years or early adult life, with no antecedent factors for epilepsy.[13,14] Febrile seizures occurred in only 2.7% of family members of the familial MTLE subjects, a frequency similar to that reported (3–4%) in the general population.[14] The nature of the aura suggested mesial temporal origin, and déjà vu and jamais vu appeared to be overrepresented in this group of patients.[14] Focal seizures with impaired awareness and especially focal to bilateral tonic-clonic seizures were infrequent, and the EEG recordings often showed no epileptiform abnormalities. Similar to sporadic *bMTLE*, the mild nature of this hereditary epilepsy may well explain why it could so often be undiagnosed. Similar families were also reported by other authors,[15] with an overrepresentation of migraine in some of them. Also, in some of these patients with familial MTLE *Hs* was later found on MRI. Similar to sporadic MTLE, there may be a remarkable intra- and interfamilial phenotypic heterogeneity with respect to history of febrile seizures, severity of the epilepsy, and presence of *Hs*.[16]

All these clinical similarities between sporadic and familial *bMTLE* are in line with the overwhelming evidence suggesting that familial MTLE is mainly a polygenic disorder with a high heritability but absence of dominant segregation that shares the genetics determinants with apparently sporadic *bMTLE* beginning in adolescence or adulthood.[7,14] Such a proposal of common polygenic and rare dominant inheritance in MTLE predicts the allelic structure of *bMTLE* to be similar to that of idiopathic generalized epilepsies.[14] Therefore, the common approaches of chromosomal linkage mapping in affected families and further in identification of candidate genes have so far not been doing well for heritable diseases like *bMTLE* that involves multiple genes and their interactions, and also the gene-environment interactions.

Indeed, no genes responsible for familial MTLE have been so far identified, and only one genetic locus has been mapped to chromosome 4q in a family with evidence of autosomal dominant inheritance.[17] Occasionally, familial MTLE can occur in the context of familial focal epilepsy with variable foci related to mutations of the mTORC1-repressor DEPDC5 (DEP domain-containing protein 5) gene.[18] It can also coexist in the context of families with dominant inheritance of febrile seizure syndromes including genetic epilepsy with febrile seizures plus. These are genetically heterogeneous with a variety of loci being reported and, in some families, the sodium channel genes SCN1A and SCN1B have been implicated.[19,20]

Very rarely, familial MTLE may also be the presenting feature of choreoacanthocytosis and may delay its diagnosis.[21] Although some polymorphisms (prodynorphin gene, Apoε gene, GABA B receptor gene, prion protein gene) have been described in MTLE patients,[8] there are no such data deriving from kindreds with familial MTLE. No families with MTLE were found to have an LGI-1 mutation, even though some had affected family members with both mesial symptoms and auditory features. This further supports the notion that familial MTLE and familial lateral TLE constitute separate genetic syndromes.

5.3 *bMTLE* and Imaging Genetics

Identifying genes in complex disorders is a major current challenge across all diseases including MTLE, and the major strategies being used to explore their molecular basis are genome-wide association studies for common genetic variants or deep sequencing for rare variants.[14] These types of approach are likely to be complemented by endophenotype studies, including brain imaging, which combines genetic assessment with multimodal neuroimaging to discover neural systems linked to genetic abnormalities or variation. The concept of endophenotype, first introduced in the psychiatric genetic literature, is extremely attractive.[22] Endophenotypes are relatively simple, stable biologic phenomena that can be measured objectively and are genetically determined. In theory, although the disorder itself will usually result from a mixture of multiple genetic and nongenetic abnormalities, the endophenotype, as a simpler biologic phenomenon, will have a more straightforward genetic and pathophysiological basis.[22]

This method may be constructively applied to all epilepsies, especially MTLE. Epilepsy, indeed, encompasses a spectrum of clinical phenotypes, defined by EEG, seizure type, and brain imaging criteria. Reducing such complex phenotypes into components, whether they are neurophysiological, biochemical, endocrine, neuroanatomical, cognitive or neuropsychological, is described as an *endophenotype* strategy or approach.[22,23] Since there is an imperfect relationship between genes and epilepsy phenotypes so that different combinations of genes (and resultant changes in neurobiology) may contribute to any complex epilepsy, it may be possible to assay the result of aberrant genes through more biologically "simple" approaches. In theory, although the disorder itself will usually result from a mixture of multiple genetic and nongenetic abnormalities, the endophenotype, as a simpler biologic phenomenon, will have a more straightforward genetic and pathophysiological basis.

There is no reliable method to select behavioral or biological markers for studies aimed at evaluating endophenotypes. Generally, a marker should be selected with respect to several criteria including feasibility and reliability of its measurement, high heritability, availability of knowledge on its underlying neurobiology/genetics, and possible relevance for the disorder under study.[23] To address that endophenotype identified in probands is highly heritable, a study design that compares prevalence rates of a marker between subjects at familial risk of the disease and those without the risk is a powerful strategy.[23] In this way, to function as an endophenotype, a marker must be found in the unaffected relatives at a higher rate than in the general population.

In the field of epilepsy, potential endophenotypes may include electroencephalographic features and abnormal responses to provocation tests such as the intermittent photic stimulation, structural, or functional imaging findings. Much work had been done on the genetics of epilepsy-related endophenotypes in the 1980s and 1990s, but success in identifying relevant disease genes had been very unsatisfactory. Novel impetus for epilepsy-related endophenotypes is now coming from advanced imaging studies including functional MRI and connectivity studies, and the endophenotypes used in this context are usually called intermediate phenotypes. The intermediate phenotype strategy is based on the assumption that gene effects at the level of the brain are a more direct effect of genetic variation than is complex behavior, and will show association in carriers of risk alleles even if the carriers show no clinical diagnostic characteristics.[24] There are two concepts of imaging genetics as an intermediate phenotype that may be applied to MTLE.[24,25] One model is that imaging genetics may allow the discovery of risk genes for MTLE; the second assumption is that imaging genetics may allow discovering reproducible effects of single gene variants on simple brain measures,[25] for example, the hippocampus volume or other subcortical regions implicated in MTLE. These two concepts have different hypotheses, but the methods to test the hypotheses are the same and include association analyses between neuroimaging data and genetic variations.[24,25]

5.3.1 Imaging Genetics of Hippocampal Morphology in MTLE

The mesolimbic system, especially the hippocampus, is well known for being involved in the pathogenesis of *bMTLE*. Notably, size and morphology of the hippocampus shows considerable variation in the general population and is estimated to be sufficiently heritable.[26] Therefore, this structure may be an attractive candidate for genetic correlation based on the approach of an intermediate phenotype/endophenotype. Accordingly, results from neuroimaging studies have illustrated that genetic factors may contribute to altered hippocampal morphology in MTLE and may thus serve as promising phenotypes in the search for TLE genes. Indeed, a recent advanced neuroimaging study has shown that first-degree relatives of MTLE-HS patients, who have never had seizures, have smaller and more asymmetric hippocampi compared with age- and sex-matched controls.[27] Conversely, none of them had *Hs* on careful qualitative and quantitative analyses, consistent with previously published data showing discordance for *Hs* in monozygotic twins.[28] The latter observation would imply that the heritability of *Hs* itself may be very low. Important, the results of the advanced neuroimaging study are largely compatible with observations from other two studies of families with MTLE. A study of 23 individuals from two German families with MTLE showed that many members had febrile seizures, and all of them had hippocampal atrophy, as did 6 of 10 asymptomatic individuals.[29] Similarly, in a series of 11 Brazilian kindreds with familial MTLE, about one-third of asymptomatic relatives had evidence of hippocampal volume loss.[30] Moreover, small hippocampus with incomplete rotation and deformation was also shown to be related to haploinsufficiency of the SOX2 gene.[31] Even though no evidence emerged that SOX2 mutation or variation could contribute commonly to MTLE or *Hs* in humans,[31] SOX2 dysfunction might also be an additional candidate mechanism for mesial temporal abnormalities associated with MTLE.

Taken together, these results support a nontrivial association between morphometric brain abnormalities of the anterior-limbic neural substrate and family history of MTLE, which may become a potential candidate as a morphological endophenotype of MTLE (Figure 5.1). Since there is good evidence that hippocampal shape and

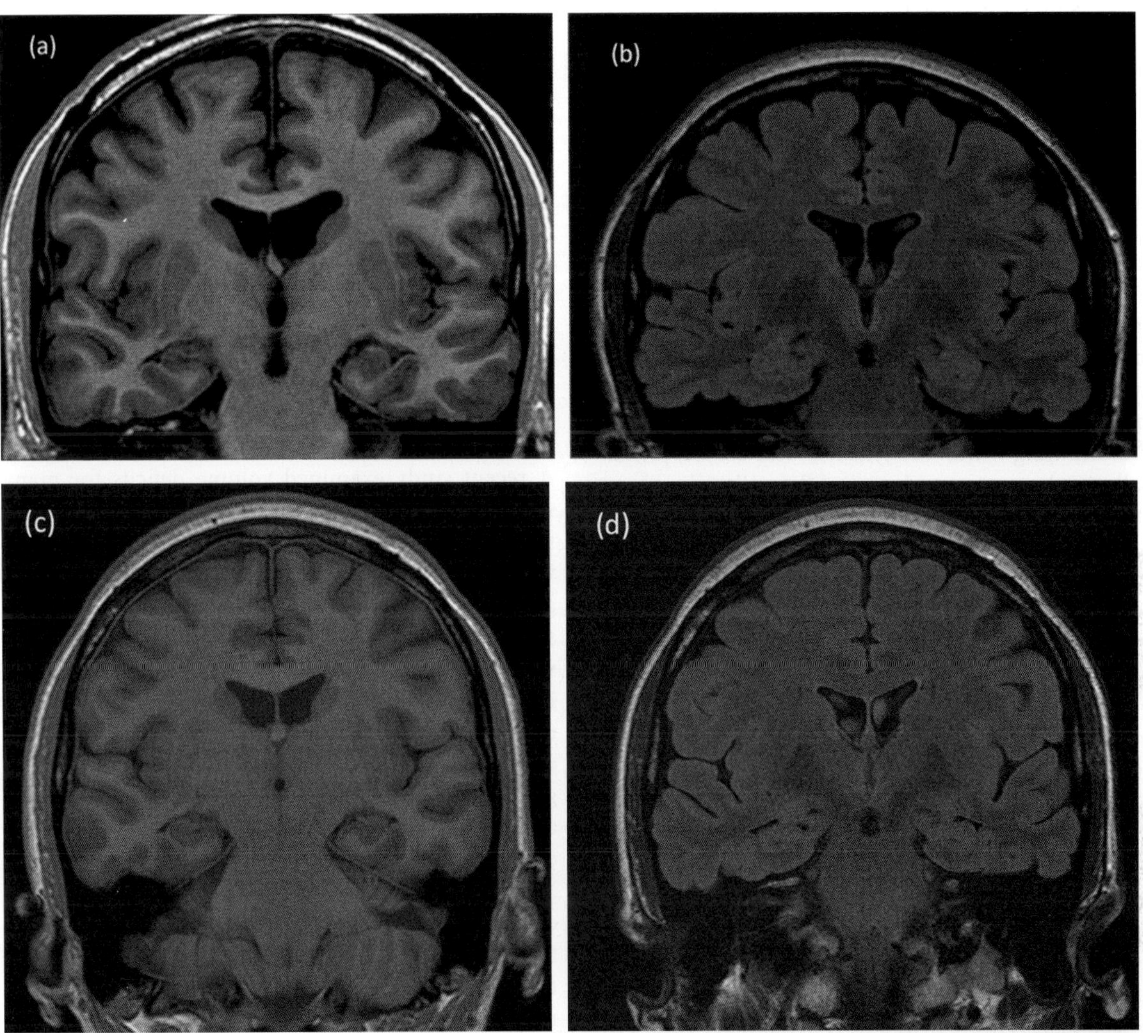

Figure 5.1. (a, b) Brain MRI of a 31-year-old woman with *bMTLE*. (a) Coronal T1-weighted image shows left deeper and vertical collateral sulcus together with the reduction of the horizontal part of the parahippocampal gyrus. (b) Coronal FLAIR image shows left hippocampal hypointensity. (c, d) Brain MRI of the woman's unaffected 59-year-old mother. (c) Coronal T1-weighted image shows signs of left hippocampal malrotation with deeper and vertical collateral sulcus together with the reduction of the horizontal part of the parahippocampal gyrus. (d) Coronal FLAIR image shows a regular signal intensity of the hippocampal formation bilaterally.

positioning abnormalities are more prevalent in MTLE patients,[32] it is tantalizing to hypothesize that, in either sporadic or familial *bMTLE*, patients may carry subtle morphologic abnormalities of the hippocampus with susceptibility for developing the disease including *Hs* in the presence of environmental or additional genetic factors, resembling complex inheritance. In this way, it is reasonable to hypothesize that hippocampal shape and positioning abnormalities may be used to help identify susceptibility alleles for *Hs*. Obviously, there is the need to determine the relationship between MRI and

histological findings in order to clarify the spectrum of hippocampal developmental abnormalities and their significance in the genesis of MTLE or *Hs*.

5.3.2 Imaging Genetics of Cortical and Subcortical Structures in MTLE

Several advanced morphometric and functional MRI studies have clearly demonstrated that extrahippocampal and extratemporal regions are involved in the epileptic network underlying MTLE, as well as that alterations of brain tissue integrity in gray and

51

white matter are found in patients with MTLE regardless of the presence of *Hs*. Cross-sectional and longitudinal structural MRI studies in drug-resistant MTLE have shown correlations with disease duration in both neocortical and mesiotemporal regions, in contrast to work in *bMTLE*, likely representing seizure-induced damage.[33] Conversely, there seems to be a resemblance of findings in benign and refractory cases with regard to cortical atrophy in sensorimotor regions. Indeed, the involvement of neocortical, especially sensorimotor, regions remote from the seizure focus was initially reported in refractory MTLE groups regardless of the presence of *Hs*.[34] Afterward, an advanced imaging study illustrated significant cortical thinning within the sensorimotor cortex, principally in the postcentral gyrus bilaterally in patients with *bMTLE* regardless of MRI signs of *Hs* as well as the side of *Hs*.[35] Thus, the sensorimotor cortex may be involved in the epileptogenic network underlying MTLE, as suggested by experimental data demonstrating that the hippocampus is associated with sensorimotor processes, especially innate processes involving control of motor responses to sensory stimuli, and that hippocampal mechanisms can directly influence locomotor activity.[36]

Likewise, a voxel-based morphometry (VBM) MRI study in patients with *bMTLE* demonstrated structural abnormalities located in the hippocampus and thalamus that were strikingly similar to previous findings in VBM studies of patients with refractory MTLE.[37] The severity and extent of volume loss seen on brain VBM was found to parallel the volume loss detected on routine MRI study, being more severe in patients with routine MRI evidence of *Hs*. In an updated series of 85 sporadic *bMTLE* patients with normal MRI, VBM analysis confirmed a reduction in gray matter volume in the thalamus, bilaterally (Figure 5.2). Since these patients with *bMTLE* had a very mild epileptic disorder with no seizures at long-term follow-up, it is tempting to hypothesize that the thalamus is primarily caught up in the epileptogenic network underlying the disease. This view is supported by the observation of no significant correlation between either the duration of the epilepsy or the age at onset and extent of brain damage in these patients with MTLE.[37]

Overall, these findings suggest that cortical and subcortical brain structures might represent attractive candidate endophenotypes reflecting particular genetic risk factors highly relevant to MTLE.

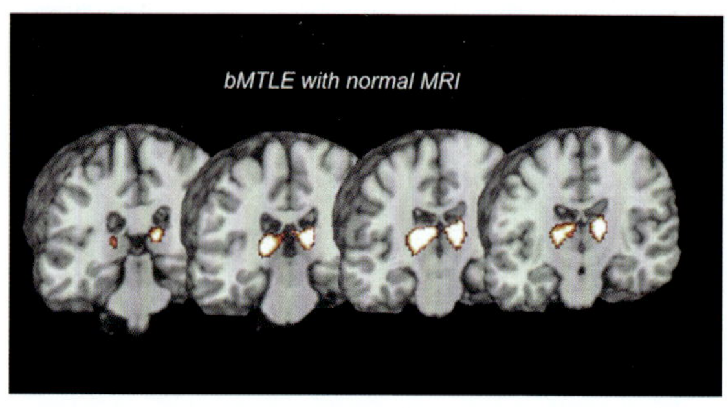

Figure 5.2. Voxel-based-analysis of the whole brain shows gray matter decrease of the bilateral thalami in a series of 85 patients with *bMTLE*.

A recent study examined volume deficits for many subcortical structures traits in unrelated patients with sporadic MTLE and their unaffected siblings with the aim of studying the suitability of these traits as endophenotypes.[38] The advanced neuroimaging study, however, failed to depict any subcortical volume deficits in the unaffected individuals, suggesting that volume deficits in subcortical structures might be largely determined by nongenetic factors in sporadic MTLE.[38] To the best of our knowledge, no study has assessed MTLE-related cortical gray matter alterations, which may represent a more suitable endophenotype in MTLE.

Interestingly, very recently a significant reduction of fractional anisotropy along the white matter of the temporal lobes in drug-resistant MTLE has been found, implying it as a valuable biological marker of refractoriness.[39] Furthermore, the evidence of both diffusion abnormalities and reduced cortical thickness of the corpus callosum only in patients with refractory MTLE suggested that differences in the distribution of such alterations might be also related to refractoriness.[40]

5.4 Conclusions and Future Directions

Brain images offer a powerful means to assess the genetic effects, so the expectation is that the combination of the genetic and neuroimaging approaches has introduced a new era of imaging genetics that will allow an unprecedented amount of novel scientific insights related to the underlying biology and genetics of epilepsies, especially MTLE. Nonetheless, since brain imaging can measure signals at over 2 million locations in the brain, and each signal is surveyed while screening the genome as well, the resulting statistical search is so vast that very large sample sizes are needed to achieve meaningful results and depict genetic variants that influence the brain.[25,26] A relatively limited sample size was also a major limit of recent attempts to elucidate the genetic architecture of complex epilepsies other than a variety of issues including phenotypic ambiguities and restricted genetic scope.[41] Most important, significantly associated single nucleotic polymorphisms have an expected effect size <0.5% of the phenotypic variance in quantitative traits, even for measures extracted from MRI.[25,26]

Therefore, only a large sample size achieved through international collaboration and new methods might overcome the difficulty in detecting a true signal of a small effect size. So, with the recent advent of global neuroimaging consortia, like the ENIGMA Consortium (http://enigma.ini.usc.edu), imaging studies are now becoming well powered to depict genetic variants that reliably affect the brain.[25,42] Another challenge is that most genetic variations found by genome wide association studies are located in regions where the functional consequences of the variants are unknown, so there is the need on turning new gene variants related to the disease into biological pathways linking associated genes, and the identification of functional annotations, some of which are cell type specific, enriched in disease associations.

Even so, imaging studies can already provide information that is not available in genetic comparison of patients and controls. Imaging can identify potential mechanisms and circuits promoting disease risk, making it easier to develop and evaluate treatments.[42] A deeper understanding of disease-modifying pathways in MTLE should also make it easier to discover disease-modifying factors that drive resilience and risk for MTLE, which could lead to clinically relevant markers of prognosis and outcome.

References

1. Manford M, Hart YM, Sander JW, Shorvon SD. National General Practice Study of Epilepsy (NGPSE): partial seizure patterns in a general population. *Neurology*. 1992;**42**:1911–7.

2. Falconer MA, Serafetinides EA, Corsellis JAN. Etiology and pathogenesis of temporal lobe epilepsy. *Arch Neurol*. 1964;**10**:233–40.

3. de Tisi J, Bell GS, Peacock JL, et al. The long-term outcome of adult epilepsy surgery, patterns of seizure remission, and relapse: a cohort study. *Lancet*. 2011;**378**:1388–95.

4. Semah F, Picot MC, Adam C, et al. Is the underlying cause of epilepsy a major prognostic factor for recurrence? *Neurology*. 1998;**51**:1256–62.

5. Spooner CG, Berkovic SF, Mitchell LA, Wrennall JA, Harvey AS. New-onset temporal lobe epilepsy in children: lesion on MRI predicts poor seizure outcome. *Neurology*. 2006;**67**:2147–53.

6. Labate A, Gambardella A, Andermann E, et al. Benign mesial temporal lobe epilepsy. *Nat Rev Neurol*. 2011;**7**: 237–40.

7. Aguglia U, Gambardella A, Le Piane E, et al. Mild non-lesional temporal lobe epilepsy. A common unrecognized disorder with onset in adulthood. *Can J Neurol Sci*. 1998;**25**:282–6.

8. Gambardella A, Manna I, Labate A, et al. GABA(B) receptor 1 polymorphism (G1465A) is associated with temporal lobe epilepsy. *Neurology*. 2003;**60**:560–3.

9. Labate A, Ventura P, Gambardella A, et al. MRI evidence of mesial temporal sclerosis in sporadic "benign" temporal lobe epilepsy. *Neurology*. 2006;**66**: 562–5.

10. Labate A, Aguglia U, Tripepi G, et al. Long-term outcome of mild mesial temporal lobe epilepsy: a prospective longitudinal cohort study. *Neurology*. 2016;**86**:1904–10.

11. Pitkänen A, Löscher W, Vezzani A, et al. Advances in the development of biomarkers for epilepsy. *Lancet Neurol*. 2016;**15**:843–56.

12. Engel J Jr, Pitkänen A, Loeb JA, et al. Epilepsy biomarkers. *Epilepsia*. 2013;**54**(S4):S61–9.

13. Berkovic SF, McIntosh A, Howell RA, Mitchell A, Sheffield LJ, Hopper JL. Familial temporal lobe epilepsy: a common disorder identified in twins. *Ann Neurol*. 1996;**40**:227–35.

14. Crompton DE, Scheffer IE, Taylor I, et al. Familial mesial temporal lobe epilepsy: a benign epilepsy syndrome showing complex inheritance. *Brain*. 2010;**133**:3221–31.

15. Gambardella A, Messina D, Le Piane E, et al. Familial temporal lobe epilepsy. Autosomal dominant inheritance in a large pedigree from southern Italy. *Epilepsy Res*. 2000;**38**:127–32.

16. Kobayashi E, D'Agostino MD, Lopes-Cendes I, et al. Hippocampal atrophy and T2 weighted signal changes in familial mesial temporal lobe epilepsy. *Neurology*. 2003;**60**:405–9.

17. Hedera P, Blair MA, Andermann E, et al. Familial mesial temporal lobe epilepsy maps to chromosome 4q13.2-q21.3. *Neurology*. 2007;**68**:2107–12.

18. Dibbens LM, de Vries B, Donatello S, et al. Mutations in DEPDC5 cause familial focal epilepsy with variable foci. *Nat Genet*. 2013;45:546–51.

19. Colosimo E, Gambardella A, Mantegazza M, et al. Electroclinical features of a family with simple febrile seizures and temporal lobe epilepsy associated with SCN1A loss-of-function mutation. *Epilepsia*. 2007;48:1691–6.

20. Scheffer IE, Harkin LA, Grinton BE, et al. Temporal lobe epilepsy and GEFS+ phenotypes associated with SCN1B mutations. *Brain*. 2007;130:100–9.

21. Al-Asmi A, Jansen AC, Badhwar A, et al. Familial temporal lobe epilepsy as a presenting feature of choreoacanthocytosis. *Epilepsia*. 2005;46:1256–63.

22. Gottesman II, Gould TD. The endophenotype concept in psychiatry: etymology and strategic intentions. *Am J Psychiatry*. 2003;160:636–45.

23. Hasler G, Drevets WC, Gould TD, Gottesman II, Manji HK. Toward constructing an endophenotype strategy for bipolar disorders. *Biol Psychiatry*. 2006;60:93–105.

24. Hashimoto R, Ohi K, Yamamori H, et al. Imaging genetics and psychiatric disorders. *Curr Mol Med*. 2015;15:168–75.

25. Medland SE, Jahanshad N, Neale BM, Thompson PM. Whole-genome analyses of whole-brain data: working within an expanded search space. *Nat Neurosci*. 2014;17:791–800.

26. Thompson PM, Cannon TD, Narr KL, et al. Genetic influences on brain structure. *Nat Neurosci*. 2001;4:1253–8.

27. Tsai MH, Pardoe HR, Perchyonok Y, et al. Etiology of hippocampal sclerosis: evidence for a predisposing familial morphologic anomaly. *Neurology*. 2013;81:144–9.

28. Jackson GD, McIntosh AM, Briellmann RS, Berkovic SF. Hippocampal sclerosis studied in identical twins. *Neurology*. 1998;51:78–84.

29. Fernández G, Effenberger O, Vinz B, et al. Hippocampal malformation as a cause of familial febrile convulsions and subsequent hippocampal sclerosis. *Neurology*. 1998;50:909–17.

30. Kobayashi E, Li LM, Lopes-Cendes I, Cendes F. Magnetic resonance imaging evidence of hippocampal sclerosis in asymptomatic, first-degree relatives of patients with familial mesial temporal lobe epilepsy. *Arch Neurol*. **2002**;59:1891–4.

31. Sisodiya SM, Ragge NK, Cavalleri GL, et al. Role of SOX2 mutations in human hippocampal malformations and epilepsy. *Epilepsia*. 2006;47:534–42.

32. Bernasconi N, Kinay D, Andermann F, Antel S, Bernasconi A. Analysis of shape and positioning of the hippocampal formation: an MRI study in patients with partial epilepsy and healthy controls. *Brain*. 2005;128:2442–52.

33. Bernhardt BC, Worsley KJ, Kim H, Evans AC, Bernasconi A, Bernasconi N. Longitudinal and cross-sectional analysis of atrophy in pharmacoresistant temporal lobe epilepsy. *Neurology*. 2009;72:1747–54.

34. Bernhardt BC, Bernasconi N, Concha L, Bernasconi A. Cortical thickness analysis in temporal lobe epilepsy: reproducibility and relation to outcome. *Neurology*. 2010;74:1776–84.

35. Labate A, Cerasa A, Aguglia U, Mumoli L, Quattrone A, Gambardella A. Neocortical thinning in "benign" mesial temporal lobe epilepsy. *Epilepsia*. 2011;52:712–7.

36. Bast T, Feldon J. Hippocampal modulation of sensorimotor processes. *Prog Neurobiol*. 2003;70:319–45.

37. Labate A, Cerasa A, Gambardella A, Aguglia U, Quattrone A. Hippocampal and thalamic atrophy in mild temporal lobe epilepsy: a VBM study. *Neurology*. 2008;71:1094–101.

38. Alhusaini S, Scanlon C, Ronan L, et al. Heritability of subcortical volumetric traits in mesial temporal lobe epilepsy. *PLOS ONE*. 2013;8:e61880.

39. Labate A, Cherubini A, Tripepi G, et al. White matter abnormalities differentiate severe from benign temporal lobe epilepsy. *Epilepsia*. 2015;56:1109–16.

40. Caligiuri ME, Labate A, Cherubini A, et al. Integrity of the corpus callosum in patients with benign temporal lobe epilepsy. *Epilepsia* 2016;57:590–6.

41. International League Against Epilepsy Consortium on Complex Epilepsies. Genetic determinants of common epilepsies: a meta-analysis of genome-wide association studies. *Lancet Neurol*. 2014;13:893–903.

42. Whelan CD, Altmann A, Botía JA, et al. Structural brain abnormalities in the common epilepsies assessed in a worldwide ENIGMA study. *Brain*. 2018;141:391–408.

Computational Neuroimaging of Epilepsy

6

Seok-Jun Hong, Min Liu, Ravnoor Gill, Edward Hogan, Neda Bernasconi, and Andrea Bernasconi

6.1 Introduction

Epilepsy, a disorder of the brain characterized by an enduring predisposition to generate epileptic seizures, affects over 50 million people worldwide. Underlying etiologies include structural, metabolic, and genetic abnormalities.[1] More than one-third of these patients suffer from seizures that are resistant to antiepileptic drugs. Consequences of recurrent seizures are severe and include an increased risk of injury, psychosocial and economic impairment, and death.[2,3] In cases of drug-resistant focal onset epilepsy, when a structural abnormality is identified, its surgical resection offers patients the chance at a complete cure.[4] Temporal lobe epilepsy related to mesiotemporal sclerosis (TLE) and extratemporal lobe epilepsies secondary to cortical developmental malformations, particularly focal cortical dysplasia (FCD), are the two most common drug-resistant epilepsy syndromes amenable to surgery,[5] accounting for 60–80% of investigations at tertiary centers. Importantly, early identification of a lesion allows timely resective surgery, limits the long-term effects of recurrent seizures and medication, and has been shown to have positive consequences on cognitive outcome and brain development.[6]

Magnetic resonance imaging (MRI) provides a unique and versatile, noninvasive method for computer-aided lesion detection and brain-wide evaluation of the epileptic process (Figure 6.1). This chapter provides an overview of advanced quantitative image analysis methods used to investigate main epilepsy syndromes: TLE, neocortical extratemporal epilepsy related to cortical dysplasias, and idiopathic generalized epilepsy.

6.2 Computational Imaging of Temporal Lobe Epilepsy

6.2.1 Analysis of Mesiotemporal Structures

In TLE, the epileptogenic network generally encompasses variable degrees of neuronal loss and astrogliosis across hippocampal subfields, the amygdala, and the entorhinal cortex, a pathological disease hallmark referred to as mesiotemporal sclerosis (MTS).[7] Based on earlier qualitative studies and recent quantitative MRI-histopathological correlation analyses of the hippocampus, it is now clearly established that cell loss and gliosis relate to atrophy and increased T2-weighted signal, respectively.[8] In many patients, MTS can be visualized on MRI as noticeable hippocampal atrophy and increased T2-weighted signal. However, the visual identification of morphological and signal characteristics of the hippocampus is highly subjective and depends heavily on the experience of the reader. For at least two decades, manual MRI volumetry has been a commonly employed quantitative technique to reliably assess mesiotemporal lobe atrophy, as it has been demonstrated to be more sensitive than visual evaluation.[9] This method, generally performed on high-resolution T1-weighted structural MR images, allows indeed for a reliable lateralization of the seizure focus. Volumetry of the entorhinal cortex, amygdala, and temporopolar region, as well as the thalamus, may play also roles in lateralization of the seizure focus.[10] Specifically, in patients with a normal-appearing hippocampal structure imaging of visual inspection of conventional contrasts, entorhinal cortex atrophy detected by volumetry provides accurate lateralization of the seizure focus in 25% of cases.[11] In sum, quantification of mesiotemporal structural damage is strongly recommended when considering epilepsy surgery in order to detect subtle atrophy or abnormal signal increases ipsilateral to the seizure focus, and to establish objectively the degree of integrity of the contralateral structures. Indeed, bilateral mesial temporal lobe atrophy raises concerns of markedly reduced chance of seizure freedom after surgery and an increased risk of memory impairment.[12]

From a practical perspective, manual segmentation of mesiotemporal lobe structures requires highly

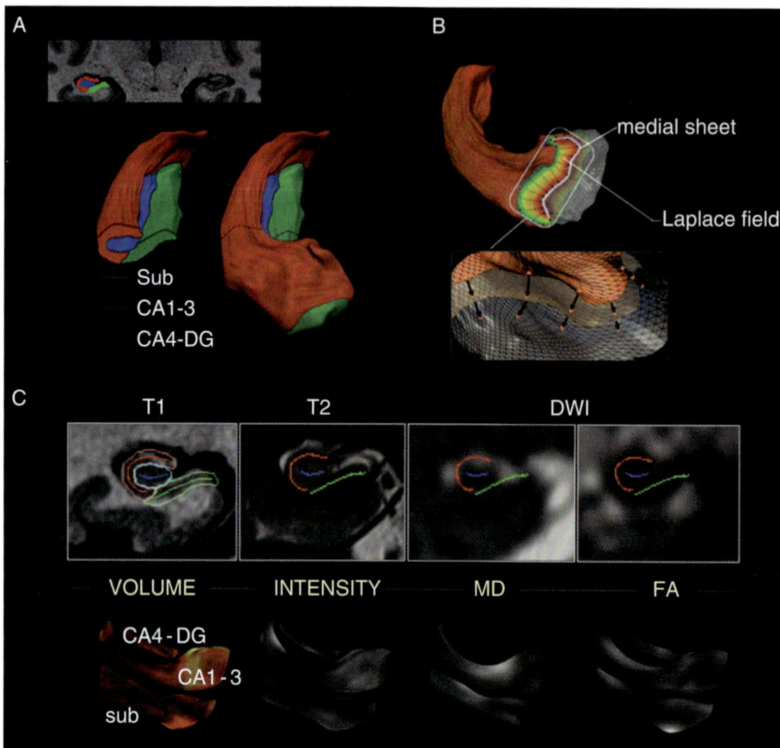

Figure 6.2. Multiparameter hippocampal subfield subface analysis.
(A) Hippocampal subfield labels are generated on high-resolution T1-weighted MRI images.
(B) Labels are then converted to surfaces and parameterized using a spherical harmonics approach, providing shape inherent subject correspondence. For each subfield surface, a medial sheet is extracted by solving a three-dimensional Hamilton-Jacobi equation from the outer hull. The sheet inherits correspondence from the parameterized hull by vertex propagation along a Laplacian field.
(C) This representation allows for dense sampling of volume (derived from T1-weighted MRI), T2-signal intensity, and diffusion-weighted MRI features (MD = mean diffusivity; FA = fractional anisotropy) with minimal partial volume effect along the entire anatomical extent of each subfield (CA1-3, CA4-dentate gyrus, and subiculum).

Overall, this work highlights the role of computational modeling and machine learning to assist clinical decision making.

6.2.2 Beyond Mesiotemporal Structures

Beside mesiotemporal damage in the hippocampus, amygdala, and entorhinal cortex, histopathological examination of postmortem and surgical specimens indicates an extended distribution of abnormalities in TLE, preferentially involving limbic structures. Findings include impaired fiber arrangement of the fimbria-fornix, neuronal loss and gliosis in the ipsilateral mediodorsal thalamic nucleus, abnormal myelination and axonal degeneration of the temporal pole, and gliotic changes of the frontal pole and orbitofrontal cortex,[29] anomalies indicative of widespread network alterations in this condition.

In addition to providing an increasingly detailed in vivo characterization of lesions, neuroimaging research has been instrumental in demonstrating network pathology in epilepsies traditionally considered as focal. In TLE, MRI morphometry has contributed to our understanding of system-level pathology by revealing widespread atrophy that extends beyond the mesial temporal into the lateral temporal and frontocentral cortices,[30,31] as well as the thalamus.[32–34] Cortical thinning is thought to be secondary to excitotoxic effects of seizure spread through thalamocortical pathways. Indeed, the location and extent of neocortical atrophy relate to degree and distribution of thalamic pathology, lending support to the concept that this deep brain structure is an important hub in the pathologic network of TLE.[32] Additional morphological alterations include increased folding complexity of the insular and cingulate regions, mainly ipsilateral to the seizure focus,[35] and decreased surface area, particularly in the temporal lobe.[36] The presence of such anomalies in the asymptomatic siblings suggests, at least in part, genetic influence.[37]

With respect to the white matter, diffusion MRI studies have shown altered fractional anisotropy and mean diffusivity suggestive of decreased axonal density and altered myelin membranes in limbic tracts, including the fornix and cingulum, and frontotemporal tracts, such as the uncinate and arcuate fasciculi, the temporo-occipital tract, the inferior longitudinal fasciculus, as well as the corpus callosum.[38,39]

Notably, a track-based segment analysis revealed a centrifugal pattern of mean diffusivity increases in major tracts carrying temporal lobe connections.[40] These anomalies likely reflect astrogliosis and microstructure derangement related to seizure activity in the vicinity of the focus, which may serve as a dynamic marker of the epileptic activity and lateralize the seizure focus, particularly in patients with unremarkable MRI. Given its proximity to the neocortex and key role in cortico-cortical connectivity, superficial white matter (SWM) is a key candidate region to evaluate the interplay between structure and function. We recently designed a surface-based method allowing for continuous sampling of diffusion parameters in this compartment,[41] a region so far neglected in TLE neuroimaging studies. Our analysis revealed ipsilateral anomalies in the temporolimbic regions that were independent from cortical thinning; conversely, they related to hippocampal volume. Interestingly, grouping patients based on histopathology showed higher load of SWM diffusion alterations in cases with cell loss and gliosis compared to those with isolated gliosis,[41] cross-validating the relationship between mesiotemporal sclerosis and SWM microstructural anomalies. Notably, the proximity of SWM alterations to the mesiotemporal lobe and close relationship to hippocampal pathology may corroborate connectivity-based models of regional susceptibility, in which regions anatomically connected to the disease epicenter may undergo most marked structural alterations.[42] Importantly, we showed that SWM damage, mediated by hippocampal atrophy, resulted in large-scale default mode anomalies,[41] a link to be targeted by future longitudinal studies in new-onset epilepsy, ideally complemented by behavioral phenotyping to understand the cognitive consequences of this causal chain.

6.3 Computational Imaging of Focal Cortical Dysplasia

Malformations of cortical development are found in 20% of specimens obtained during epilepsy surgery. Among them, FCD is the most common, accounting for 71% of cases.[5] In FCD Type I, an elusive pathological entity, microcolumns are typified by more than eight vertically aligned neurons, predominantly located in layers 3 and 4.[43] Compared to healthy cortices, regions harboring microcolumns may present with reduced cell size, increased neuronal density, as well as a tendency for decreased cortical thickness.[44] Notably, as the microcolumnar pattern resembles the unit of the radial lineage model of corticogenesis, its presence suggests a remnant of perturbed events affecting the final, postmigrational neuronal positioning.[45] Currently, Type I dysplasias cannot be seen on MRI. Indeed, the strictly intracortical character of FCD Type I lesions together with the lack of large-scale cytological anomalies has so far precluded a clear definition of their signature even on high-resolution 3T MRI.[46] Moving toward ultra-high-field 7T MRI offers the possibility to increase spatial resolution to the range of 350–500 μm, a scale at which cortical layers may be visualized in vivo.[47] Paralleling increases in resolution, quantitative contrasts provide time-efficient measurements of longitudinal T1 and transverse T2 relaxation times, which may distinguish cortical myelin and iron,[48] two determinants of cortical laminar architecture. Case reports have shown encouraging increased sensitivity for the detection of subtle FCD Type II lesions,[49] setting the conceptual basis for the use in Type I lesions.

Cortical dysplasias Type II are typified by both disorganization of layers and cytological anomalies,[43] including dysmorphic neurons and balloon cells. Ectopic neurons, and hypertrophic astrocytes may be present in the juxtacortical white matter, which may also show demyelination. On MRI, FCD Type II lesions are typically characterized by visible increased cortical thickness, blurred gray-white matter interface, and increased T2-weighted signal, particularly apparent on FLAIR images.[9] Notably, these lesions may vary greatly in size, ranging from those that affect extended cortical territories to a few voxels only. The transmantle sign, a funnel-shaped hyperintensity extending from the FCD to the lateral ventricle and thought to represent the footprint of disrupted neuronal migration along radial glial processes may help detecting such smaller lesions. In addition, subtle dysplasias may be associated with unusual gyral patterns and are preferentially located at a bottom of an abnormally deep sulcus.[50] As a result, conventional radiological assessment, even when carried out by expert observers, often fails to detect such lesions.[9] This clinical challenge has motivated the development of computer-aided methods aimed at objectively detecting FCD lesions by means of tissue modeling, detailed in the next section.

6.3.1 Automated Detection of FCD Type II Lesions

While voxel-based morphometry was originally developed to quantify group differences in gray matter density derived from T1-weighted MRI, several studies have used this approach to identify "excessive" gray matter associated with FCD in individual patients. Applying an arbitrary threshold (e.g., >1 *SD* above the mean gray matter concentration in healthy controls) as indicative of abnormality, areas of increased density colocalize with lesional voxels in 63–86% of cases.[51] Notably, the marked hyperintensity of some lesions leads to gray matter misclassification, reducing considerably the sensitivity of the technique.[51] Moreover, while this method may give diagnostic confidence in detecting medium to large lesions, its yield in small FCD has not been demonstrated. The analysis of MRI intensities or quantitative MRI contrasts, such as T2 relaxometry, double inversion recovery, and magnetization transfer ratio imaging, are also amenable to voxel-based comparison. These approaches have a sensitivity of 87–100% in detecting obvious FCD lesions,[9] but have low yield in cases with negative conventional MRI.

Our group pioneered the design of computer-based algorithms modeling the distinctive characteristics of FCD Type II.[52,53] Initial methods targeted cortical thickening and blurring of the GM-WM boundary on T1-weighted MRI, enhancing visual detection rates by 30% relative to standard evaluation. Other groups have replicated our techniques with similar results.[54–56] Notably, sensitivity and specificity may be hampered by the fact that various maps are inspected visually. In addition, the yield and diagnostic confidence depend heavily on the reader's ability to integrate the diverse and complex information embedded in the various maps, an expertise that only few centers have developed so far. Limited generalizability also stems from the fact that such approaches have been validated using only large lesions easily recognized by routine radiology.[55] Another source of difficulty may arise from the paucity or lack of localizing clinical and EEG findings to direct the search for FCD, particularly in patients with seizures originating in the frontal lobe, the preferential location of this malformation. To overcome these challenges, over the years, we have developed a series of increasingly sophisticated algorithms for automatic FCD detection, operating on 1.5 T MRI data, which

initially relied on voxel-based texture analysis combined with a Bayesian classifier.[57] These tools were validated with mid- to large-sized lesions visible on routine radiological inspection; nevertheless, classification failed in up to 20% of cases. Such performance is unsatisfactory given current referral patterns to epilepsy surgery centers, with an increasing number of patients with nondiagnostic clinical MRI, even at 3.0 Tesla. Indeed, the absence of a visible lesion is one of the greatest challenges in epilepsy surgery and has led to an increase in invasive EEG studies with implanted intracranial electrodes. Yet, without informed, image-guided implantation, even with widespread coverage, EEG sampling errors may occur in up to 40% of cases; consequently the target cannot be defined and the outcome of surgery, if considered, is poorer.[1] In these patients, lesions are subtle, with morphological characteristics that may differ only slightly from normal tissue.

As an alternative approach to voxel-based techniques, surface-based methods preserve cortical topology and allow quantifying sulco-gyral anomalies, at times the only sign of dysgenesis. We thus recently opted for a surface-based framework and combined various morphological (cortical thickness, curvature, and sulcal depth), intensity features (relative intensity and intensity gradients), microstructure (mean diffusivity and fractional anisotropy), and function, taking advantage of their covariance to unveil subthreshold tissue properties not readily identified by a single modality. To characterize in vivo MRI signatures of FCD Type IIA and Type IIB, we have designed a multisurface approach that systematically sampled intra- and subcortical lesional features.[58] In addition, geodesic distance mapping quantified the same features in the lesion perimeter. Logistic regression assessed the relationship between MRI and histology, while supervised pattern learning was used for individualized subtype prediction. In this comprehensive analysis, we demonstrated that FCD Type IIB is characterized by abnormal morphology, intensity, diffusivity, and connectivity across all surfaces, while Type IIA lesions present only with increased FLAIR signal and reduced diffusion anisotropy close to the gray-white matter interface. Similar to lesional patterns, perilesional anomalies in all domains were more marked in Type IIB, extending up to 16 mm from the manual lesion label. Importantly, structural MRI markers correlated with categorical histological characteristics. Finally, a profile-based classifier predicted

FCD subtypes with equal sensitivity of 85%, while maintaining a high specificity of 94% against healthy and disease controls. This work demonstrated that image processing applied to widely available MRI contrasts has the ability to dissociate FCD subtypes at a mesoscopic level. Integrating in vivo staging of pathological traits with automated lesion detection may be clinically relevant, as it could conceivably assist emerging approaches, such as minimally invasive thermal ablation, which do not supply tissue specimen. It may also help guiding and monitoring novel lesion-specific drug treatments, such as mTOR inhibitors.

We recently designed a surface-based two-step classifier for automated FCD detection of "MRI-negative FCD.[59] This procedure, which relies on a linear discriminant classification scheme of multiple features derived from T1-weighted MRI, first recognizes vertices with the highest detection rate and subsequently removes false positive clusters using tissue texture characteristics and spatial priors (Figure 6.3A). Given that all subjects were initially diagnosed as MRI-negative, this fully automated approach yielding 74% sensitivity in healthy and disease controls offers a substantial gain in sensitivity over standard radiological assessment (Figure 6.3B). When applying our classifier trained on 3.0 Tesla images to an independent dataset of patients with histologically proven FCD acquired at 1.5 Tesla, we maintained high sensitivity (71%), supporting generalizability. Notably, using T1 and FLAIR intensities, we subsequently achieved even higher sensitivity (83%) and specificity (92%).[60]

6.3.2 Assessing Brain-Wide Integrity in Cortical Dysplasias

Because of the crucial role in defining the surgical target, MRI studies in FCD have been primarily dedicated to lesion detection in single patients, as discussed in the previous section. Aside from a few case reports describing gray matter and white anomalies,[61] whole-brain cohort-specific structural brain anomalies have been only rarely assessed. To fill this knowledge gap, we recently examined systematically whole-brain MRI morphology in dysplasia-related frontal lobe epilepsy[46] and found that, relative to controls, patients with FCD Type I display multilobar cortical atrophy that is most marked in ipsilateral frontal cortices. Conversely, in Type II FCD, anomalies beyond the primary lesions are typified by multilobar frontocentral cortical thickening. Cortical folding also diverged, with increased complexity in prefrontal cortices in Type I and decreases in Type II. We hypothesized that cortical thickening in Type II may indicate delayed pruning, while a thin cortex in Type I likely results from combined effects of seizure excitotoxicity and the primary malformation. In addition, group-level patterns successfully guided automated subtype classification (Type I: 100%, Type II: 96%), seizure focus lateralization (Type I: 92%, Type II: 86%), and outcome prediction (Type I: 92%, Type II: 82%), thereby demonstrating clinical relevance.

We assessed large-scale brain organization across the entire spectrum of prevalent cortical malformations.[62] Based on experimental evidence suggesting that distributed effects of focal insults are modulated by stages of brain development, we postulated differential patterns of network anomalies across subtypes of malformations. Graph theoretical analysis of structural covariance networks indicated a consistent rearrangement toward a regularized architecture characterized by increased path length and clustering, as well as disrupted rich-club topology, overall suggestive of inefficient global and excessive local connectivity. Notably, we observed a gradual shift in network reconfigurations across subgroups, with only subtle changes in FCD Type II, moderate effects in heterotopia, and maximal effects in polymicrogyria (Figure 6.4). Analysis of resting-state functional connectivity also revealed gradual network changes, with most marked rearrangement in polymicrogyria; contrary to findings in the structural domain, however, functional architecture was characterized by decreases in both local and global parameters. These findings support the concept that time of insult during corticogenesis impacts the severity of topological network reconfiguration. Specifically, late-stage malformations, typified by polymicrogyria, may disrupt the formation of large-scale cortico-cortical networks and thus lead to a more profound impact on whole-brain organization than early stage disturbances of predominantly radial migration patterns observed in cortical dysplasia Type II, which likely affect a relatively confined cortical territory.

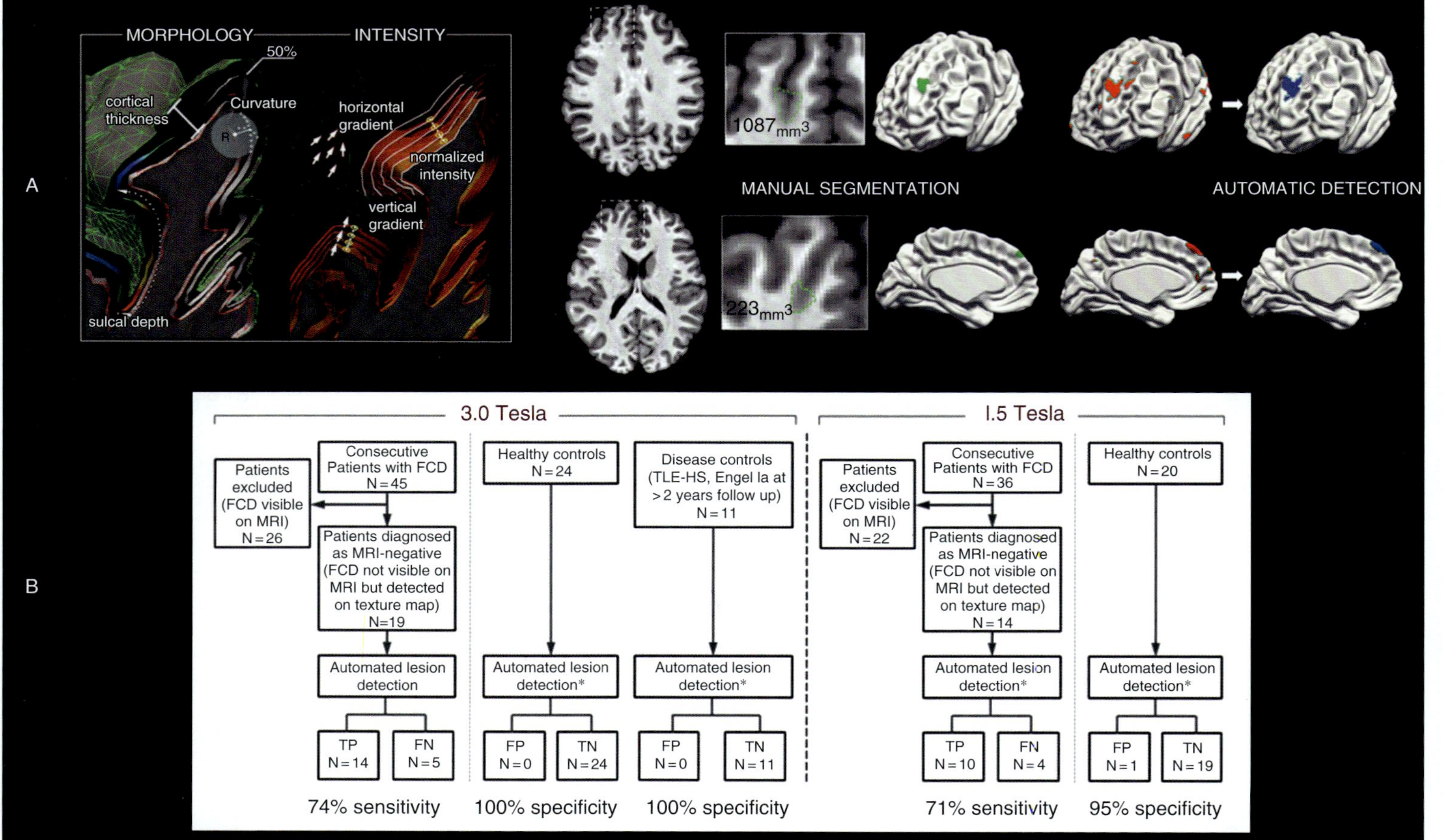

Figure 6.3. Automated detection of focal cortical dysplasia (FCD) Type II. (A) After surface-based extraction of morphology and intensity-based features modeling the *in vivo* characteristics of FCD, a vertex-wise classification identifies putative lesions (red), whereas the subsequent cluster-wise classification discards false positives except the cluster (blue) colocalizing with the manual label (green). The axial T1-weighted MRI sections show the region containing the FCD (dashed square). The magnified panel displays two manually segmented FCD labels (dotted squares) and their volume; the label is projected onto a surface template. (B) The flow diagram and study design of the only classifier with Class II evidence for diagnostic accuracy are shown. Sources of spectrum bias were ruled out by evaluating the specificity of the algorithm against healthy individuals and clinically well-characterized disease controls. To minimize incorporation bias, the classifier trained on 3.0 Tesla data was tested on an independent cohort of patients and controls examined at 1.5 Tesla.

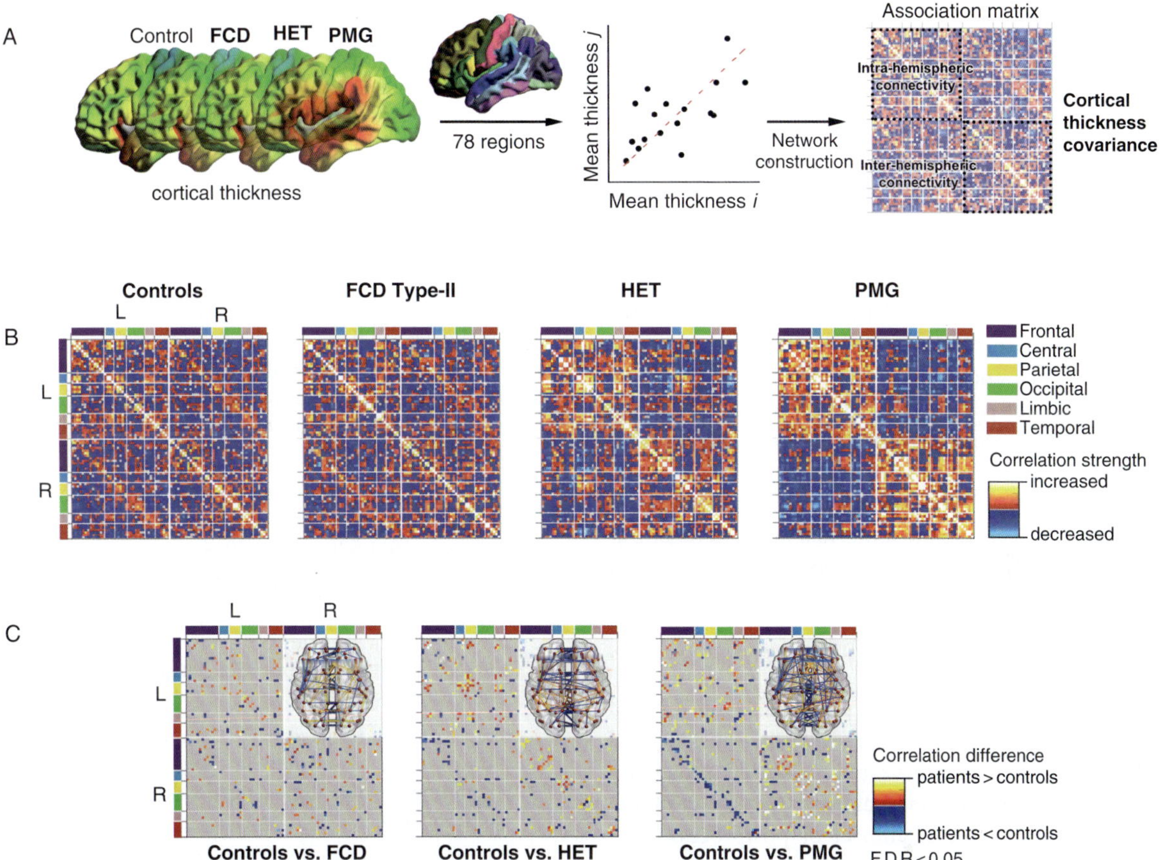

Figure 6.4. Structural network alterations across the spectrum of malformations of cortical development. (A) Covariance networks based on cortical thickness correlations in healthy controls ($n = 82$), and patients with malformations of cortical development ($n = 154$), including focal cortical dysplasia Type II (FCD-II), heterotopia (HET), and polymicrogyria (PMG). The structural covariance matrix is obtained by calculating the cross-correlation coefficient of mean cortical thickness between 78 pairs of regions i and j across subjects. (B) For each cohort (i.e., controls, FCD-II, HET, and PMG), a group-level matrix is obtained. For display purposes, parcels are color coded according to brain regions listed beside the matrices. The color bar indicates the correlation strength. (C) Significant group differences in interregional correlations corrected for multiple comparisons at FDR < 0.05. Increase/decrease in patients relative to controls are shown in red/blue in the matrices and corresponding network graphs. In (B) and (C), L and R refer to left and right hemispheres, respectively.

6.4 Computational Imaging of Idiopathic Generalized Epilepsy

Idiopathic generalized epilepsies (IGE) form a group of electroclinical syndromes characterized by absence seizures, myoclonic jerks, and generalized tonic-clonic seizures (alone or in varying combinations) in association with generalized discharges of spikes and waves in the presence of a normal EEG background. A handful of postmortem observations have reported heterotopic neurons in the white matter, an indistinct cortical boundary and dyslamination;[63] nevertheless, these findings have not been reproduced.[64] Advanced quantitative image postprocessing of structural MRI has identified subtle thalamic and cortical changes.[65,66] Notably, however, findings from manual MRI volumetry of the thalamus have been somewhat heterogeneous: initial studies were negative,[67] while more recent data have shown atrophy[68] or both atrophy and hypertrophy within a single cohort.[65] Similarly, whole-brain, lobe-wise volumetry has provided uncertain results.[69] These results may reflect not only the heterogeneity of structural anomalies across subsyndromes, but also interrater variability of manual segmentations. Indeed, evidence from VBM analyses showed consistent reduction in thalamic[70] and neocortical gray matter density.[71] Using cortical thickness measures and thalamic volumetry, we showed bilateral thalamic atrophy corresponding to known anatomical connections from sacrificial tracer studies and widespread cortical thinning that was most prominent in frontocentral areas, with a prevalence of up to 40%. In patients, thalamocortical network correlations increased in frontocentral and parietal regions, but decreased in limbic areas. Importantly, patients with poorly controlled seizures showed an even faster progression of atrophy of these neocortical regions, suggesting that neocortical atrophy in IGE is likely the consequence of generalized seizure activity inducing thalamocortical network remodeling.[66] Similar distributions of neocortical atrophy have been reported in juvenile myoclonic epilepsy.[72] One study demonstrated widespread changes in mean curvature, a surrogate marker of abnormal brain development, suggesting that structural anomalies in juvenile myoclonic epilepsy extend beyond the thalamus and frontal lobes.[73] Studies employing structural and functional MRI also found abnormal network organization, with structure-function decoupling,[74] although findings were only partially reproduced.[75]

6.5 Conclusion

By revealing subtle lesions that previously eluded visual inspection, quantitative image processing of structural MRI has clearly demonstrated increased sensitivity compared to traditional visual inspection. Overall, the most significant clinical impact of postprocessing is that cases with pharmacoresistant seizures considered MRI-negative at first have been increasingly become MRI-positive, thereby offering the life-changing benefits of epilepsy surgery to more patients. Aside from diagnostics, ongoing developments in computational imaging, combined with machine learning, lend us unprecedented opportunities to design novel biomarkers to stage and monitor the disease, and ultimately provide personalized therapeutic interventions.

References

1. Fisher RS, Cross JH, D'Souza C, et al. Instruction manual for the ILAE 2017 operational classification of seizure types. *Epilepsia.* 2017;**58**:531–42.

2. Keezer MR, Sisodiya SM, Sander JW. Comorbidities of epilepsy: current concepts and future perspectives. *Lancet Neurol.* 2016;**15**:106–15.

3. Devinsky O, Hesdorffer DC, Thurman DJ, et al. Sudden unexpected death in epilepsy: epidemiology, mechanisms, and prevention. *Lancet Neurol.* 2016;**15**: 1075–88.

4. Tellez-Zenteno JF, Hernandez Ronquillo L, Moien-Afshari F, et al. Surgical outcomes in lesional and non-lesional epilepsy: a systematic review and meta-analysis. *Epilepsy Res.* 2010;**89**:310–8.

5. Blumcke I, Spreafico R, Haaker G, et al. Histopathological findings in brain tissue obtained during epilepsy surgery. *N Engl J Med.* 2017;**377**: 1648–56.

6. Skirrow C, Cross JH, Cormack F, et al. Long-term intellectual outcome after temporal lobe surgery in childhood. *Neurology.* 2011;**76**:1330–7.

7. Thom M. Review: hippocampal sclerosis in epilepsy: a neuropathology review. *Neuropathol Appl Neurobiol.* 2014;**40**:520–43.

8. Goubran M, Bernhardt BC, Cantor-Rivera D, et al. In vivo MRI signatures of hippocampal subfield pathology in intractable epilepsy. *Hum Brain Mapp.* 2016;**37**:1103–19.

9. Bernasconi A, Bernasconi N. Unveiling epileptogenic lesions: the contribution of image processing. *Epilepsia.* 2011;**52**(suppl 4):20–4.

10. Bernasconi N, Bernasconi A, Caramanos Z, et al. Mesial temporal damage in temporal lobe epilepsy: a volumetric MRI study of the hippocampus, amygdala and parahippocampal region. *Brain.* 2003;**126**:462–9.

11. Bernasconi N, Bernasconi A, Caramanos Z, et al. Entorhinal cortex atrophy in epilepsy patients exhibiting normal hippocampal volumes. *Neurology.* 2001;**56**:1335–9.

12. McLachlan RS, Pigott S, Tellez-Zenteno JF, et al. Bilateral hippocampal stimulation for intractable temporal lobe epilepsy: impact on seizures and memory. *Epilepsia.* 2010;**51**:304–7.

13. Iglesias JE, Augustinack JC, Nguyen K, et al. A computational atlas of the hippocampal formation using ex vivo, ultra-high resolution MRI: application to adaptive segmentation of in vivo MRI. *NeuroImage.* 2015;**115**:117–37.

14. Coupe P, Manjon JV, Fonov V, et al. Nonlocal patch-based label fusion for hippocampus segmentation. *Med Image Comput Comput Assist Interv.* 2010;**13**:129–36.

15. Bernasconi N, Kinay D, Andermann F, et al. Analysis of shape and positioning of the hippocampal formation: an MRI study in patients with partial epilepsy and healthy controls. *Brain.* 2005;**128**:2442–52.

16. Kim H, Mansi T, Bernasconi N, et al. Surface-based multi-template automated hippocampal segmentation: application to temporal lobe epilepsy. *Med Image Anal.* 2012;**16**:1445–55.

17. Kulaga-Yoskovitz J, Bernhardt BC, Hong SJ, et al. Multi-contrast submillimetric 3 Tesla hippocampal subfield segmentation protocol and dataset. *Sci Data.* 2015;**2**:150059.

18. Sone D, Sato N, Maikusa N, et al. Automated subfield volumetric analysis of hippocampus in temporal lobe epilepsy using high-resolution T2-weighted MR imaging. *NeuroImage Clin.* 2016;**12**:57–64.

19. Caldairou B, Bernhardt BC, Kulaga-Yoskovitz J, et al. A surface patch-based segmentation method for hippocampal subfields. In: Ourselin S, Joskowicz L, Sabuncu MR, et al., eds. *Medical Image Computing and Computer-Assisted Intervention—MICCAI. 2016: 19th International Conference, Athens, Greece, October 17–21, 2016, Proceedings*, pt. 2. Cham: Springer; 2016:379–87.

20. Pipitone J, Park MT, Winterburn J, et al. Multi-atlas segmentation of the whole hippocampus and subfields using multiple automatically generated templates. *NeuroImage.* 2014;**101**:494–512.

21. Wolz R, Aljabar P, Hajnal JV, et al. LEAP: learning embeddings for atlas propagation. *NeuroImage.* 2010;**49**:1316–25.

22. Yushkevich PA, Pluta JB, Wang H, et al. Automated volumetry and regional thickness analysis of hippocampal subfields and medial temporal cortical structures in mild cognitive impairment. *Hum Brain Mapp.* 2015;**36**:258–87.

23. Santyr BG, Goubran M, Lau JC, et al. Investigation of hippocampal substructures in focal temporal lobe epilepsy with and without hippocampal sclerosis at 7T. *J Magn Reson Imaging.* 2017;**45**:1359–70.

24. Kim H, Bernhardt BC, Kulaga-Yoskovitz J, et al. Multivariate hippocampal subfield analysis of local MRI intensity and volume: application to temporal lobe epilepsy. *Med Image Comput Comput Assist Interv.* 2014;**17**:170–8.

25. Bernhardt BC, Bernasconi A, Liu M, et al. The spectrum of structural and functional imaging abnormalities in temporal lobe epilepsy. *Ann Neurol.* 2016;**80**:142–53.

26. Bernhardt BC, Hong SJ, Bernasconi A, et al. Magnetic resonance imaging pattern learning in temporal lobe epilepsy: classification and prognostics. *Ann Neurol.* 2015;**77**:436–46.

27. Maccotta L, Moseley ED, Benzinger TL, et al. Beyond the CA1 subfield: local hippocampal shape changes in MRI-negative temporal lobe epilepsy. *Epilepsia.* 2015;**56**:780–8.

28. Kim H, Mansi T, Bernasconi A, et al. Vertex-wise shape analysis of the hippocampus: disentangling positional differences from volume changes. *Med Image Comput Comput Assist Interv.* 2011;**14**:352–9.

29. Thom M, Bertram EH. Temporal lobe epilepsy. *Handbook Clin Neurol.* 2012;**107**:225–40.

30. Bernasconi N, Duchesne S, Janke A, et al. Whole-brain voxel-based statistical analysis of gray matter and white matter in temporal lobe epilepsy. *NeuroImage.* 2004;**23**:717–23.

31. Bernhardt BC, Bernasconi N, Concha L, et al. Cortical thickness analysis in temporal lobe epilepsy: reproducibility and relation to outcome. *Neurology.* 2010;**74**:1776–84.

32. Bernhardt BC, Bernasconi N, Kim H, et al. Mapping thalamocortical network pathology in temporal lobe epilepsy. *Neurology.* 2012;**78**:129–36.

33. Keller SS, Richardson MP, Schoene-Bake JC, et al. Thalamotemporal alteration and postoperative seizures in temporal lobe epilepsy. *Ann Neurol.* 2015;**77**:760–74.

34. Caciagli L, Bernhardt BC, Hong SJ, et al. Functional network alterations and their structural substrate in drug-resistant epilepsy. *Front Neurosci.* 2014;**8**:411.

35. Voets NL, Bernhardt BC, Kim H, et al. Increased temporolimbic cortical folding complexity in temporal lobe epilepsy. *Neurology.* 2011;**76**:138–44.

36. Alhusaini S, Whelan CD, Doherty CP, et al. Temporal cortex morphology in mesial temporal lobe epilepsy patients and their asymptomatic siblings. *Cereb Cortex.* 2016;**26**:1234–41.

37. Alhusaini S, Whelan CD, Sisodiya SM, et al. Quantitative magnetic resonance imaging traits as endophenotypes for genetic mapping in epilepsy. *NeuroImage Clin.* 2016;**12**:526–34.

38. Ahmadi ME, Hagler DJ Jr, McDonald CR, et al. Side matters: diffusion tensor imaging tractography in left and right temporal lobe epilepsy. *AJNR Am J Neuroradiol.* 2009;**30**:1740–7.

39. Bonilha L, Edwards JC, Kinsman SL, et al. Extrahippocampal gray matter loss and hippocampal deafferentation in patients with temporal lobe epilepsy. *Epilepsia.* 2010;**51**.

40. Concha L, Kim H, Bernasconi A, et al. Spatial patterns of water diffusion along white matter tracts in temporal lobe epilepsy. *Neurology.* 2012;**79**:455–62.

41. Liu M, Bernhardt BC, Hong SJ, et al. The superficial white matter in temporal lobe epilepsy: a key link between structural and functional network disruptions. *Brain.* 2016;**139**:2431–40.

42. Fornito A, Zalesky A, Breakspear M. The connectomics of brain disorders. *Nat Rev Neurosci.* 2015;**16**:159–72.

43. Blumcke I, Thom M, Aronica E, et al. The clinicopathologic spectrum of focal cortical dysplasias: a consensus classification proposed by an ad hoc Task Force of the ILAE Diagnostic Methods Commission. *Epilepsia.* 2011;**52**:158–74.

44. Muhlebner A, Coras R, Kobow K, et al. Neuropathologic measurements in focal cortical dysplasias: validation of the ILAE. 2011 classification system and diagnostic implications for MRI. *Acta Neuropathol.* 2012;**123**:259–72.

45. Barkovich AJ, Guerrini R, Kuzniecky RI, et al. A developmental and genetic classification for malformations of cortical development: update. 2012. *Brain.* 2012;**135**:1348–69.

46. Hong SJ, Bernhardt BC, Schrader DS, et al. Whole-brain MRI phenotyping in dysplasia-related frontal lobe epilepsy. *Neurology.* 2016;**86**:643–50.

47. Trampel R, Bazin PL, Pine K, et al. In-vivo magnetic resonance imaging (MRI) of laminae in the human cortex. *NeuroImage.* 2017. doi:10.1016/j.neuroimage.2017.09.037.

48. Stuber C, Morawski M, Schafer A, et al. Myelin and iron concentration in the human brain: a quantitative study of MRI contrast. *NeuroImage.* 2014;**93** (pt 1):95–106.

49. De Ciantis A, Barba C, Tassi L, et al. 7T MRI in focal epilepsy with unrevealing conventional field strength imaging. *Epilepsia.* 2016;**57**:445–54.

50. Besson P, Andermann F, Dubeau F, et al. Small focal cortical dysplasia lesions are located at the bottom of a deep sulcus. *Brain.* 2008;**131**:3246–55.

51. Colliot O, Antel SB, Naessens VB, et al. In vivo profiling of focal cortical dysplasia on high-resolution MRI with computational models. *Epilepsia.* 2006;**47**:134–42.

52. Bernasconi A, Antel SB, Collins DL, et al. Texture analysis and morphological processing of magnetic resonance imaging assist detection of focal cortical dysplasia in extra-temporal partial epilepsy. *Ann Neurol.* 2001;**49**:770–5.

53. Antel SB, Bernasconi A, Bernasconi N, et al. Computational models of MRI characteristics of focal cortical dysplasia improve lesion detection. *NeuroImage.* 2002;**17**:1755–60.

54. Huppertz HJ, Grimm C, Fauser S, et al. Enhanced visualization of blurred gray-white matter junctions in focal cortical dysplasia by voxel-based 3D MRI analysis. *Epilepsy Res.* 2005;**67**:35–50.

55. Huppertz HJ, Kurthen M, Kassubek J. Voxel-based 3D MRI analysis for the detection of epileptogenic lesions at single subject level. *Epilepsia.* 2009;**50**:155–6.

56. Bien CG, Szinay M, Wagner J, et al. Characteristics and surgical outcomes of patients with refractory magnetic resonance imaging-negative epilepsies. *Arch Neurol.* 2009;**66**:1491–9.

57. Antel SB, Collins DL, Bernasconi N, et al. Automated detection of focal cortical dysplasia lesions using computational models of their MRI characteristics and texture analysis. *NeuroImage.* 2003;**19**:1748–59.

58. Hong SJ, Bernhardt BC, Caldairou B, et al. Multimodal MRI profiling of focal cortical dysplasia type II. *Neurology.* 2017;**88**:734–42.

59. Hong SJ, Kim H, Schrader D, et al. Automated detection of cortical dysplasia type II in MRI-negative epilepsy. *Neurology.* 2014;**83**:48–55.

60. Gill RS, Hong S-J, Fadaie F, et al. Automated detection of epileptogenic cortical malformations using multimodal MRI. In: Cardoso MJ, Arbel T, Carneiro G, et al., eds. *Deep Learning in Medical Image Analysis and Multimodal Learning for Clinical Decision Support: Third International Workshop Proceedings.* Cham: Springer; 2017: 349–56.

61. Bonilha L, Montenegro MA, Rorden C, et al. Voxel-based morphometry reveals excess gray matter concentration in patients with focal cortical dysplasia. *Epilepsia.* 2006;**47**:908–15.

62. Hong SJ, Bernhardt BC, Gill RS, et al. The spectrum of structural and functional network alterations in malformations of cortical development. *Brain.* 2017;**140**:2133–43.

63. Meencke HJ, Janz D. Neuropathological findings in primary generalised epilepsy: a study of eight cases. *Epilepsia.* 1984;**25**:8–21.

64. Opeskin K, Kalnins RM, Halliday G, et al. Idiopathic generalized epilepsy: lack of significant microdysgenesis. *Neurology.* 2000;**55**:1101–6.

65. Betting LE, Li LM, Lopes-Cendes I, et al. Correlation between quantitative EEG and MRI in idiopathic generalized epilepsy. *Hum Brain Mapp.* 2010;**31**:1327–38.

66. Bernhardt BC, Rozen DA, Worsley KJ, et al. Thalamo-cortical network pathology in idiopathic generalized epilepsy: insights from MRI-based morphometric correlation analysis. *NeuroImage.* 2009;**46**:373–81.

67. Natsume J, Bernasconi N, Andermann F, et al. MRI volumetry of the thalamus in temporal, extratemporal, and idiopathic generalized epilepsy. *Neurology.* 2003;**60**:1296–300.

68. Ciumas C, Savic I. Structural changes in patients with primary generalized tonic and clonic seizures. *Neurology.* 2006;**67**:683–6.

69. Caplan R, Levitt J, Siddarth P, et al. Frontal and temporal volumes in childhood absence epilepsy. *Epilepsia.* 2009;**50**:2466–72.

70. Chan CH, Briellmann RS, Pell GS, et al. Thalamic atrophy in childhood absence epilepsy. *Epilepsia.* 2006;**47**:399–405.

71. O'Muircheartaigh J, Vollmar C, Barker GJ, et al. Focal structural changes and cognitive dysfunction in juvenile myoclonic epilepsy. *Neurology.* 2011;76:34–40.

72. Tae WS, Kim SH, Joo EY, et al. Cortical thickness abnormality in juvenile myoclonic epilepsy. *J Neurol.* 2008;255:561–6.

73. Ronan L, Alhusaini S, Scanlon C, et al. Widespread cortical morphologic changes in juvenile myoclonic epilepsy: evidence from structural MRI. *Epilepsia.* 2012;53: 651–8.

74. Zhang Z, Liao W, Chen H, et al. Altered functional-structural coupling of large-scale brain networks in idiopathic generalized epilepsy. *Brain.* 2011;134:2912–28.

75. Liao W, Zhang Z, Mantini D, et al. Relationship between large-scale functional and structural covariance networks in idiopathic generalized epilepsy. *Brain Connect.* 2013;3:240–54.

Chapter 7

Imaging White Matter Pathology in Epilepsy

Min Liu, Luis Concha, Boris C. Bernhardt, Neda Bernasconi, and Andrea Bernasconi

7.1 Introduction

While the study of the gray matter has been the main focus of epilepsy research, observations of white matter (WM) atrophy and gliosis date back to early seminal histopathological assessment of surgically resected tissue in patients with temporal lobe epilepsy.[1] WM may subserve seizure propagation, forming an integral part of the epileptogenic network. Recent advances in MRI techniques and quantitative postprocessing algorithms have provided unprecedented tools to examine the microstructure and architecture of this compartment. While early studies based on manual segmentation have provided volume measurements of this compartment, more recent computational approaches integrating various imaging contrasts, such as T2 relaxometry, diffusion tractography, and myelin imaging, lend biologically meaningful indices of its microstructural properties. In the past two decades, a large body of work has indeed revealed novel information regarding the distribution and nature of WM pathology in epilepsy. Together with gray matter findings, these results consolidate the view that focal epilepsy is a system-level disorder.

7.2 Voxel-Based Techniques

Early MRI studies of patients with temporal lobe epilepsy (TLE) noted WM atrophy alongside hippocampal anomalies during visual inspection.[2] Such observations motivated quantitative analysis using manual volumetry. Similar to hippocampal volumetry, manual tracing requires expert knowledge of anatomy and provides a single summary statistic of the queried structure. In particular, its application to the fimbria-fornix, the major projection of the hippocampus in TLE, revealed significant atrophy concordant with the side of hippocampal atrophy.[3] While the time-consuming nature of the procedure has confined its application to small WM structures, automated methods enabled efficient quantitation of lobar and global volumes. Indeed, several studies have identified marked reduction of the WM lobar volume in the temporal, frontal, and parietal lobes mainly ipsilateral to the side of seizure focus.[4] The degree of WM loss was associated with disease duration[4] and the severity of cognitive impairment, including (but not limited to) processing speed and full-scale and verbal IQ.[5] Voxel-based morphometry studies in TLE have shown WM reduction mostly colocalized with GM abnormalities.[6] Particularly, our group found WM loss in the temporopolar, entorhinal, and perirhinal areas ipsilateral to the seizure focus, suggesting a disconnection process involving preferentially frontolimbic pathways.[7]

WM pathology has also been studied using T2 relaxometry, a quantitative analysis of T2-weighted signal. Region-of-interest analysis and voxel-based statistical comparisons reported group-level increases in the anterior temporal lobe WM in TLE patients.[8] We showed that while hippocampal T2 relaxometry provides a correct lateralization of the seizure,[9] prolongation of WM T2-relaxation time in the temporal stem occurs in a rather bilateral and symmetric fashion.[10] Altogether, these early studies set the basis for large-scale extrahippocampal WM disruption in TLE.

7.3 Diffusion-Weighted MRI

Despite relatively coarse spatial resolution (typically 2 mm), diffusion-weighted MRI and its analytical extensions, particularly diffusion tensor imaging (DTI), exploit the diffusion of water molecules to probe tissue microstructure, which is especially useful in coherently organized tissue, such as the WM.[11] While its sensitivity and specificity for the identification of the primary epileptogenic lesion are modest, this technique has revealed WM anomalies in several forms of epilepsy, particularly TLE.[12]

The most often reported diffusion MRI metric is related to diffusion anisotropy (e.g., fractional anisotropy, FA), which tends to be high in regions where densely packed axons have very similar orientations. Reduced diffusion anisotropy at the expense of increased radial diffusivity is interpreted as reduced axonal density and alterations of myelin architecture. Histological examination of the fornix and the temporal pole WM structures derived from TLE surgical specimens have confirmed these anomalies.[13,14] Small caliber axons, in particular, are reduced in number in TLE patients.[15] Although perhaps to a lesser degree, other histological features, such as WM gliosis, may also play a role in altered water diffusion metrics. Notably, orientational coherence of axons is reduced by the infiltration of oligodendrocytes, which are more common in regions of WM hyperintensity, and may reflect an attempt to compensate for progressive myelin loss.[16] Independently of gliosis, the number of heterotopic neurons, commonly seen in the superficial WM, is higher in TLE patients than in controls;[17] however, their density and overall dimension appear to be insufficient to modulate water diffusion profiles.

7.3.1 Diffusion Tensor Imaging of Deep White Matter Tracts

Diffusion data can be processed through tractography algorithms,[18] which reconstruct fiber pathways running along plausible trajectories in voxel space. While somewhat challenged in regions of crossing fibers and gray-white matter interface, tractography can generate consistent results for a number of deep WM tracts.[19] Probabilistic tract atlases further enable automated tract labeling by providing a standard template for fiber location and orientation.[20] The virtually reconstructed trajectories have shown good correspondence with postmortem dissection, and have been cross-validated by comparative sacrificial tracing studies in nonhuman primates.[21] Diffusion characteristics can be also assessed throughout the brain without a priori hypothesis. Tract-based spatial statistics (TBSS) allow projecting individual FA values onto a group-wise WM skeleton, so that the structural core is reasonably aligned even in the presence of slight misregistration across subjects;[22] furthermore, sensitivity is increased due to fewer number of voxelwise comparisons.

In TLE, diffusion abnormalities are evident in limbic structures,[23] including the fornix and cingulum,[24] but also in the temporal lobe association tracts, such as the arcuate and uncinate fasciculi.[25] Although changes are more marked ipsilateral to the seizure focus, there is a considerable degree of bilateral involvement in a number of tracts. Tracked-based segmental analysis of these tracts has revealed centrifugal pattern of mean diffusivity increases, where the degree of anomalies gradually tapered off as the tracts exited the temporal lobe.[26] Outside the temporal lobe, diffusion anomalies are found in the corpus callosum and the frontoparietal pathways. Studies on the effect of hemispheric laterality on diffusion parameters have yielded conflicting results, with some reports showing more alterations in right TLE,[27] while others showed the opposite.[28] These discrepancies could be due to differences in the TLE cohorts with respect to the severity of hippocampal damage, precipitating factors, or disease duration. Several factors appear to influence the degree and extent of WM diffusion abnormalities, yet the largest effect appears to be related to the presence of mesial temporal sclerosis. Patients with severe mesiotemporal lobe sclerosis have more widespread and severe diffusion abnormalities than those with subtle pathology.[27] Diffusion anomalies have also been shown to be positively correlated to longer disease duration.[29] Combining TBSS with resting-state functional MRI, we have highlighted the relationship between temporal lobe WM anomalies and reduced functional connectivity of the default mode network,[30] an assembly of regions thought to play a key role in internally generated cognition, memory, and future planning.

Electron microscopy of TLE fornix specimens obtained at surgery revealed considerably more damage in patients with severe hippocampal sclerosis, consisting of reduced axonal density, and myelin abnormalities.[13] Similar characteristics were noted bilaterally in the TLE fornix of autopsy specimens.[15] Axonal degeneration is expected in the fornix ipsilateral to the focus, and although a fraction of axons crosses the midline toward the contralateral hippocampus, ipsilateral hippocampal pathology may not be the only factor leading to bilateral abnormalities of the fornix and other WM fascicles. The fact that diffusion anomalies may be present soon after diagnosis[31] is suggestive of a likely contribution of neurodevelopmental processes. Recently, a multifactorial model of WM damage in TLE was proposed, in which the underlying pathology and effects of seizures play a significant role in comparison to genetic factors.[32]

Postoperative diffusion MRI studies have revealed two distinct profiles that may inform of specific histopathological changes. Comparing post- to presurgical values, several studies have found decreased FA and increased MD in the fimbria fornix, parahippocampal cingulum, uncinate fasciculus, and inferior fronto-occipital fasciculus ipsilateral to the epileptic focus.[33]

Alterations of diffusion parameters of WM fascicles directly affected by surgical resections are likely due to Wallerian degeneration. Indeed, acute reduction of diffusion anisotropy at the expense of reduced longitudinal diffusivity[34] is reflective of axonal fragmentation or beading; conversely chronic increased radial diffusivity likely relates to reduced axonal density and myelin degradation.[35] WM distant from the resection, including the corona radiata and external capsules, showed increased FA.[28,33] While originally deemed as plastic phenomena and functional adaptations, the paradoxical increase of FA in ipsilateral tracts is likely due to the degeneration of a single fiber population in regions of fiber crossing. Conversely, increased anisotropy in the hemisphere contralateral to surgery may in fact be related to plastic changes in language networks.[28]

Compared to the large body of literature in TLE, fewer studies have addressed WM alterations in focal cortical dysplasia (FCD). Voxel-based analysis has revealed reduced anisotropy and increased diffusivity, subjacent to the MRI-visible lesion or diffusely across both hemispheres.[36,37] Aside from the direct effects of the structural pathology, reduced anisotropy may be related to abnormal myelination or the presence of ectopic neurons. Conversely, increased diffusivity has been interpreted as a defect of neurogenesis or cell loss resulting in increased extracellular space. Only a few studies to date have used fiber tractography in FCD. An early study showed both a reduction of the subcortical fibers and reduced connection between the subcortex and deep WM.[38]

7.3.2 Surface-Based Analysis of the Superficial White Matter

Most DTI studies have focused on assessing the integrity of deep WM bundles. The superficial WM immediately subjacent to the cortical mantle has been largely neglected. Given its anatomical proximity to neocortical networks and role in maintaining long-range connectivity, the analysis of this compartment may provide novel insights in the pathophysiology of various epilepsy syndromes. The complex fiber trajectories of the less densely myelinated superficial WM challenge conventional diffusion tractography and TBSS.[39] Conversely, surface-based approaches allow vertex-wise sampling of diffusion parameters at multiple WM depths. In TLE, diffusion alterations (increased diffusivity and decreased anisotropy) of the superficial WM are found primarily in ipsilateral limbic regions[40] (Figure 7.1). In extratemporal epilepsy related to FCD Type II, subtle diffusion anomalies are found both at the lesional site and in the normal appearing perilesional WM[41] (Figure 7.2).

7.3.3 Diffusion Tensor Imaging: Relation to Cognition

Cognitive abilities are dependent on the orchestrated action of several brain regions interconnected by association and commissural fibers, many of which are affected in epilepsy. It is thus not surprising that TLE patients may have cognitive deficits beyond temporal lobe functions.[42] Diffusion metrics of the superior longitudinal fasciculus, cingulum, and temporal lobe WM correlate with working memory performance.[43] Similarly, diffusion anisotropy of the uncinate, arcuate, and inferior fronto-occipital fascicles, as well as temporal cingulum positively correlate to verbal memory.[44] Moreover, alterations of diffusion parameters in the temporal lobe WM are associated with delayed and immediate memory performances.[45,46] Preliminary data in patients with both idiopathic generalized and focal epilepsy show WM diffusion alterations already at disease onset,[31] which were shown to correlate with executive functions.[47]

Tractography has shown potential clinical utility in presurgical mapping of the language network with the goal of avoiding deficit. In TLE, diffusion abnormalities in the arcuate, uncinate, and inferior fronto-occipital fasciculi have been shown to be associated with impaired language performance.[48] Left TLE patients who exhibited atypical language activations in functional MRI presented reduced FA asymmetry in connecting WM pathways,[49] and the structural disruption was correlated with decreased functional connectivity between frontal and temporal cortices.[50] Notably, a greater asymmetry in pathways connecting to eloquent cortices correlated with worse decline in naming function after anterior temporal lobe resection also in the dominant hemisphere.[51] WM diffusion

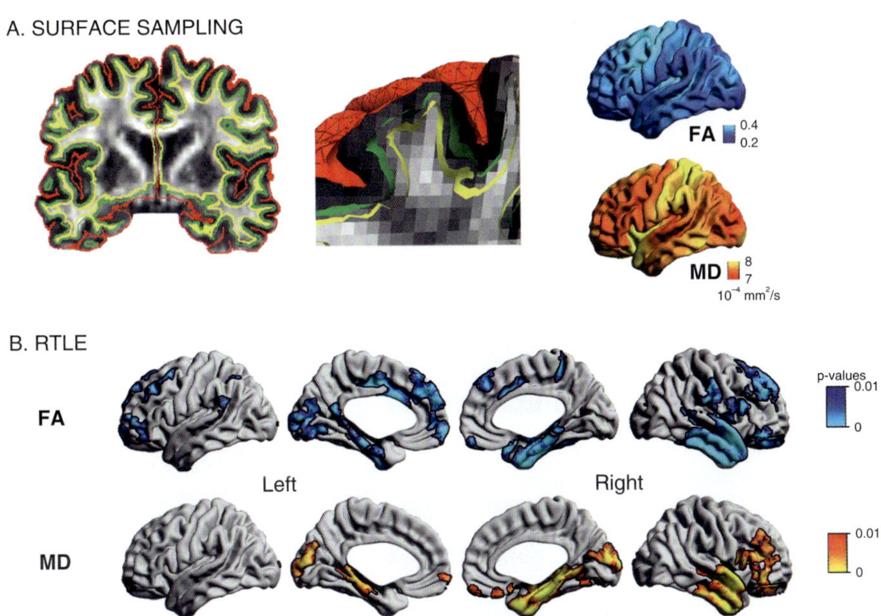

Figure 7.1. Surface-based mapping of superficial white matter (SWM) diffusion in TLE. (A) After generating the inner (WM-GM, green) and outer (GM-CSF, red) cortical surfaces, we computed a Laplacian potential field between the WM-GM interface and the ventricular walls to guide placement of a surface running 2 mm below the WM-GM boundary (SWM, yellow). Fractional anisotropy (FA) and mean diffusivity (MD) were sampled on this surface. (B) SWM diffusion anomalies in right (RTLE) patients relative to controls showing predominantly ipsilateral temporal and limbic anomalies. Similar anomalies are found in left TLE (not shown). Findings were corrected for multiple comparisons thresholded at $P_{FWE} < .05$ using random field theory for nonisotropic images (cluster threshold $p < .01$).

characteristics may thus provide valuable information to predict which patients are at higher risk of developing cognitive deficits. So far however, most studies have been group-based; future efforts should assess the predictive value of this technique at an individual level. Using FA asymmetry of the arcuate fasciculus, language could be lateralized in 83% of patients, with increases up to 96% when combining information from functional MRI and handedness.[52] Additionally, if progressive WM abnormalities are secondary to ongoing seizures, then appropriate treatment could halt further cognitive deterioration. Another application of tractography has been in surgical planning to avoid functional deficit. More recently, WM damage in TLE patients has been shown to be a factor associated to poor postsurgical outcome. A study using tractography and automated fiber quantification found that diffusion abnormalities in the ipsilateral dorsal fornix and in the contralateral parahippocampal bundle contribute to persistent post-operative seizures.[53]

7.4 Analytical Approaches to the White Matter beyond Tensor Model

The tensor model has provided extremely valuable information regarding WM microstructure in epilepsy, yet DTI has a number of limitations. First and foremost is its incapability of resolving fiber crossing, which may result in erroneous estimations of diffusion anisotropy. Second, the tensor model assumes that diffusion is Gaussian, while in reality the complex intracellular and extracellular environment causes the diffusion of water molecules to deviate considerably from this pattern; this oversimplification prevents precise description of tissue microstructure. Many alternative analytical methods have been proposed in the last decade that extend the capabilities of diffusion MRI and are beginning to be used for the study of epilepsy as acquisition requirements are relaxed to the point that they can be applied in clinical populations. For example, diffusion kurtosis imaging, an extension of tensor model to measure the degree to which diffusion deviates from Gaussian behavior, i.e., mean kurtosis, yields complementary information regarding diffusion heterogeneity.[54] In TLE, in addition to revealing more marked diffusion profiles along ipsilateral WM fibers,[55,56] this technique has shown abnormalities in both gray and white matter extending to regions not detected by conventional DTI measures. Diffusion spectrum imaging, a method querying complex distributions of intravoxel fiber orientation, has revealed that reduced FA in the mesiotemporal lobes in TLE patients likely results from the reduction in intracellular diffusion related to neurite (i.e., dendrites and axonal) loss, whereas reduced FA in extratemporal regions was confounded

71

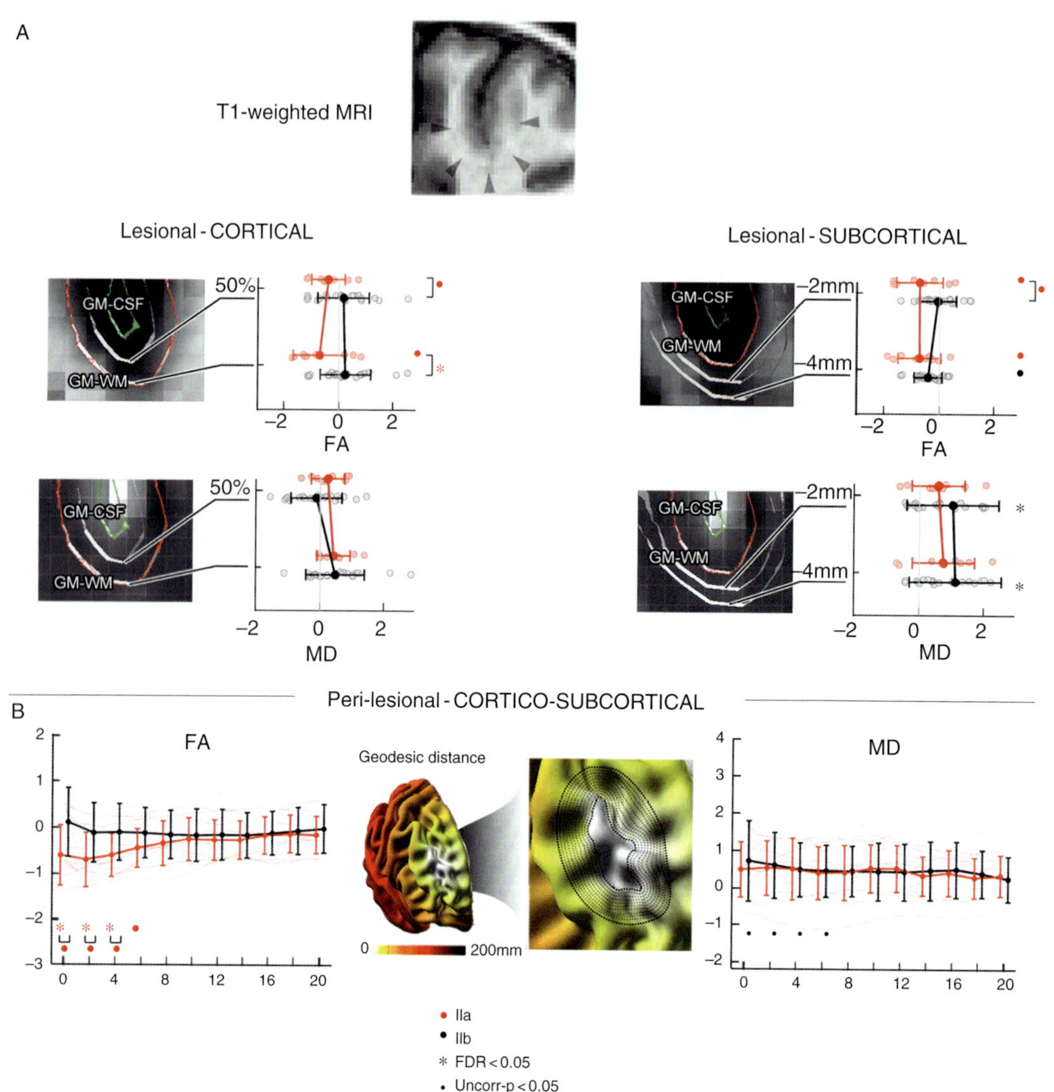

Figure 7.2. Multisurface profiling of diffusion MRI (dMRI) parameters in focal cortical dysplasia. (A) For lesion profiling, individual patients (normalized with respect to corresponding regions in healthy controls) are plotted as a function of intracortical and subcortical level, separately for Type IIA (red dot) and Type IIB (black dot); zero reference line indicates the mean of controls; mean values and standard deviations in patients are shown as horizontal lines, color coded by patient group. An example case is with arrowheads pointing to the lesion on the T1-weighted MRI. Overall, Type IIB lesions display increased subcortical mean diffusivity (MD), while IIA are associated with decreased fractional anisotropy (FA) at the gray matter (GM)-white matter (WM) interface and subcortical WM. (B) For distance-based profiling, normalized features are averaged across cortical and subcortical surfaces, and plotted relative to the geodesic distance from the primary lesion in steps of 4 mm. Subcortical diffusion alterations extend up to 16 mm, with anomalies being more marked in Type IIB. Conversely, reduced cortical FA, specific to Type IIA, extended up to 6 mm outside the lesion. Asterisks indicate significant differences with controls and between cohort contrasts after correction of multiple comparisons (FDR < .05); small dots indicate uncorrected findings at $p < .05$.

by fiber orientation changes, potentially reflecting disorganized fiber orientation and packing.[57] Restricted spectrum imaging is based on a multicompartment model that quantifies the degree to which diffusion heterogeneity within a voxel is driven by intra-axonal water versus extracellular diffusion, while accounting for fiber orientation. This multishell acquisition method has detected more widespread reduction of neurite density than anisotropy.[58]

Fixel-based analysis is a diffusion-weighted MRI reconstruction technique that combines the measurements of fiber cross-sectional area, a measure of morphology, with microstructural information of fiber density,[59] thereby providing a sensitive marker of intra-axonal volume. Fixel-based morphometry has assisted in identifying WM atrophy secondary to repeated focal seizures and global insults in TLE patients.[32] Neurite orientation dispersion and density imaging, commonly referred to as NODDI,[60] is another advanced diffusion imaging reconstruction technique based on a multishell acquisition protocol that allows an estimation of intra- and extracellular volume fractions in order to measure neurites morphology. This technique has the advantage to model both gray and white matter, a desired ability in the investigation of epilepsy, yet so far not exploited. As more complex models are introduced to probe individual tissue components, histologic-MRI correlational studies may help validate the biological underpinning of the novel metrics.

7.5 White Matter Connectome

Growing evidence of extensive structural and functional abnormalities in epilepsy has emphasized the notion that epilepsy is a disorder affecting distributed neural networks. Recent methodological advances in graph theoretical analysis lend tools to characterize topological aspects of interconnected regions.[61] In this mathematical formalism, the brain connectome is modeled using a parcellation scheme of gray matter regions, i.e., nodes. Importantly, WM pathways reconstructed by in vivo diffusion tractography provide a good approximation of the actual anatomical connectivity, forming the edges between nodes. Notably, however, there is so far no consensus on metrics of connectivity strength, with some studies adopting parameters based on deterministic tractography (including FA, number of streamlines, and fiber lengths), while others using fiber connections density derived from probabilistic tractography.

Most DTI-based network studies have focused on focal epilepsy, particularly TLE. Collectively, findings have argued for a reorganization of macrolevel topological properties. In local circuits, on the other hand, increased integration was found within limbic and default mode networks, even though fiber connectivity among these regions was reduced.[62] This dichotomy of changes between large-scale topology and individual fiber level illustrates the advantage of network analyses in addressing how regions interrelate as a whole, which is not possible via querying the integrity of a specific connection, and may potentially provide useful diagnostic and prognostic information for the assessment of individual patients.[63] Indeed, combined with machine learning algorithms, presurgical connectome data have shown the ability to predict surgery outcome in 70% of TLE patients, an accuracy similar to "expert-based" clinical decision.[64] Moreover, the presurgical connection strengths between ipsilateral temporal and extratemporal regions have been able to predict seizure-free surgical outcome in TLE patients with performances far beyond that of presurgical clinical data.[55] Specifically, compared to seizure-free patients, presurgical connectomes in non-seizure-free patients exhibited higher integration within ipsilateral temporal lobe subnetwork.[61]

In light of work in healthy populations that utilizes structural connectomes to simulate functional dynamics,[65] recent advances in network diffusion science offer new models to simulate the communication process between network nodes and have provided preliminary evidence for the spreading pattern of neuronal damage from the seizure focus at a group level in TLE.[66] Emerging data suggest that computational models of seizure spread and epileptogenicity based on structural connectomes[67] may also help localizing surgical targets[68] and possibly improve the prediction of postsurgical seizure outcome.[69] In patients with IGE, network analysis based on diffusion MRI and resting state functional MRI data has identified altered topology in regions that had previously been hypothesized to play a role in the epileptogenesis of this condition, including the mesial frontal cortex, putamen, and thalamus.[70,71]

To date, WM organization in cohorts with validated FCD has been seldom studied. Our recent study employing gray matter structural covariance and functional connectivity analysis suggests marked network rearrangement particularly in FCD I.[72] In patients with frontal lobe epilepsy and suspected subtle dysplasia, overall reductions of global and nodal efficiency have been observed, as well as network segregation. Given the macro-scale analyses, it is likely that some connectome findings may relate to cognitive difficulties commonly seen in epileptic populations. Indeed, a structural connectivity study reported that patients with severe cognitive impairment may have more marked alterations in network topology compared with healthy individuals as well as patients with only little cognitive impairment.[73]

7.6 Conclusion

Epilepsy is a network disorder and the WM is the substrate connecting its various components. A proper conceptualization of this condition entails equal attention to both the gray and white matter at the site of the seizure focus and at a distance from it.[74] Novel in vivo diffusion-weighted MRI acquisition and reconstruction techniques, but also emerging imaging methods assessing myelin content,[75] provide an increasingly comprehensive view of this pivotal compartment and its relationship to neuronal processes. They have successfully characterized pathology and mapped its regional distribution. Although initial studies have suggested the relevance of single markers of WM integrity to clinical outcomes, a better understanding of this relationship demands combined analysis of descriptors of gray and white matter microstructure, connectome-level features, and lesional markers. In this context, machine learning is the method of choice to extract critical features, patterns, and relationships from such high-dimensional datasets that might otherwise be missed. Moreover, paralleling the move toward big data analyses in neuroscience, epilepsy research would benefit from data-sharing initiatives to validate results and address new hypotheses.

References

1. Falconer MA, Serafetinides EA, Corsellis JAN. Etiology and pathogenesis of temporal lobe epilepsy. *Arch Neurol.* 1964;**10**:233–48.

2. Meiners LC, Witkamp TD, de Kort GA, et al. Relevance of temporal lobe white matter changes in hippocampal sclerosis. Magnetic resonance imaging and histology. *Invest Radiol.* 1999;**34**:38–45.

3. Kuzniecky R, Bilir E, Gilliam F, et al. Quantitative MRI in temporal lobe epilepsy: evidence for fornix atrophy. *Neurology.* 1999;**53**:496–501.

4. Seidenberg M, Kelly KG, Parrish J, et al. Ipsilateral and contralateral MRI volumetric abnormalities in chronic unilateral temporal lobe epilepsy and their clinical correlates. *Epilepsia.* 2005;**46**:420–30.

5. Hermann B, Seidenberg M, Bell B, et al. Extratemporal quantitative MR volumetrics and neuropsychological status in temporal lobe epilepsy. *J Int Neuropsychol Soc.* 2003;**9**:353–62.

6. McMillan AB, Hermann BP, Johnson SC, et al. Voxel-based morphometry of unilateral temporal lobe epilepsy reveals abnormalities in cerebral white matter. *NeuroImage.* 2004;**23**:167–74.

7. Bernasconi N, Duchesne S, Janke A, et al. Whole-brain voxel-based statistical analysis of gray matter and white matter in temporal lobe epilepsy. *NeuroImage.* 2004;**23**:717–23.

8. Pell GS, Briellmann RS, Waites AB, et al. Voxel-based relaxometry: a new approach for analysis of T2 relaxometry changes in epilepsy. *NeuroImage.* 2004;**21**:707–13.

9. Bernasconi A, Bernasconi N, Caramanos Z, et al. T2 relaxometry can lateralize mesial temporal lobe epilepsy in patients with normal MRI. *NeuroImage.* 2000;**12**:739–46.

10. Townsend TN, Bernasconi N, Pike GB, et al. Quantitative analysis of temporal lobe white matter T2 relaxation time in temporal lobe epilepsy. *NeuroImage.* 2004;**23**:318–24.

11. Concha L. A macroscopic view of microstructure: using diffusion-weighted images to infer damage, repair, and plasticity of white matter. *Neuroscience.* 2014;**276**:14–28.

12. Slinger G, Sinke MR, Braun KP, et al. White matter abnormalities at a regional and voxel level in focal and generalized epilepsy: A systematic review and meta-analysis. *NeuroImage Clin.* 2016;**12**:902–9.

13. Concha L, Livy DJ, Beaulieu C, et al. In vivo diffusion tensor imaging and histopathology of the fimbria-fornix in temporal lobe epilepsy. *J Neurosci.* 2010;**30**:996–1002.

14. Garbelli R, Milesi G, Medici V, et al. Blurring in patients with temporal lobe epilepsy: clinical, high-field imaging and ultrastructural study. *Brain.* 2012;**135**:2337–49.

15. Ozdogmus O, Cavdar S, Ersoy Y, et al. A preliminary study, using electron and light-microscopic methods, of axon numbers in the fornix in autopsies of patients with temporal lobe epilepsy. *Anat Sci Int.* 2009;**84**:2–6.

16. Stefanits H, Czech T, Pataraia E, et al. Prominent oligodendroglial response in surgical specimens of patients with temporal lobe epilepsy. *Clin Neuropathol.* 2012;**31**:409–17.

17. Thom M, Holton JL, D'Arrigo C, et al. Microdysgenesis with abnormal cortical myelinated fibres in temporal lobe epilepsy: a histopathological study with calbindin D-28-K immunohistochemistry. *Neuropathol Appl Neurobiol.* 2000;**26**:251–7.

18. Behrens TE, Johansen-Berg H, Woolrich MW, et al. Non-invasive mapping of connections between human thalamus and cortex using diffusion imaging. *Nat Neurosci.* 2003;**6**:750–7.

19. Wakana S, Caprihan A, Panzenboeck MM, et al. Reproducibility of quantitative tractography methods applied to cerebral white matter. *NeuroImage.* 2007;**36**:630–44.

20. Hagler DJ Jr, Ahmadi ME, Kuperman J, et al. Automated white-matter tractography using a probabilistic diffusion tensor atlas: Application to temporal lobe epilepsy. *Hum Brain Mapp.* 2009;**30**:1535–47.

21. Dauguet J, Peled S, Berezovskii V, et al. Comparison of fiber tracts derived from in-vivo DTI tractography with 3D histological neural tract tracer reconstruction on a macaque brain. *NeuroImage.* 2007;**37**:530–8.

22. Smith SM, Jenkinson M, Johansen-Berg H, et al. Tract-based spatial statistics: voxelwise analysis of multi-subject diffusion data. *NeuroImage.* 2006;**31**:1487–505.

23. Thivard L, Lehericy S, Krainik A, et al. Diffusion tensor imaging in medial temporal lobe epilepsy with hippocampal sclerosis. *NeuroImage.* 2005;**28**:682–90.

24. Concha L, Beaulieu C, Gross DW. Bilateral limbic diffusion abnormalities in unilateral temporal lobe epilepsy. *Ann Neurol.* 2005;**57**:188–96.

25. Otte WM, Dijkhuizen RM, van Meer MP, et al. Characterization of functional and structural integrity in experimental focal epilepsy: reduced network efficiency coincides with white matter changes. *PLOS ONE.* 2012;**7**:e39078.

26. Concha L, Kim H, Bernasconi A, et al. Spatial patterns of water diffusion along white matter tracts in temporal lobe epilepsy. *Neurology.* 2012;**79**:455–62.

27. Liu M, Concha L, Lebel C, et al. Mesial temporal sclerosis is linked with more widespread white matter changes in temporal lobe epilepsy. *NeuroImage Clin.* 2012;**1**:99–105.

28. Pustina D, Doucet G, Sperling M, et al. Increased microstructural white matter correlations in left, but not right, temporal lobe epilepsy. *Hum Brain Mapp.* 2015;**36**:85–98.

29. Keller SS, Schoene-Bake JC, Gerdes JS, et al. Concomitant fractional anisotropy and volumetric abnormalities in temporal lobe epilepsy: cross-sectional evidence for progressive neurologic injury. *PLOS ONE.* 2012;**7**:e46791.

30. Voets NL, Beckmann CF, Cole DM, et al. Structural substrates for resting network disruption in temporal lobe epilepsy. *Brain.* 2012;**135**:2350–7.

31. Hutchinson E, Pulsipher D, Dabbs K, et al. Children with new-onset epilepsy exhibit diffusion abnormalities in cerebral white matter in the absence of volumetric differences. *Epilepsy Res.* 2010;**88**:208–14.

32. Vaughan DN, Raffelt D, Curwood E, et al. Tract-specific atrophy in focal epilepsy: Disease, genetics, or seizures? *Ann Neurol.* 2017;**81**:240–50.

33. Winston GP, Stretton J, Sidhu MK, et al. Progressive white matter changes following anterior temporal lobe resection for epilepsy. *NeuroImage Clin.* 2014;**4**:190–200.

34. Liu M, Gross DW, Wheatley BM, et al. The acute phase of Wallerian degeneration: longitudinal diffusion tensor imaging of the fornix following temporal lobe surgery. *NeuroImage.* 2013;**74**:128–39.

35. Concha L, Gross DW, Wheatley BM, et al. Diffusion tensor imaging of time-dependent axonal and myelin degradation after corpus callosotomy in epilepsy patients. *NeuroImage.* 2006;**32**:1090–9.

36. Eriksson SH, Rugg-Gunn FJ, Symms MR, et al. Diffusion tensor imaging in patients with epilepsy and malformations of cortical development. *Brain.* 2001;**124**:617–26.

37. Fonseca Vde C, Yasuda CL, Tedeschi GG, et al. White matter abnormalities in patients with focal cortical dysplasia revealed by diffusion tensor imaging analysis in a voxelwise approach. *Front Neurol.* 2012;**3**:121.

38. Lee SK, Kim DI, Mori S, et al. Diffusion tensor MRI visualizes decreased subcortical fiber connectivity in focal cortical dysplasia. *NeuroImage.* 2004;**22**:1826–9.

39. Reveley C, Seth AK, Pierpaoli C, et al. Superficial white matter fiber systems impede detection of long-range cortical connections in diffusion MR tractography. *Proc Natl Acad Sci USA.* 2015;**112**:E2820–8.

40. Liu M, Bernhardt BC, Hong SJ, et al. The superficial white matter in temporal lobe epilepsy: a key link between structural and functional network disruptions. *Brain.* 2016;**139**:2431–40.

41. Hong SJ, Bernhardt BC, Caldairou B, et al. Multimodal MRI profiling of focal cortical dysplasia type II. *Neurology.* 2017;**88**:734–42.

42. Hermann B, Seidenberg M, Lee EJ, et al. Cognitive phenotypes in temporal lobe epilepsy. *J Int Neuropsychol Soc.* 2007;**13**:12–20.

43. Winston GP, Stretton J, Sidhu MK, et al. Structural correlates of impaired working memory in hippocampal sclerosis. *Epilepsia.* 2013;**54**:1143–53.

44. McDonald CR, Leyden KM, Hagler DJ, et al. White matter microstructure complements morphometry for predicting verbal memory in epilepsy. *Cortex.* 2014;**58**:139–50.

45. Yogarajah M, Powell HW, Parker GJ, et al. Tractography of the parahippocampal gyrus and material specific memory impairment in unilateral temporal lobe epilepsy. *NeuroImage.* 2008;**40**:1755–64.

46. Riley JD, Franklin DL, Choi V, et al. Altered white matter integrity in temporal lobe epilepsy: association with cognitive and clinical profiles. *Epilepsia.* 2010;**51**:536–45.

47. Ekmekci B, Bulut HT, Gumustas F, et al. The relationship between white matter abnormalities and cognitive functions in new-onset juvenile myoclonic epilepsy. *Epilepsy Behav.* 2016;**62**:166–70.

48. McDonald CR, Ahmadi ME, Hagler DJ, et al. Diffusion tensor imaging correlates of memory and language impairments in temporal lobe epilepsy. *Neurology.* 2008;**71**:1869–76.

49. Powell HW, Parker GJ, Alexander DC, et al. Abnormalities of language networks in temporal lobe epilepsy. *NeuroImage.* 2007;**36**:209–21.

50. Takaya S, Liu H, Greve DN, et al. Altered anterior-posterior connectivity through the arcuate fasciculus in temporal lobe epilepsy. *Hum Brain Mapp.* 2016;**37**:4425–38.

51. Powell HW, Parker GJ, Alexander DC, et al. Imaging language pathways predicts postoperative naming deficits. *J Neurol Neurosurg Psychiatry.* 2008;**79**:327–30.

52. Ellmore TM, Beauchamp MS, Breier JI, et al. Temporal lobe white matter asymmetry and language laterality in epilepsy patients. *NeuroImage.* 2010;**49**:2033–44.

53. Keller SS, Glenn GR, Weber B, et al. Preoperative automated fibre quantification predicts postoperative seizure outcome in temporal lobe epilepsy. *Brain.* 2017;**140**:68–82.

54. Wu EX, Cheung MM. MR diffusion kurtosis imaging for neural tissue characterization. *NMR Biomed.* 2010;**23**:836–48.

55. Bonilha L, Jensen JH, Baker N, et al. The brain connectome as a personalized biomarker of seizure outcomes after temporal lobectomy. *Neurology.* 2015;**84**:1846–53.

56. Glenn GR, Jensen JH, Helpern JA, et al. Epilepsy-related cytoarchitectonic abnormalities along white matter pathways. *J Neurol Neurosurg Psychiatry.* 2016;**87**:930–6.

57. Lemkaddem A, Daducci A, Kunz N, et al. Connectivity and tissue microstructural alterations in right and left temporal lobe epilepsy revealed by diffusion spectrum imaging. *NeuroImage Clin.* 2014;**5**:349–58.

58. Loi RQ, Leyden KM, Balachandra A, et al. Restriction spectrum imaging reveals decreased neurite density in patients with temporal lobe epilepsy. *Epilepsia.* 2016;**57**:1897–906.

59. Raffelt DA, Tournier JD, Smith RE, et al. Investigating white matter fibre density and morphology using fixel-based analysis. *NeuroImage.* 2017;**144**:58–73.

60. Zhang H, Schneider T, Wheeler-Kingshott CA, et al. NODDI: practical in vivo neurite orientation dispersion and density imaging of the human brain. *NeuroImage.* 2012;**61**:1000–16.

61. Bernhardt BC, Hong S, Bernasconi A, et al. Imaging structural and functional brain networks in temporal lobe epilepsy. *Front Hum Neurosci.* 2013;**7**:624.

62. Bonilha L, Nesland T, Martz GU, et al. Medial temporal lobe epilepsy is associated with neuronal fibre loss and paradoxical increase in structural connectivity of limbic structures. *J Neurol Neurosurg Psychiatry.* 2012;**83**:903–9.

63. Gleichgerrcht E, Kocher M, Bonilha L. Connectomics and graph theory analyses: novel insights into network abnormalities in epilepsy. *Epilepsia.* 2015;**56**:1660–8.

64. Munsell BC, Wee CY, Keller SS, et al. Evaluation of machine learning algorithms for treatment outcome prediction in patients with epilepsy based on structural connectome data. *NeuroImage.* 2015;**118**:219–30.

65. Misic B, Betzel RF, Nematzadeh A, et al. Cooperative and competitive spreading dynamics on the human connectome. *Neuron.* 2015;**86**:1518–29.

66. Abdelnour F, Mueller S, Raj A. Relating cortical atrophy in temporal lobe epilepsy with graph diffusion-based network models. *PLOS Comput Biol.* 2015;**11**:e1004564.

67. Besson P, Bandt SK, Proix T, et al. Anatomic consistencies across epilepsies: a stereotactic-EEG informed high-resolution structural connectivity study. *Brain.* 2017;**140**:2639–52.

68. Hutchings F, Han CE, Keller SS, et al. Predicting surgery targets in temporal lobe epilepsy through structural connectome based simulations. *PLOS Comput Biol.* 2015;**11**:e1004642.

69. Proix T, Bartolomei F, Guye M, et al. Individual brain structure and modelling predict seizure propagation. *Brain.* 2017;**140**:641–54.

70. Vollmar C, O'Muircheartaigh J, Symms MR, et al. Altered microstructural connectivity in juvenile myoclonic epilepsy: the missing link. *Neurology.* 2012;**78**:1555–9.

71. Zhang Z, Liao W, Chen H, et al. Altered functional-structural coupling of large-scale brain networks in idiopathic generalized epilepsy. *Brain.* 2011;**134**:2912–28.

72. Hong SJ, Bernhardt BC, Gill RS, et al. The spectrum of structural and functional network alterations in malformations of cortical development. *Brain.* 2017;**140**:2133–43.

73. Vaessen MJ, Jansen JF, Vlooswijk MC, et al. White matter network abnormalities are associated with cognitive decline in chronic epilepsy. *Cereb Cortex.* 2012;**22**:2139–47.

74. Bernasconi A. Connectome-based models of the epileptogenic network: a step towards epileptomics? *Brain.* 2017;**140**:2525–7.

75. Bernhardt BC, Fadaie F, Vos de Wael R, et al. Preferential susceptibility of limbic cortices to microstructural damage in temporal lobe epilepsy: a quantitative T1 mapping study. *NeuroImage.* 2017. doi:10.1016/j.neuroimage.2017.06.002.

Chapter

8

Network Modeling of Epilepsy Using Structural and Functional MRI

Lorenzo Caciagli, Boris C. Bernhardt, Andrea Bernasconi, and Neda Bernasconi

8.1 Introduction

Since the first classification attempts, major advancements in our understanding of human epileptogenesis have posed notable challenges to the conventional models of "focal" and "generalized" epilepsies. A wealth of observations from experimental paradigms, animal models, and studies in humans indicates that specific cortical and subcortical networks are involved in the generation of focal and generalized seizures, suggesting that epilepsies may be conceptualized as disorders of neural networks.[1,2]

8.2 Epilepsy as a Network Disorder

In temporal lobe epilepsy (TLE), which has been long considered the prototypical focal epilepsy syndrome, studies employing intracranial electroencephalography (EEG) demonstrated that seizure activity involves a widespread set of regions that consists of mesiotemporal, neocortical, as well as subcortical structures.[1] Advancements in functional and structural magnetic resonance imaging (MRI) techniques have further revealed complex connectional derangements.[3,4] Initial evidence of widespread structural and functional reconfigurations is also emerging for epilepsies secondary to cortical malformations, particularly focal cortical dysplasia.[4] Moreover, several studies suggest that abnormalities outside the lesional boundaries may negatively impact the outcome of epilepsy surgery, which is still suboptimal in up to 40% of candidates despite rigorous selection.[5–8] Collectively, these findings have prompted a major conceptual shift from the conventional interpretation of focal epilepsies, and emphasize the importance of a network approach to adequately capture the neurobiology of these disorders.

In idiopathic generalized epilepsies (IGE), it is widely accepted that an aberrant thalamocortical interaction plays a major role in the generation of spike-wave discharges.[9] Several lines of evidence from experimental and animal models indicate that while some cortical regions may play a pivotal role in seizure generation, others would be relatively uninvolved.[10] EEG and magnetoencephalography (MEG) analyses focusing on the initiation of human absence seizures demonstrated the predominant involvement of focal cortical areas, mostly represented by mesial frontal and orbitofrontal cortices.[11–13] In addition, imaging studies reported a rather localized pattern of structural and functional abnormalities in several IGE syndromes.[2,14] Findings indicate that generalized epilepsies might be better understood as arising from the pathological activity of neuronal networks encompassing specific corticosubcortical structures. This is also reflected in the recently formulated concept of *system epilepsies*, which would view some of the IGE syndromes as the pathologic expression of specific neural systems, normally subserving identifiable functions in the healthy brain.[15]

Hence, adopting a network perspective seems nowadays compelling to fully capture the complexity of human epilepsies. In this regard, a well-established role is attributed to noninvasive imaging techniques, given their ability to probe connectivity in vivo at multiple scales and with complementary modalities. In the following paragraphs, we discuss the underpinnings of network analysis in epilepsy from an imaging standpoint, focusing on the most widely employed MRI methods. We first summarize the evidence obtained for focal epilepsy syndromes, placing particular emphasis on TLE and extratemporal epilepsy secondary to focal cortical dysplasia (FCD). A subsequent paragraph details findings in patients with IGE. Finally, we outline the relevance of network modeling techniques for clinical decision making.

8.3 Network Modeling Using Functional and Structural MRI

At its simplest, a network can be envisioned as a collection of items with pairwise relationships. This concept can be translated at multiple levels; the brain as a whole can be considered a hierarchically organized network, partitioned into mutually interconnected units responsible for information processing spanning from local circuits to broad functional areas.[16] The shift to a network perspective has assumed a particular relevance in epilepsy, since structures that are part of an epileptogenic network are thought to be involved in the generation and expression of seizures, and to the maintenance of the disorder.[1]

Structural connectivity refers to direct anatomical associations between brain regions.[17] The two major MRI techniques employed to map structural networks in humans are diffusion-weighted imaging and covariance analysis (or morphometric correlations) (Figure 8.1). Diffusion imaging provides voxel-wise information about the magnitude and directionality of water diffusion and is utilized to assess the microstructural integrity of the white matter. The use of tractography algorithms further allows reconstructing fiber pathways along plausible diffusion trajectories.[18,19] Despite limitations in signal disambiguation for voxels containing intersecting fibers, results of diffusion imaging analyses show a high degree of consistency and reproducibility and have been cross-validated against tract-tracing studies conducted on animal models.[20–23] An alternative technique to generate structural network representations of the human brain is MRI covariance analysis. This framework infers networks from interregional correlations of structural markers, such as gray matter volume or cortical thickness. The existence of a network link between two regions is derived from a high correlation of morphological markers between them, while a low correlation speaks against a link. Covariance patterns of structural markers may reflect trophic and/or signaling interactions among distant areas,[24] exhibit high correspondence with maturational networks,[25] and overlap with networks derived via diffusion imaging and resting-state fMRI.[26–28] Differences in structural measures also appear to covary within assemblies of brain regions belonging to well characterized functional systems, such as those subserving visual, auditory, motor, language, and other cognitive functions.[24,29,30] Thus, covariance patterns seem to reflect brain connectivity,

and may aid in unveiling interregional pathological influences in the context of brain disorders.

Functional connectivity refers to statistical associations between spatially distributed neurophysiological time series.[31] Functional connectivity measures can be derived from a variety of signal sources, spanning from electrophysiological techniques, such as EEG and MEG, to imaging methods, the most prominent of which is represented by functional MRI (fMRI). In fMRI, changes in blood-oxygenation-level-dependent (BOLD) signal are utilized to infer neuronal activity under a neurovascular coupling model. While the temporal resolution of fMRI is generally low (order of seconds), its spatial resolution falls in the millimeter range and whole-brain coverage is permitted. A versatile tool is represented by resting-state fMRI paradigms, during which subjects lie still in the scanner without performing any tasks (Figure 8.1). Acquiring resting-state fMRI datasets exhibits several advantages compared with task-based paradigms, including the possibility to assess multiple regions simultaneously, as well as reduced demand for patients with impaired ability to engage in tasks.[32] Analysis of such sequences has led to the identification of brain networks, which show coherent fluctuations in their intrinsic, spontaneous activity. Resting-state networks display a high degree of reproducibility across subjects, as well as robust correspondence with task-involved systems.[33] Hence, resting-state fMRI can be utilized to probe intrinsic functional networks and assess disease-specific connectional derangements.[34]

In recent years, graph theory has emerged as a unique framework to characterize network organizational properties at a whole-brain level.[35] Graph theory formalizes a network as a collection of *nodes*, corresponding to brain regions, interconnected by pairwise *edges*, derived from structural and functional connectivity estimates (Figure 8.1). Nodes can often be clustered into modules, which show a dense internal connectivity but a relative segregation from the rest of the network. Centrality-based metrics allow identifying *hubs*, i.e., nodes with a high degree of connections to the rest of the network and prominent roles in network dynamics. Another set of measures addresses the efficiency of local and global information transmission, such as the clustering coefficient, which quantifies connection density within the local environment subnetwork surrounding a node, and path length, which describes the average number of connections between any pairs of nodes. In healthy individuals, whole-brain

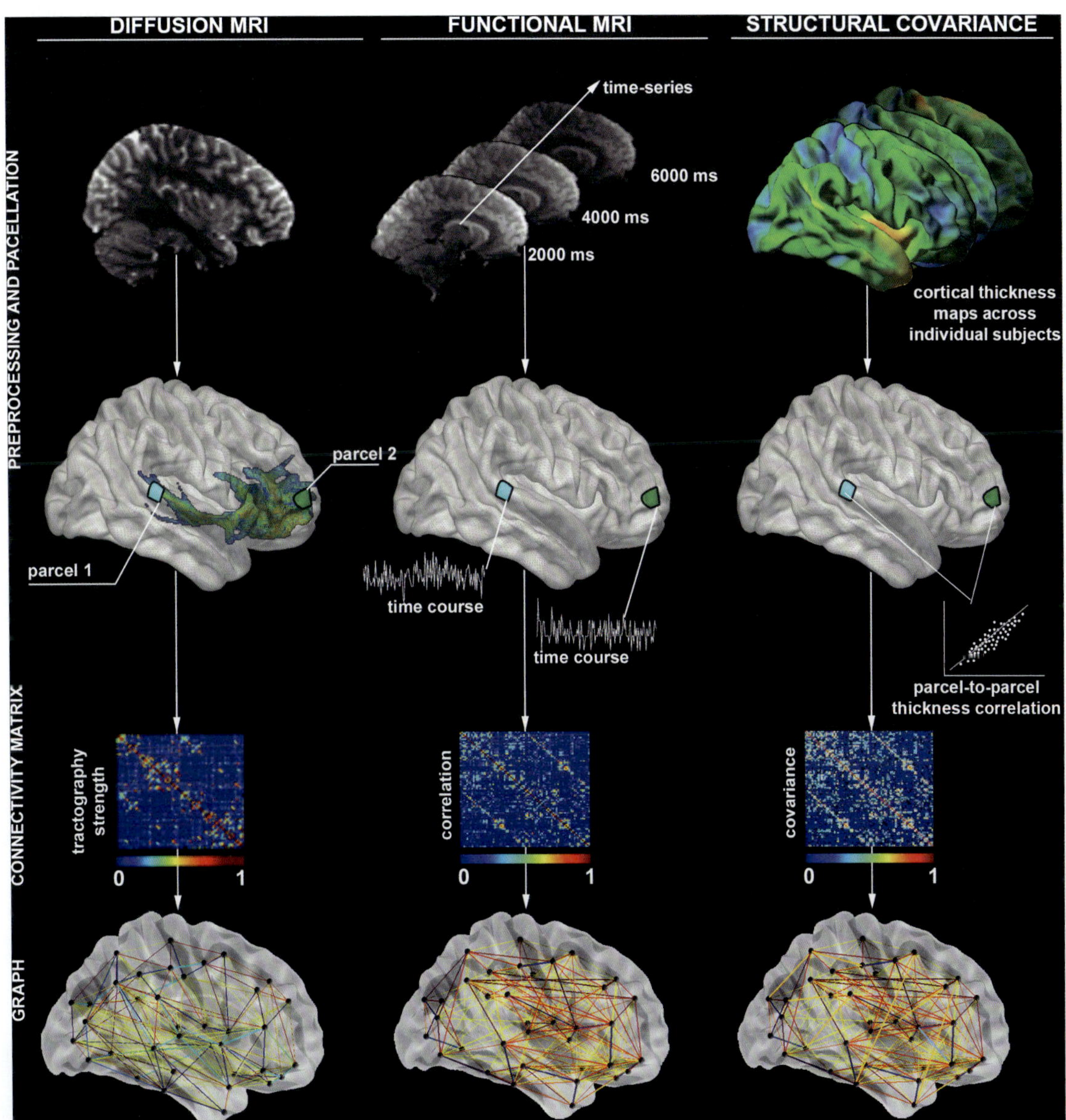

Figure 8.1. MRI methods to model networks in the human brain. The passages leading to derive brain networks through different MRI modalities (diffusion MRI, functional MRI, and structural MRI covariance) are here summarized. Within each modality, data are usually preprocessed and analyzed in parcellated anatomical space. Tractography algorithms permit reconstructing fiber pathways between parcels along plausible diffusion trajectories. Resting-state fMRI and structural MRI covariance rely on cross-correlations between seed and target regions. A connectome-based approach implies the iterative assessment of pairwise associations between parcels, and leads to derive connectivity matrices. An alternative representation of a connectivity matrix is a graph, in which nodes correspond to brain areas and edges represent connections. Adapted from Bernhardt et al., *Epilepsy Behav* 2015 with permission.

topological properties exhibit small world attributes.[36] Combining high levels of clustering with overall short paths, a small world architecture represents an efficient arrangement to achieve segregation and integration of information processing while minimizing wiring costs.[37] Graph theoretical analysis provides the unique opportunity to assess high-order properties of functional and structural networks, with the possibility to span from rather localized domains to a whole-brain level.

8.4 Evidence for Extensive Network Disruptions in Temporal Lobe Epilepsy

Studies probing the integrity of structural and functional networks in TLE have found evidence of widespread abnormalities affecting temporolimbic circuits as well as several large-scale networks, along with striking reconfigurations of whole-brain organizational properties (Figure 8.2). Morphometric correlation analyses of cortical thickness and gray matter volume revealed decreased structural coordination between mesiotemporal regions and an extensive assembly of neocortical areas, including lateral temporal neocortices[38,39] as well as prefrontal, frontocentral, occipitotemporal, and cingulate cortices.[38–40] Covariance of thalamic atrophy with cortical thickness of mesiotemporal,[41] frontocentral, and lateral temporal neocortices[42] has been detected, pointing to a prominent involvement of the thalamus in the pathologic network of TLE. Considering the underlying white matter, abnormal diffusion parameters have been observed for several temporolimbic tracts both ipsilateral and contralateral to the seizure focus, including fornix, cingulum, and uncinate fasciculus.[43–47] Moreover, reduced fractional anisotropy has been documented for several extratemporal bundles, including the superior and inferior longitudinal fascicles, occipitofrontal fascicle, external and internal capsules, and corpus callosum.[44,45,47–49] Diffusion derangements also encompass specific connections arising from the thalamus, such as thalamo-mesiotemporal fibers, ipsilateral anterior thalamic radiation, and tracts linking ipsilateral thalamus with the precentral gyrus.[8,48,50] Mean diffusivity changes seem to be less extensive, and display a progressive reversal as a function of the anatomical distance from the epileptogenic focus.[46] Collectively, these abundant diffusion abnormalities imply a striking reconfiguration of white matter architecture in TLE, with changes being more prominent ipsilateral to the seizure focus, but also involving the contralateral hemisphere. More pronounced and diffuse disruptions in connectivity measures have been

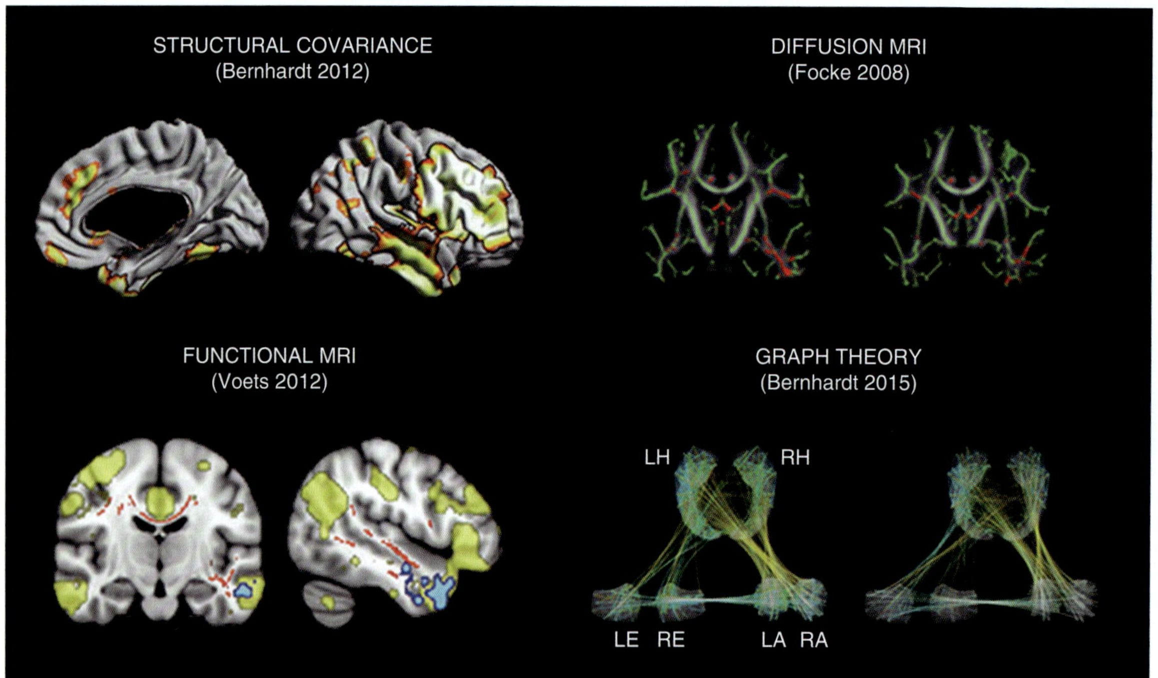

Figure 8.2. Network abnormalities in TLE. Exemplary findings of network alterations in TLE compared with controls are here displayed. Top left panel shows areas with aberrant cortical thickness correlations with medial thalamic volume (adapted from Bernhardt et al., *Neurology* 2012 with permission). Top right panel displays fiber tracts with reduced fractional anisotropy (adapted from Focke et al., *NeuroImage* 2008 with permission). Bottom left panel shows abnormal fMRI time series (yellow) within a spatial component closely overlapping with the default mode network (adapted from Voets et al., *Brain* 2012 with permission). Bottom right panel details results of a graph-theoretical analysis on interregional covariance of surface-based metrics (LH/RH = left/right hippocampus; LE/RE = left/right entorhinal cortex; LA/RA = left/right amygdala). Covariance increases/decreases are depicted in green/yellow (adapted from Bernhardt et al., *Cereb Cortex* 2015 with permission).

reported for left than right TLE.[44,48,51] In addition, recent analyses indicated more extensive reduction of fiber density and cross-sectional area in TLE patients with hippocampal sclerosis compared with those not exhibiting global decreases in hippocampal volume.[47]

Extensive anomalies have also been inferred from functional connectivity measures. Resting-state fMRI studies found impaired connectivity estimates for mesiotemporal structures ipsilateral to the seizure focus, mostly involving links between anterior and posterior hippocampus, and between anterior hippocampus and entorhinal cortex.[52,53] Reduced functional connectivity was additionally detected between ipsilateral and contralateral hippocampus, ipsilateral and contralateral insula, and between epileptogenic mesiotemporal structures and bilateral lateral temporal neocortices.[54–56] Impaired ipsilateral functional integration may coexist with enhanced connectivity in contralateral mesiotemporal networks.[52] It is suggested that more marked connectional derangements may take place in left than right TLE, both ipsilateral and contralateral to the seizure focus.[54] At a whole-brain level, bilaterally impaired functional connectivity has been consistently detected for areas pertaining to the default mode network (DMN), which is traditionally composed of mesiotemporal lobes and mesial prefrontal, lateral, and midline parietal areas.[57–61] DMN activity is enhanced during task-free periods, and this network is thought to play a major role in internally directed activities, such as memory, future planning, and mind wandering.[62] Interictal spike-correlated dysfunctions have also been mapped in regions belonging to the DMN by studies employing EEG-fMRI.[63,64] It remains poorly understood whether DMN integrity may be differentially affected according to the side of the epileptogenic focus. Hippocampal decoupling from anterior and posterior DMN nodes, however, appeared more prominent in patients with histology-confirmed hippocampal sclerosis, while being relatively less evident in those with gliosis only.[65] Connectional derangements in TLE have also been documented for other large-scale functional systems, including sensory-motor, attentional, episodic memory, working memory, and language networks; altered functional integration has been furthermore described between mesiotemporal and subcortical structures, such as thalamus and cerebellum.[58–60,66–71] Disruption of thalamocortical functional connectivity were shown as ipsilateral in patients with focal seizures only, while involving both ipsilateral and contralateral structures in patients with additional secondarily generalized seizures.[72] EEG-fMRI studies have indicated that activations correlated with temporal lobe spikes encompass a widespread ipsilateral network, most frequently extending to anterior hippocampus, amygdala, insula, superior temporal gyrus, basal ganglia, cerebellum, midcingulate as well as piriform cortex.[73,74] In view of its unique connectivity profile, the piriform cortex has been recently pinpointed as a common hub within networks involved in seizure generation in focal epilepsies, including TLE.[75]

Although neuroimaging-derived structural and functional abnormalities in TLE show considerable overlap, relatively few multimodal imaging studies have directly addressed cross-domain relationships. It is documented that decreased network integration of the hippocampus could be partially explained by estimates of its gray matter density.[60,76] More detailed structural analyses, however, suggest that the magnitude of hippocampal T2 signal changes, followed by hippocampal atrophy, may relate to the extent of its functional disconnection from anterior and posterior DMN epicenters.[65] Moreover, disrupted functional connectivity between mesiotemporal structures and neocortical targets, including regions belonging to the DMN and sensory-motor networks, was also associated with altered diffusion parameters of the interconnecting white matter tracts.[60,61] In a recent study, abnormal functional amplitude metrics of midline and lateral default mode areas were shown to be mediated by microstructural abnormalities of the temporolimbic superficial white matter.[77] Globally, these findings suggest that disruptions in morphological and architectural features may account for derangements in intrinsic functional connectivity in TLE.

Graph-theoretical studies on diffusion MRI datasets are suggestive of profound rearrangements within ipsilateral and contralateral mesiotemporal lobe subnetworks,[8,51,78] which seem to account for a shift toward a more regularized topology.[78] Analyses of functional data also found evidence of deranged limbic nodal topology[55,79] and detected changes indicative of compensatory reorganization of the contralateral mesiotemporal subnetwork.[55] While aberrant topological properties were shown to maximally affect hippocampus, thalamus, and anteromesial prefrontal cortex in TLE with hippocampal sclerosis, patients with normal hippocampal volumetry may only exhibit altered connectional properties of the ipsilateral temporal neocortex.[80] Pronounced abnormalities have also been detailed by graph-theoretical studies addressing whole-brain network properties.[81] An analysis on resting-state fMRI networks in bilateral TLE revealed decreased clustering and path length, along with a major redistribution of network hubs, which would indicate a random network arrangement.[79] More recent resting-state fMRI work on unilateral TLE found increased clustering and path length, pointing to a more regular topology.[82] A regularization of whole-brain network topology as well as pronounced shifts in the distribution of hubs and modularity was also reported by graph-theory studies on structural MRI datasets, such as cortical thickness or gray matter volume correlations[83,84] and diffusion MRI metrics,[85,86] and by graph-theoretical analyses on EEG-derived networks.[87,88] Evidence pointing to a regularization of whole-brain networks also came from a recent meta-analysis, which included electrophysiology and imaging graph-theory studies on focal epilepsy cohorts.[89] Graph-theoretical studies also indicated reduced coupling between structural and functional networks, which may be partially modulated by disease duration.[90] Connectome alterations appeared to be more pervasive in left TLE than right TLE patients.[51,84]

8.5 Emerging Evidence for Disturbed Connectivity in Extratemporal Epilepsies

Extratemporal epilepsy related to cortical malformations, particularly FCD, is a major cause of drug-resistant seizures. In recent years, advances in MRI processing have led to substantial improvements in lesion detection.[91] In parallel, several studies have unveiled the existence of morphological and architectural abnormalities, encompassing gray matter density, cortical thickness, sulcal depth, diffusion parameters, and local functional markers in areas at a distance from the dysplastic cortex.[92–97] These findings have prompted further attempts to assess the integrity of structural and functional networks in epilepsy secondary to FCD.

Diffusion MRI studies in patients with FCD-related frontal lobe epilepsy (FLE) found bilateral reductions in fractional anisotropy, encompassing the superior longitudinal fasciculus, uncinate fasciculus, cingulum, as well as corpus callosum.[92,98–100]

Bilateral diffusion abnormalities involving the corpus callosum and several interlobar fiber tracts are also described for pediatric FLE patients with non-diagnostic MRI.[101,102] Widespread diffusivity increases are also reported for bilateral frontal, temporal, and parietal lobes and corpus callosum.[99,103] Only a few fMRI studies probed the integrity of functional networks in homogeneous populations of patients with FCD. Functional connectivity analyses seeded from dysplastic cortices revealed complex rearrangements of network properties, involving variable combinations of hyper- and hypoconnectivity.[104] There is also evidence of higher signal variability of local resting-state fMRI measures in a sample of patients with extratemporal epilepsy, half of whom with FCDs.[105] Striking reconfigurations of language and memory networks, with evidence of intra- and interhemispheric redistribution of function, have been detailed for FLE and mixed focal epilepsy patients.[106–109] In adult and pediatric populations with FLE and suspected dysplasia, resting-state fMRI analyses have displayed extensive connectional derangements, involving language,[110] working memory,[111,112] sensory-motor networks,[113] as well as DMN, attentional, visual, and auditory networks.[114] Focusing on patients with malformations of cortical development or pure dysplasia cohorts, EEG-fMRI studies have shown widely distributed spike-correlated BOLD signal changes, frequently involving cortical and subcortical areas at a distance from the seizure onset zone.[73,115–117] These studies are indicative of widespread epileptogenic networks, which may also imply increased functional connectivity between the epileptogenic region and remote brain areas, possibly with patient-specific connectional profiles.[118] A graph-theoretical study conducted on resting-state fMRI data in adult patients with nonlesional focal epilepsy found evidence for a reduced global network efficiency as well as decreased clustering coefficient, suggestive of a network randomization.[119] Conversely, metrics indicative of a more regularized network architecture were reported by resting-state fMRI analyses on a mixed adult sample of FCD-related and MRI-negative extratemporal epilepsy,[120] and in a small group of adults with polymicrogyria.[121] Results of DTI and resting-state fMRI analyses in pediatric populations with nonlesional FLE are also compatible with a more regularized network topology.[122,123]

8.6 Network Disturbances in Idiopathic Generalized Epilepsies

Challenging conventional beliefs, several studies carried out in the last decade have provided compelling evidence that structural abnormalities are indeed present in patients with IGE, and most frequently encompass thalamus and frontal cortical areas.[2,124] In parallel, the pivotal role of activity changes in the thalamus and diffuse cortical areas during generalized spike-wave discharges was documented by EEG-fMRI analyses.[125] As a result, a wealth of investigations has recently assessed the integrity of structural and functional networks in IGE, with prominent attention to the interplay between thalamus and neocortical areas (Figure 8.3).

In a patient cohort of IGE with generalized tonic-clonic seizures (IGE-GTCS), one of the first studies to assess the integrity of thalamocortical networks using MRI covariance reported enhanced thalamic structural coupling with frontocentral, parietal, and lateral temporal cortices, along with connectivity reductions between thalamus and limbic areas.[126] Studies employing diffusion MRI have largely focused on juvenile myoclonic epilepsy (JME), and consistently indicate impaired thalamo-frontocortical connectivity, particularly in the frontal lobe.[127–130] White matter abnormalities in JME, however, also encompass associative fiber tracts, such as the superior longitudinal fasciculus, the uncinate fasciculus,[131,132] and segments of the corpus callosum connecting prefrontal cortices, supplementary motor areas (SMA), and posterior cingulate cortices.[130,133–135] Vollmar and colleagues recently described a complex pattern of diffusion derangements in JME, including increased structural connectivity of pre-SMA with motor cortices and descending motor pathways, and reduced connectivity between pre-SMA and prefrontal/frontopolar areas, which may represent a correlate of the impaired frontal lobe functions observed in this syndrome. In addition, enhanced connectivity between SMA and the occipital lobe as well as lateral temporal neocortices was also described.[136] A graph theoretical analysis on a diffusion MRI dataset indicated

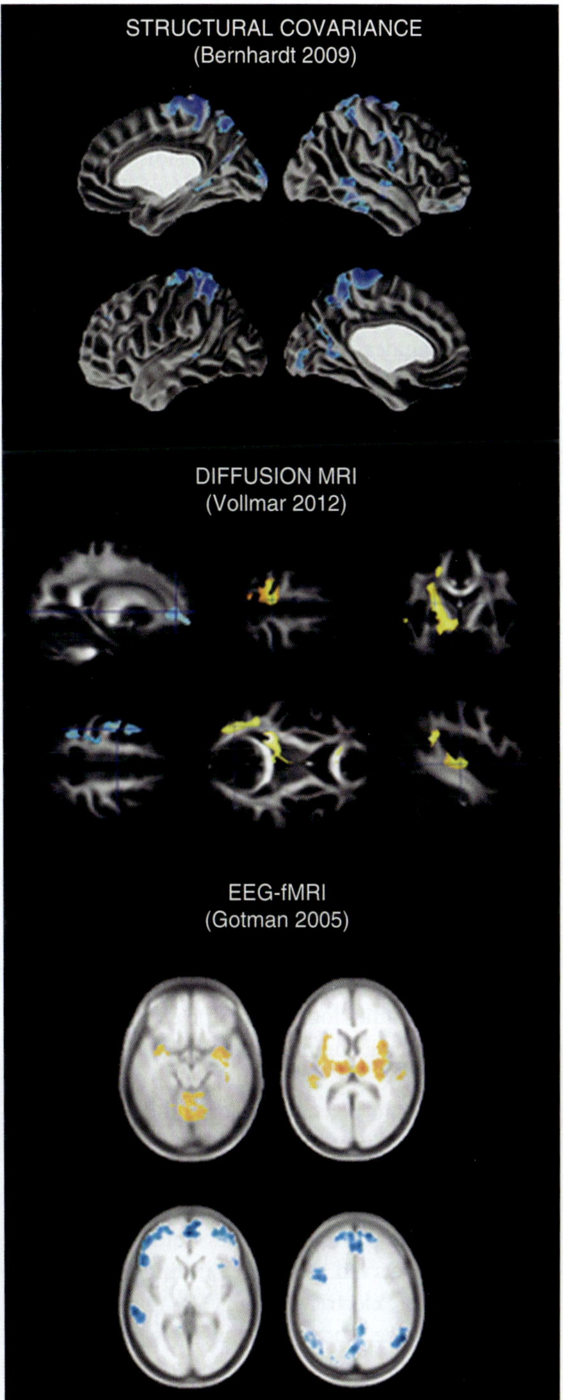

Figure 8.3. Network abnormalities in IGE. Exemplary findings of network alterations in IGE compared with controls are here depicted. Top panel highlights regions with increased cortical thickness correlations with thalamic volume in patients with IGE-GTCS (adapted from Bernhardt et al., *NeuroImage* 2009 with permission). In the middle panel, increases (warm colors) and decreases (cold colors) in DTI connectivity of the pre-SMA/SMA are displayed in the

Caption for Figure 8.3. (cont.)

upper/lower row (adapted from Vollmar et al., *Neurology* 2012 with permission). Lower panel displays EEG-correlated fMRI activations (warm colors) and deactivations (cold colors) in a mixed IGE cohort (adapted from Gotman et al., with permission, copyright (2005) National Academy of Sciences, USA).

increased structural connectivity in a subnetwork including primary motor cortex, parietal lobe areas, right hippocampus, and subcortical structures such as putamen and cerebellum.[137] Collectively, these findings are suggestive of complex reconfigurations affecting white matter networks in JME. Diffusion imaging studies on mixed IGE cohorts also found evidence of disrupted white matter microstructure, involving anterior thalamocortical fibers, corticospinal tracts, and several cortico-cortical projections;[138,139] a subgroup analysis failed to detect major differences between JME patients and the remainder mixed IGE group.[138] Preliminary evidence of diffusion abnormalities extending to cortico-cortical and corticosubcortical fibers is also available for childhood absence epilepsy (CAE).[140,141]

Studies employing EEG-fMRI have led to major improvements in our understanding of functional networks implicated in generalized spike-wave discharges and absence seizures. Discharge-correlated BOLD activations are consistently observed in thalamus and frontal lobes, along with deactivations in a collection of areas pertaining to the DMN.[125,142–146] Other analyses additionally emphasized the contribution of deactivations in basal ganglia, such as the caudate.[147–149] Thalamocortical BOLD activations are hypothesized as underlying the generation of the epileptic paroxysms, in line with evidence from experimental and animal models. On the other hand, perturbation of the brain's default state is thought to arise as an indirect consequence of these discharges,[125] although more recent evidence points to a permissive role of precuneal activity, which would act as the initiator of spike-wave discharges.[150] Studies focusing on the dynamic changes taking place during absence seizures revealed that cortical activity within a set of frontal and parietal regions precedes thalamic activation,[151,152] and that frontal activation may exhibit a causal relationship to subsequent thalamic changes.[153] While BOLD dynamic changes exhibit high variability across subjects, consistent within-subject results across multiple epileptic events were observed.[152] These findings are in line with results of electrical source mapping studies, which emphasize focal patterns of ictal onsets, mostly involving mesial frontal and orbitofrontal cortices,[11,154] and would revive the concept of a "cortical focus" as the initiator of generalized spike-wave discharges.[10] More recent EEG-fMRI work, however, identified activity changes within multiple distinct subnetworks during absence seizures, and suggests that early discharge-associated changes may

involve an extended network encompassing medial parietal, lateral occipital, frontopolar areas as well as subcortical structures.[155]

Resting-state fMRI assessments have reported a widespread set of connectional derangements involving the DMN in CAE, IGE-GTCS, and mixed IGE cohorts.[156–159] Pairwise functional connectivity between DMN nodes is usually reported as being decreased. On the other hand, enhanced connectivity has been observed between mesial prefrontal cortex and regions implicated in "task-positive" cortical regions, including the superior parietal lobule and the intraparietal sulcus, and between medial parieto-occipital cortices and primary sensory networks, suggesting a reduced segregation of the DMN.[157,160] Increased interhemispheric connectivity is described for orbitofrontal cortices in CAE,[161] and for anterior cingulate and mesial prefrontal cortices in IGE-GTCS.[162] Additional abnormalities in CAE include decreased functional connectivity between anterior insular/frontal opercular and mesial prefrontal cortices,[163] which may indicate impaired functioning of the attentional network in this syndrome, as well as abnormal connectional profiles between thalamus and diffuse cortical areas.[160] The numerous assessments of thalamocortical functional connectivity patterns, however, do not exhibit homogeneous results. In patients with IGE-GTCS, decreased resting-state functional connectivity was described between the medial dorsal thalamic nucleus and bilateral orbitofrontal cortices, amygdalae, and basal ganglia.[164] Studies on mixed IGE groups identified increased functional connectivity between posterior thalamus and mesial prefrontal/anterior cingulate cortices.[165] Opposite results for the same cortical targets were instead obtained when seeding from anteromedial thalamic subsections,[166] which seem to exhibit reduced amplitude of low-frequency fluctuations.[167] Connectivity between thalamus and posteromedial default mode areas, such as the posterior cingulate cortex, was uniformly reported as decreased.[165,166] Finally, a recent study assessing a heterogeneous IGE population documented increased functional connectivity measures for a majority of the analyzed thalamocortical connections, including thalamotemporal, prefrontal-thalamic, occipitothalamic, as well as motor and premotor thalamic networks.[168]

Studies assessing patients with JME and their unaffected siblings revealed enhanced functional connectivity between motor and prefrontal cortices, which accounts for the coactivation of the motor system during cognitive tasks described in this syndrome and represents a disease endophenotype.[169,170] Such functional connectional patterns are also mirrored by increased structural connectivity of the underlying white matter tracts in JME, as reported by diffusion MRI studies.[136] Additional alterations in JME include decreased functional connectivity between thalamus and the SMA, which parallels abnormal diffusivity parameters of the interconnecting white matter projections.[129]

Graph theoretical findings in IGE cohorts are, so far, relatively heterogeneous. A diffusion MRI study on children with CAE reported whole-brain topological rearrangement, indicated by reduced clustering coefficient and increased path length,[141] while a resting-state fMRI analysis indicated aberrant centrality measures of network hubs, including fronto-temporoparietal DMN nodes as well as thalamus.[171] Conversely, no major differences between adult patients with JME and controls were found for whole-brain diffusion MRI-derived graph theoretical metrics.[137] In IGE-GTCS, combined graph-theory approaches on resting-state fMRI and diffusion MRI datasets indicated disruptions in whole-brain network topology, with a shift toward a random network configuration in both domains.[172] Across both imaging modalities, major derangements in nodal characteristics[172] were particularly evident for key corticosubcortical network hubs, pointing to an impaired "rich club" organization.[173] There was also evidence for decoupling between structural and functional connectivity networks, which was shown to be negatively associated with epilepsy duration and may thus underlie disease progression.[172]

8.7 Relevance of Network Analysis for Clinical Purposes

In epilepsy, the majority of studies have employed MRI-derived connectivity measures to describe functional and structural correlates of disease processes. Increasing evidence indicates that network analyses may also represent promising tools to inform clinical practice, especially with respect to the management of focal epilepsy. Several studies have assessed the potential of resting-state fMRI and diffusion MRI analyses to lateralize the seizure focus. Patterns of thalamotemporal functional connectivity were shown to categorize left and right TLE patients with high sensitivity and specificity,[53,174]

whereas connectivity increases in contralateral limbic structures may aid the identification of the nonepileptic temporal lobe.[175] A lateralization index derived from the analysis of local functional markers, such as the amplitude of the low-frequency fluctuations (ALFF), seems to effectively distinguish left from right TLE.[176] Successful detection of the seizure focus could also be attained by contrasting the local functional connectivity profiles within lesional areas to that of contralateral corresponding regions.[177–179] Implementing functional topology metrics within an automated algorithm resulted in prospective prediction of focus laterality with similar accuracy rates compared to video-EEG, and with a significantly improved yield compared to expert qualitative MRI inspection.[180]

Initial evidence indicates that structural and functional network modeling may provide valuable indications regarding disease monitoring and prognosis. A longitudinal analysis based on cortical thickness correlations showed an increase in path length over time, suggesting that serial analyses of network topology metrics might capture changes potentially related to disease progression.[83] Graph-theoretical analyses in a sample of TLE patients who underwent epilepsy surgery revealed increased network resilience metrics in individuals rendered seizure-free compared to those with poor outcome.[181] Moreover, a longitudinal comparison of presurgical with postsurgical connectome-level metrics documented a reorganization of connections involving mesial prefrontal cortices and temporoparietal junctions, which appeared more pronounced in seizure-free patients.[181] In a diffusion MRI study in TLE, disrupted organizational properties within the ipsilateral temporal subnetwork and increased structural connectivity between a set of ipsilateral and contralateral temporoparietal areas were associated with poor postsurgical seizure control.[8] Recent analyses in TLE have also emphasized the relevance of thalamic connectivity. Bilateral thalamic atrophy and abnormal DTI-derived thalamotemporal probabilistic paths were distinctive features of patients with persistent seizures after surgery compared with those rendered seizure-free.[50] In a resting-state fMRI analysis, patients with poor postsurgical seizure outcome presented with higher values of centrality ("hubness") metrics within ipsilateral and contralateral thalami, along with enhanced connectivity of both thalami with bilateral cortical targets.[182] Computing whole-brain functional connectivity from seizure foci identified with EEG-fMRI, Negishi and colleagues found that high functional connectivity within the epileptogenic hemisphere may be a predictor of successful seizure control after surgery.[183] The potential of interhemispheric functional connectivity differences to distinguish patients with good and suboptimal surgical outcome is also reported by other resting-state fMRI work.[184] Recent studies have attempted to identify connectome-based biomarkers to predict outcome at the individual level. Embedding preoperative white matter architectural features into classification algorithms led to a fairly accurate prediction of postsurgical outcomes in TLE,[185,186] which displayed higher sensitivity and specificity than a combined scored derived from clinical characteristics.[186] Machine-learning models based on thalamic connectional properties, as derived by resting-state fMRI graph-theoretical analysis, were recently shown to provide a more accurate prediction of postsurgical outcome than algorithms based on clinical variables.[182]

Network analysis may prove useful for deriving functional or structural biomarkers of cognitive functions. Resting-state functional connectivity maps obtained from seeds placed in the inferior frontal cortices predicted hemispheric dominance for language in TLE.[187] Functional connectivity between areas belonging to frontotemporal language networks was demonstrated to be associated with verbal IQ measures[188] and with scores in word fluency and text reading tasks.[110] Several studies have also found imaging correlates of verbal and nonverbal episodic memory. In TLE, functional connectivity disruptions between mesiotemporal lobes and posterior DMN areas may underlie impairments in episodic memory performance.[76,189,190] Enhanced functional connectivity between the epileptogenic mesiotemporal lobe and DMN regions, however, is elsewhere reported to correlate with worse memory abilities.[191,192] Recruitment of contralateral mesiotemporal and DMN structures might instead relate to improved memory performance.[52,191,192] Measures of bilateral functional connectivity between the thalamus and neocortical areas, including prefrontal cortex and supramarginal gyrus, may be associated with working memory scores.[71] Of note, diffusion abnormalities in specific fiber tracts, such as uncinate, arcuate, inferior longitudinal fascicles and cingulum, were shown to correlate with memory and language abilities,[193–195] and may be used to predict scores in verbal memory tests.[196] In populations of adult and pediatric localization-related epilepsy patients, graph theoretical metrics derived from fMRI[119,122] and

diffusion MRI datasets[197] displayed the ability to mirror cognitive performance across several domains.

Connectivity metrics may assist the prediction of neurocognitive outcome after epilepsy surgery. Postsurgical episodic memory performance was shown to correlate negatively with functional connectivity between the epileptogenic hippocampus and precunei, and positively with connectional profiles between the contralateral hippocampus and precunei.[189] A combination of diffusion MRI, resting-state as well as task-based fMRI was recently shown to accurately predict verbal fluency outcomes following anterior temporal lobe surgery.[198] Moreover, an analysis on resting-state fMRI data points to the usefulness of regional graph theory measures in informing predictive models of postsurgical cognitive outcome across multiple domains, including language, attention, working memory, and verbal episodic memory.[199] Interestingly, enhanced whole-brain integration of the contralateral unaffected hippocampus also predicted a better cognitive outcome,[199] adding further proof that recruitment of contralateral areas may represent a compensatory phenomenon favoring memory abilities.

8.8 Conclusions

Nowadays, a network perspective is essential to the understanding of the development, progression, and management of epilepsy. Due to recent advancements in noninvasive imaging, networks can be mapped at multiple levels, spanning from local and interregional connectivity to whole-brain topological attributes, thus providing a window into the complex pattern of disease-specific effects. Functional and structural MRI analyses have contributed to substantial improvements in our understanding of brain connectivity and have challenged the originally proposed dichotomy between focal and generalized syndromes. In focal epilepsies, there is evidence for organizational derangements at a whole-brain level. In IGE, although reconfigurations of white matter architecture and shifts in network topology are described, several studies do not support a homogeneous whole-brain involvement, and rather emphasize the role of distinct networks. It is anticipated that the rich set of connectivity metrics and topological properties can serve both diagnostics and prognostics in individual patients. Caution is required before their clinical implementation, which will need to be preceded by validation on large, multicenter cohorts with assessments of reliability, sensitivity, and specificity, and formulation of guidelines pertaining to data acquisition and analysis.

References

1. Spencer SS. Neural networks in human epilepsy: evidence of and implications for treatment. *Epilepsia.* 2002;43:219–27.

2. Richardson MP. Large scale brain models of epilepsy: dynamics meets connectomics. *J Neurol Neurosurg Psychiatry.* 2012;83:1238–48.

3. Bernhardt BC, Hong S, Bernasconi A, Bernasconi N. Imaging structural and functional brain networks in temporal lobe epilepsy. *Front Hum Neurosci.* 2013;7:624.

4. Caciagli L, Bernhardt BC, Hong SJ, Bernasconi A, Bernasconi N. Functional network alterations and their structural substrate in drug-resistant epilepsy. *Front Neurosci.* 2014;8:411.

5. Tellez-Zenteno JF, Dhar R, Wiebe S. Long-term seizure outcomes following epilepsy surgery: a systematic review and meta-analysis. *Brain.* 2005;128:1188–98.

6. de Tisi J, Bell GS, Peacock JL, et al. The long-term outcome of adult epilepsy surgery, patterns of seizure remission, and relapse: a cohort study. *Lancet.* 2011;378:1388–95.

7. Bernhardt BC, Bernasconi N, Concha L, Bernasconi A. Cortical thickness analysis in temporal lobe epilepsy: reproducibility and relation to outcome. *Neurology.* 2010;74:1776–84.

8. Bonilha L, Helpern JA, Sainju R, et al. Presurgical connectome and postsurgical seizure control in temporal lobe epilepsy. *Neurology.* 2013;81: 1704–10.

9. Blumenfeld H. Cellular and network mechanisms of spike-wave seizures. *Epilepsia.* 2005;46 (suppl 9):21–33.

10. Meeren HK, Pijn JP, Van Luijtelaar EL, Coenen AM, Lopes da Silva FH. Cortical focus drives widespread corticothalamic networks during spontaneous absence seizures in rats. *J Neurosci.* 2002;22:1480–95.

11. Holmes MD, Brown M, Tucker DM. Are "generalized" seizures truly generalized? Evidence of localized mesial frontal and frontopolar discharges in absence. *Epilepsia.* 2004;45:1568–79.

12. Carney PW, Jackson GD. Insights into the mechanisms of absence seizure generation provided by EEG with functional MRI. *Front Neurol.* 2014;5:162.

13. Amor F, Baillet S, Navarro V, Adam C, Martinerie J, Quyen Mle V. Cortical local and long-range synchronization interplay in human absence seizure initiation. *NeuroImage.* 2009;45:950–62.

14. Wolf P, Yacubian EM, Avanzini G, et al. Juvenile myoclonic epilepsy: a system disorder of the brain. *Epilepsy Res.* 2015;114:2–12.

15. Avanzini G, Manganotti P, Meletti S, et al. The system epilepsies: a pathophysiological hypothesis. *Epilepsia.* 2012;53:771–8.

16. Power JD, Fair DA, Schlaggar BL, Petersen SE. The development of human functional brain networks. *Neuron.* 2010;67:735–48.

17. Sporns O, Tononi G, Kotter R. The human connectome: a structural description of the human brain. *PLOS Comput Biol.* 2005;1:e42.

18. Mori S, Crain BJ, Chacko VP, van Zijl PC. Three-dimensional tracking of axonal projections in the brain by magnetic resonance imaging. *Ann Neurol.* 1999;45: 265–9.

19. Behrens TE, Johansen-Berg H, Woolrich MW, et al. Non-invasive mapping of connections between human thalamus and cortex using diffusion imaging. *Nat Neurosci.* 2003;6:750–7.

20. Le Bihan D, Mangin JF, Poupon C, et al. Diffusion tensor imaging: concepts and applications. *J Magn Reson Imaging.* 2001;13:534–46.

21. Conturo TE, Lori NF, Cull TS, et al. Tracking neuronal fiber pathways in the living human brain. *Proc Natl Acad Sci USA.* 1999;96:10422–7.

22. Knosche TR, Anwander A, Liptrot M, Dyrby TB. Validation of tractography: Comparison with manganese tracing. *Hum Brain Mapp.* 2015;36: 4116–34.

23. Jones DK, Knosche TR, Turner R. White matter integrity, fiber count, and other fallacies: the do's and don'ts of diffusion MRI. *NeuroImage.* 2013;73:239–54.

24. Alexander-Bloch A, Giedd JN, Bullmore E. Imaging structural co-variance between human brain regions. *Nat Rev Neurosci.* 2013;14:322–36.

25. Alexander-Bloch A, Raznahan A, Bullmore E, Giedd J. The convergence of maturational change and structural covariance in human cortical networks. *J Neurosci.* 2013;33:2889–99.

26. Gong G, He Y, Chen ZJ, Evans AC. Convergence and divergence of thickness correlations with diffusion connections across the human cerebral cortex. *NeuroImage.* 2012;59:1239–48.

27. Kelly C, Biswal BB, Craddock RC, Castellanos FX, Milham MP. Characterizing variation in the functional connectome: promise and pitfalls. *Trends Cogn Sci.* 2012;16:181–8.

28. Clos M, Rottschy C, Laird AR, Fox PT, Eickhoff SB. Comparison of structural covariance with functional connectivity approaches exemplified by an investigation of the left anterior insula. *NeuroImage.* 2014;99:269–80.

29. Lerch JP, Worsley K, Shaw WP, et al. Mapping anatomical correlations across cerebral cortex

(MACACC) using cortical thickness from MRI. *NeuroImage*. 2006;**31**:993–1003.

30. Zielinski BA, Gennatas ED, Zhou J, Seeley WW. Network-level structural covariance in the developing brain. *Proc Natl Acad Sci USA*. 2010;**107**:18191–6.

31. Friston KJ. Functional and effective connectivity: a review. *Brain Connect*. 2011;**1**:13–36.

32. Cabral J, Kringelbach ML, Deco G. Exploring the network dynamics underlying brain activity during rest. *Prog Neurobiol*. 2014;**114**:102–31.

33. Damoiseaux JS, Rombouts SA, Barkhof F, et al. Consistent resting-state networks across healthy subjects. *Proc Natl Acad Sci USA*. 2006;**103**:13848–53.

34. Centeno M, Carmichael DW. Network connectivity in epilepsy: resting state fMRI and EEG-fMRI contributions. *Front Neurol*. 2014;**5**:93.

35. Bullmore E, Sporns O. Complex brain networks: graph theoretical analysis of structural and functional systems. *Nat Rev Neurosci*. 2009;**10**:186–98.

36. Watts DJ, Strogatz SH. Collective dynamics of "small-world" networks. *Nature*. 1998;**393**:440–2.

37. Bullmore E, Sporns O. The economy of brain network organization. *Nat Rev Neurosci*. 2012;**13**:336–49.

38. Bernhardt BC, Worsley KJ, Besson P, et al. Mapping limbic network organization in temporal lobe epilepsy using morphometric correlations: insights on the relation between mesiotemporal connectivity and cortical atrophy. *NeuroImage*. 2008;**42**:515–24.

39. Bonilha L, Rorden C, Halford JJ, et al. Asymmetrical extra-hippocampal grey matter loss related to hippocampal atrophy in patients with medial temporal lobe epilepsy. *J Neurol Neurosurg Psychiatry*. 2007;**78**: 286–94.

40. Mueller SG, Laxer KD, Barakos J, Ian C, Garcia P, Weiner MW. Widespread neocortical abnormalities in temporal lobe epilepsy with and without mesial sclerosis. *NeuroImage*. 2009;**46**:353–9.

41. Mueller SG, Laxer KD, Barakos J, et al. Involvement of the thalamocortical network in TLE with and without mesiotemporal sclerosis. *Epilepsia*. 2010;**51**:1436–45.

42. Bernhardt BC, Bernasconi N, Kim H, Bernasconi A. Mapping thalamocortical network pathology in temporal lobe epilepsy. *Neurology*. 2012;**78**:129–36.

43. Concha L, Beaulieu C, Collins DL, Gross DW. White-matter diffusion abnormalities in temporal-lobe epilepsy with and without mesial temporal sclerosis. *J Neurol Neurosurg Psychiatry*. 2009;**80**:312–9.

44. Focke NK, Yogarajah M, Bonelli SB, Bartlett PA, Symms MR, Duncan JS. Voxel-based diffusion tensor imaging in patients with mesial temporal lobe epilepsy and hippocampal sclerosis. *NeuroImage*. 2008;**40**: 728–37.

45. Thivard L, Lehericy S, Krainik A, et al. Diffusion tensor imaging in medial temporal lobe epilepsy with hippocampal sclerosis. *NeuroImage*. 2005;**28**:682–90.

46. Concha L, Kim H, Bernasconi A, Bernhardt BC, Bernasconi N. Spatial patterns of water diffusion along white matter tracts in temporal lobe epilepsy. *Neurology*. 2012;**79**:455–62.

47. Vaughan DN, Raffelt D, Curwood E, et al. Tract-specific atrophy in focal epilepsy: disease, genetics, or seizures? *Ann Neurol*. 2017;**81**:240–50.

48. Ahmadi ME, Hagler DJ Jr, McDonald CR, et al. Side matters: diffusion tensor imaging tractography in left and right temporal lobe epilepsy. *AJNR Am J Neuroradiol*. 2009;**30**:1740–7.

49. Takaya S, Liu H, Greve DN, et al. Altered anterior-posterior connectivity through the arcuate fasciculus in temporal lobe epilepsy. *Hum Brain Mapp*. 2016;**37**:4425–38.

50. Keller SS, Richardson MP, Schoene-Bake JC, et al. Thalamotemporal alteration and postoperative seizures in temporal lobe epilepsy. *Ann Neurol*. 2015;**77**:760–74.

51. Besson P, Dinkelacker V, Valabregue R, et al. Structural connectivity differences in left and right temporal lobe epilepsy. *NeuroImage*. 2014;**100**:135–44.

52. Bettus G, Guedj E, Joyeux F, et al. Decreased basal fMRI functional connectivity in epileptogenic networks and contralateral compensatory mechanisms. *Hum Brain Mapp*. 2009;**30**:1580–91.

53. Barron DS, Fox PT, Pardoe H, et al. Thalamic functional connectivity predicts seizure laterality in individual TLE patients: application of a biomarker development strategy. *NeuroImage Clin*. 2015;**7**: 273–80.

54. Pereira FR, Alessio A, Sercheli MS, et al. Asymmetrical hippocampal connectivity in mesial temporal lobe epilepsy: evidence from resting state fMRI. *BMC Neurosci*. 2010;**11**:66.

55. Chiang S, Stern JM, Engel J Jr, Levin HS, Haneef Z. Differences in graph theory functional connectivity in left and right temporal lobe epilepsy. *Epilepsy Res*. 2014;**108**:1770–81.

56. Maccotta L, He BJ, Snyder AZ, et al. Impaired and facilitated functional networks in temporal lobe epilepsy. *NeuroImage Clin*. 2013;**2**:862–72.

57. Zhang Z, Lu G, Zhong Y, et al. Altered spontaneous neuronal activity of the default-mode network in mesial temporal lobe epilepsy. *Brain Res*. 2010;**1323**: 152–60.

58. Pittau F, Grova C, Moeller F, Dubeau F, Gotman J. Patterns of altered functional connectivity in mesial temporal lobe epilepsy. *Epilepsia*. 2012;**53**: 1013–23.

59. Haneef Z, Lenartowicz A, Yeh HJ, Levin HS, Engel J Jr, Stern JM. Functional connectivity of hippocampal networks in temporal lobe epilepsy. *Epilepsia*. 2014;**55**: 137–45.

60. Voets NL, Beckmann CF, Cole DM, Hong S, Bernasconi A, Bernasconi N. Structural substrates for resting network disruption in temporal lobe epilepsy. *Brain*. 2012;**135**:2350–7.

61. Liao W, Zhang Z, Pan Z, et al. Default mode network abnormalities in mesial temporal lobe epilepsy: a study combining fMRI and DTI. *Hum Brain Mapp*. 2011;**32**: 883–95.

62. Andrews-Hanna JR, Smallwood J, Spreng RN. The default network and self-generated thought: component processes, dynamic control, and clinical relevance. *Ann N Y Acad Sci*. 2014;**1316**:29–52.

63. Laufs H, Hamandi K, Salek-Haddadi A, Kleinschmidt AK, Duncan JS, Lemieux L. Temporal lobe interictal epileptic discharges affect cerebral activity in "default mode" brain regions. *Hum Brain Mapp*. 2007;**28**:1023–32.

64. Kobayashi E, Bagshaw AP, Benar CG, et al. Temporal and extratemporal BOLD responses to temporal lobe interictal spikes. *Epilepsia*. 2006;**47**:343–54.

65. Bernhardt BC, Bernasconi A, Liu M, et al. The spectrum of structural and functional imaging abnormalities in temporal lobe epilepsy. *Ann Neurol*. 2016;**80**:142–53.

66. Zhang Z, Lu G, Zhong Y, et al. Impaired attention network in temporal lobe epilepsy: a resting FMRI study. *Neurosci Lett*. 2009;**458**:97–101.

67. Waites AB, Briellmann RS, Saling MM, Abbott DF, Jackson GD. Functional connectivity networks are disrupted in left temporal lobe epilepsy. *Ann Neurol*. 2006;**59**:335–43.

68. Lv ZX, Huang DH, Ye W, Chen ZR, Huang WL, Zheng JO. Alteration of functional connectivity within visuospatial working memory-related brain network in patients with right temporal lobe epilepsy: a resting-state fMRI study. *Epilepsy Behav*. 2014;**35**:64–71.

69. Stretton J, Winston GP, Sidhu M, et al. Disrupted segregation of working memory networks in temporal lobe epilepsy. *NeuroImage Clin*. 2013;**2**: 273–81.

70. Sidhu MK, Stretton J, Winston GP, et al. A functional magnetic resonance imaging study mapping the episodic memory encoding network in temporal lobe epilepsy. *Brain*. 2013;**136**:1868–88.

71. Voets NL, Menke RA, Jbabdi S, et al. Thalamo-cortical disruption contributes to short-term memory deficits in patients with medial temporal lobe damage. *Cereb Cortex*. 2015;**25**:4584–95.

72. He X, Doucet GE, Sperling M, Sharan A, Tracy JI. Reduced thalamocortical functional connectivity in temporal lobe epilepsy. *Epilepsia*. 2015;**56**:1571–9.

73. Fahoum F, Lopes R, Pittau F, Dubeau F, Gotman J. Widespread epileptic networks in focal epilepsies: EEG-fMRI study. *Epilepsia*. 2012;**53**:1618–27.

74. Kobayashi E, Grova C, Tyvaert L, Dubeau F, Gotman J. Structures involved at the time of temporal lobe spikes revealed by interindividual group analysis of EEG/fMRI data. *Epilepsia*. 2009;**50**:2549–56.

75. Laufs H, Richardson MP, Salek-Haddadi A, et al. Converging PET and fMRI evidence for a common area involved in human focal epilepsies. *Neurology*. 2011;**77**:904–10.

76. McCormick C, Protzner AB, Barnett AJ, Cohn M, Valiante TA, McAndrews MP. Linking DMN connectivity to episodic memory capacity: what can we learn from patients with medial temporal lobe damage? *NeuroImage Clin*. 2014;**5**:188–96.

77. Liu M, Bernhardt BC, Hong SJ, Caldairou B, Bernasconi A, Bernasconi N. The superficial white matter in temporal lobe epilepsy: a key link between structural and functional network disruptions. *Brain*. 2016;**139**:2431–40.

78. Bernhardt BC, Bernasconi N, Hong SJ, Dery S, Bernasconi A. Subregional mesiotemporal network topology is altered in temporal lobe epilepsy. *Cereb Cortex*. 2016;**26**:3237–48.

79. Liao W, Zhang Z, Pan Z, et al. Altered functional connectivity and small-world in mesial temporal lobe epilepsy. *PLOS ONE*. 2010;**5**:e8525.

80. Vaughan DN, Rayner G, Tailby C, Jackson GD. MRI-negative temporal lobe epilepsy: a network disorder of neocortical connectivity. *Neurology*. 2016;**87**:1934–42.

81. Bernhardt BC, Bonilha L, Gross DW. Network analysis for a network disorder: the emerging role of graph theory in the study of epilepsy. *Epilepsy Behav*. 2015;**50**: 162–70.

82. Wang J, Qiu S, Xu Y, et al. Graph theoretical analysis reveals disrupted topological properties of whole brain functional networks in temporal lobe epilepsy. *Clin Neurophysiol*. 2014;**125**:1744–56.

83. Bernhardt BC, Chen Z, He Y, Evans AC, Bernasconi N. Graph-theoretical analysis reveals disrupted small-world organization of cortical thickness correlation networks in temporal lobe epilepsy. *Cereb Cortex*. 2011;**21**:2147–57.

84. Yasuda CL, Chen Z, Beltramini GC, et al. Aberrant topological patterns of brain structural network in temporal lobe epilepsy. *Epilepsia*. 2015;**56**:1992–2002.

85. Bonilha L, Nesland T, Martz GU, et al. Medial temporal lobe epilepsy is associated with neuronal fibre loss and paradoxical increase in structural connectivity

of limbic structures. *J Neurol Neurosurg Psychiatry*. 2012;**83**:903–9.

86. Liu M, Chen Z, Beaulieu C, Gross DW. Disrupted anatomic white matter network in left mesial temporal lobe epilepsy. *Epilepsia*. 2014;**55**:674–82.

87. Bartolomei F, Bettus G, Stam CJ, Guye M. Interictal network properties in mesial temporal lobe epilepsy: a graph theoretical study from intracerebral recordings. *Clin Neurophysiol*. 2013;**124**:2345–53.

88. Ponten SC, Bartolomei F, Stam CJ. Small-world networks and epilepsy: graph theoretical analysis of intracerebrally recorded mesial temporal lobe seizures. *Clin Neurophysiol*. 2007;**118**:918–27.

89. van Diessen E, Zweiphenning WJ, Jansen FE, Stam CJ, Braun KP, Otte WM. Brain network organization in focal epilepsy: a systematic review and meta-analysis. *PLOS ONE*. 2014;**9**:e114606.

90. Chiang S, Stern JM, Engel J Jr, Haneef Z. Structural-functional coupling changes in temporal lobe epilepsy. *Brain Res*. 2015;**1616**:45–57.

91. Bernasconi A, Bernasconi N, Bernhardt BC, Schrader D. Advances in MRI for "cryptogenic" epilepsies. *Nat Rev Neurol*. 2011;**7**:99–108.

92. Lee SK, Kim DI, Mori S, et al. Diffusion tensor MRI visualizes decreased subcortical fiber connectivity in focal cortical dysplasia. *NeuroImage*. 2004;**22**:1826–9.

93. Bonilha L, Montenegro MA, Rorden C, et al. Voxel-based morphometry reveals excess gray matter concentration in patients with focal cortical dysplasia. *Epilepsia*. 2006;**47**:908–15.

94. Besson P, Andermann F, Dubeau F, Bernasconi A. Small focal cortical dysplasia lesions are located at the bottom of a deep sulcus. *Brain*. 2008;**131**:3246–55.

95. Hong SJ, Kim H, Schrader D, Bernasconi N, Bernhardt BC, Bernasconi A. Automated detection of cortical dysplasia type II in MRI-negative epilepsy. *Neurology*. 2014;**83**:48–55.

96. Hong SJ, Bernhardt BC, Schrader DS, Bernasconi N, Bernasconi A. Whole-brain MRI phenotyping in dysplasia-related frontal lobe epilepsy. *Neurology*. 2016;**86**:643–50.

97. Hong SJ, Bernhardt BC, Caldairou B, et al. Multimodal MRI profiling of focal cortical dysplasia type II. *Neurology*. 2017;**88**:734–42.

98. Fonseca Vde C, Yasuda CL, Tedeschi GG, Betting LE, Cendes F. White matter abnormalities in patients with focal cortical dysplasia revealed by diffusion tensor imaging analysis in a voxelwise approach. *Front Neurol*. 2012;**3**:121.

99. Campos BM, Coan AC, Beltramini GC, et al. White matter abnormalities associate with type and localization of focal epileptogenic lesions. *Epilepsia*. 2015;**56**:125–32.

100. O'Dwyer R, Wehner T, LaPresto E, et al. Differences in corpus callosum volume and diffusivity between temporal and frontal lobe epilepsy. *Epilepsy Behav*. 2010;**19**:376–82.

101. Widjaja E, Kis A, Go C, Snead OC III, Smith ML. Bilateral white matter abnormality in children with frontal lobe epilepsy. *Epilepsy Res*. 2014;**108**:289–94.

102. Braakman HM, Vaessen MJ, Jansen JF, et al. Pediatric frontal lobe epilepsy: white matter abnormalities and cognitive impairment. *Acta Neurol Scand*. 2014;**129**:252–62.

103. Guye M, Ranjeva JP, Bartolomei F, et al. What is the significance of interictal water diffusion changes in frontal lobe epilepsies? *NeuroImage*. 2007;**35**:28–37.

104. Besseling RM, Jansen JF, de Louw AJ, et al. Abnormal profiles of local functional connectivity proximal to focal cortical dysplasias. *PLOS ONE*. 2016;**11**:e0166022.

105. Pedersen M, Omidvarnia A, Curwood EK, Walz JM, Rayner G, Jackson GD. The dynamics of functional connectivity in neocortical focal epilepsy. *NeuroImage Clin*. 2017;**15**:209–14.

106. Centeno M, Vollmar C, O'Muircheartaigh J, et al. Memory in frontal lobe epilepsy: an fMRI study. *Epilepsia*. 2012;**53**:1756–64.

107. Gaillard WD, Berl MM, Moore EN, et al. Atypical language in lesional and nonlesional complex partial epilepsy. *Neurology*. 2007;**69**:1761–71.

108. Janszky J, Ebner A, Kruse B, et al. Functional organization of the brain with malformations of cortical development. *Ann Neurol*. 2003;**53**:759–67.

109. Mbwana J, Berl MM, Ritzl EK, et al. Limitations to plasticity of language network reorganization in localization related epilepsy. *Brain*. 2009;**132**:347–56.

110. Vlooswijk MC, Jansen JF, Majoie HJ, et al. Functional connectivity and language impairment in cryptogenic localization-related epilepsy. *Neurology*. 2010;**75**:395–402.

111. Vlooswijk MC, Jansen JF, Jeukens CR, et al. Memory processes and prefrontal network dysfunction in cryptogenic epilepsy. *Epilepsia*. 2011;**52**:1467–75.

112. Braakman HM, Vaessen MJ, Jansen JF, et al. Frontal lobe connectivity and cognitive impairment in pediatric frontal lobe epilepsy. *Epilepsia*. 2013;**54**:446–54.

113. Woodward KE, Gaxiola-Valdez I, Goodyear BG, Federico P. Frontal lobe epilepsy alters functional connections within the brain's motor network: a resting-state fMRI study. *Brain Connect*. 2014;**4**:91–9.

114. Widjaja E, Zamyadi M, Raybaud C, Snead OC, Smith ML. Abnormal functional network connectivity among resting-state networks in children with frontal lobe epilepsy. *AJNR Am J Neuroradiol.* 2013;34:2386–92.

115. Federico P, Archer JS, Abbott DF, Jackson GD. Cortical/subcortical BOLD changes associated with epileptic discharges: an EEG-fMRI study at 3 T. *Neurology.* 2005;64:1125–30.

116. Tyvaert L, Chassagnon S, Sadikot A, LeVan P, Dubeau F, Gotman J. Thalamic nuclei activity in idiopathic generalized epilepsy: an EEG-fMRI study. *Neurology.* 2009;73:2018–22.

117. Thornton R, Vulliemoz S, Rodionov R, et al. Epileptic networks in focal cortical dysplasia revealed using electroencephalography-functional magnetic resonance imaging. *Ann Neurol.* 2011;70:822–37.

118. Luo C, An D, Yao D, Gotman J. Patient-specific connectivity pattern of epileptic network in frontal lobe epilepsy. *NeuroImage Clin.* 2014;4:668–75.

119. Vlooswijk MC, Vaessen MJ, Jansen JF, et al. Loss of network efficiency associated with cognitive decline in chronic epilepsy. *Neurology.* 2011;77:938–44.

120. Pedersen M, Omidvarnia AH, Walz JM, Jackson GD. Increased segregation of brain networks in focal epilepsy: an fMRI graph theory finding. *NeuroImage Clin.* 2015;8:536–42.

121. Sethi M, Pedersen M, Jackson GD. Polymicrogyric cortex may predispose to seizures via abnormal network topology: an fMRI connectomics study. *Epilepsia.* 2016;57:e64–8.

122. Vaessen MJ, Braakman HM, Heerink JS, et al. Abnormal modular organization of functional networks in cognitively impaired children with frontal lobe epilepsy. *Cereb Cortex.* 2013;23:1997–2006.

123. Vaessen MJ, Jansen JF, Braakman HM, et al. Functional and structural network impairment in childhood frontal lobe epilepsy. *PLOS ONE.* 2014;9: e90068.

124. Koepp MJ, Woermann F, Savic I, Wandschneider B. Juvenile myoclonic epilepsy–neuroimaging findings. *Epilepsy Behav.* 2013;28(suppl 1):S40–4.

125. Gotman J, Grova C, Bagshaw A, Kobayashi E, Aghakhani Y, Dubeau F. Generalized epileptic discharges show thalamocortical activation and suspension of the default state of the brain. *Proc Natl Acad Sci USA.* 2005;102:15236–40.

126. Bernhardt BC, Rozen DA, Worsley KJ, Evans AC, Bernasconi N, Bernasconi A. Thalamo-cortical network pathology in idiopathic generalized epilepsy: insights from MRI-based morphometric correlation analysis. *NeuroImage.* 2009;46:373–81.

127. Deppe M, Kellinghaus C, Duning T, et al. Nerve fiber impairment of anterior thalamocortical circuitry in juvenile myoclonic epilepsy. *Neurology.* 2008;71: 1981–5.

128. Keller SS, Ahrens T, Mohammadi S, et al. Microstructural and volumetric abnormalities of the putamen in juvenile myoclonic epilepsy. *Epilepsia.* 2011;52:1715–24.

129. O'Muircheartaigh J, Vollmar C, Barker GJ, et al. Abnormal thalamocortical structural and functional connectivity in juvenile myoclonic epilepsy. *Brain.* 2012;135:3635–44.

130. Gong J, Chang X, Jiang S, et al. Microstructural alterations of white matter in juvenile myoclonic epilepsy. *Epilepsy Res.* 2017;135:1–8.

131. Liu M, Concha L, Beaulieu C, Gross DW. Distinct white matter abnormalities in different idiopathic generalized epilepsy syndromes. *Epilepsia.* 2011;52: 2267–75.

132. Kim JH, Suh SI, Park SY, et al. Microstructural white matter abnormality and frontal cognitive dysfunctions in juvenile myoclonic epilepsy. *Epilepsia.* 2012;53:1371–8.

133. Vulliemoz S, Vollmar C, Koepp MJ, et al. Connectivity of the supplementary motor area in juvenile myoclonic epilepsy and frontal lobe epilepsy. *Epilepsia.* 2011;52:507–14.

134. O'Muircheartaigh J, Vollmar C, Barker GJ, et al. Focal structural changes and cognitive dysfunction in juvenile myoclonic epilepsy. *Neurology.* 2011;76:34–40.

135. Kim SH, Lim SC, Kim W, et al. Extrafrontal structural changes in juvenile myoclonic epilepsy: a topographic analysis of combined structural and microstructural brain imaging. *Seizure.* 2015;30:124–31.

136. Vollmar C, O'Muircheartaigh J, Symms MR, et al. Altered microstructural connectivity in juvenile myoclonic epilepsy: the missing link. *Neurology.* 2012;78:1555–9.

137. Caeyenberghs K, Powell HW, Thomas RH, et al. Hyperconnectivity in juvenile myoclonic epilepsy: a network analysis. *NeuroImage Clin.* 2015;7: 98–104.

138. Focke NK, Diederich C, Helms G, Nitsche MA, Lerche H, Paulus W. Idiopathic-generalized epilepsy shows profound white matter diffusion-tensor imaging alterations. *Hum Brain Mapp.* 2014;35: 3332–42.

139. Lee CY, Tabesh A, Spampinato MV, Helpern JA, Jensen JH, Bonilha L. Diffusional kurtosis imaging reveals a distinctive pattern of microstructural alternations in idiopathic generalized epilepsy. *Acta Neurol Scand.* 2014;130:148–55.

140. Yang T, Guo Z, Luo C, et al. White matter impairment in the basal ganglia-thalamocortical circuit of drug-naive childhood absence epilepsy. *Epilepsy Res.* 2012;**99**:267–73.

141. Xue K, Luo C, Zhang D, et al. Diffusion tensor tractography reveals disrupted structural connectivity in childhood absence epilepsy. *Epilepsy Res.* 2014;**108**:125–38.

142. Aghakhani Y, Bagshaw AP, Benar CG, et al. fMRI activation during spike and wave discharges in idiopathic generalized epilepsy. *Brain.* 2004;**127**:1127–44.

143. Hamandi K, Salek-Haddadi A, Laufs H, et al. EEG-fMRI of idiopathic and secondarily generalized epilepsies. *NeuroImage.* 2006;**31**:1700–10.

144. Salek-Haddadi A, Lemieux L, Merschhemke M, Friston KJ, Duncan JS, Fish DR. Functional magnetic resonance imaging of human absence seizures. *Ann Neurol.* 2003;**53**:663–7.

145. Benuzzi F, Mirandola L, Pugnaghi M, et al. Increased cortical BOLD signal anticipates generalized spike and wave discharges in adolescents and adults with idiopathic generalized epilepsies. *Epilepsia.* 2012;**53**:622–30.

146. Guo JN, Kim R, Chen Y, et al. Impaired consciousness in patients with absence seizures investigated by functional MRI, EEG, and behavioural measures: a cross-sectional study. *Lancet Neurol.* 2016;**15**:1336–45.

147. Moeller F, Siebner HR, Wolff S, et al. Changes in activity of striato-thalamo-cortical network precede generalized spike wave discharges. *NeuroImage.* 2008;**39**:1839–49.

148. Carney PW, Masterton RA, Harvey AS, Scheffer IE, Berkovic SF, Jackson GD. The core network in absence epilepsy. Differences in cortical and thalamic BOLD response. *Neurology.* 2010;**75**:904–11.

149. Dong L, Luo C, Zhu Y, et al. Complex discharge-affecting networks in juvenile myoclonic epilepsy: a simultaneous EEG-fMRI study. *Hum Brain Mapp.* 2016;**37**:3515–29.

150. Vaudano AE, Laufs H, Kiebel SJ, et al. Causal hierarchy within the thalamo-cortical network in spike and wave discharges. *PLOS ONE.* 2009;**4**:e6475.

151. Bai X, Vestal M, Berman R, et al. Dynamic time course of typical childhood absence seizures: EEG, behavior, and functional magnetic resonance imaging. *J Neurosci.* 2010;**30**:5884–93.

152. Moeller F, LeVan P, Muhle H, et al. Absence seizures: individual patterns revealed by EEG-fMRI. *Epilepsia.* 2010;**51**:2000–10.

153. Szaflarski JP, DiFrancesco M, Hirschauer T, et al. Cortical and subcortical contributions to absence seizure onset examined with EEG/fMRI. *Epilepsy Behav.* 2010;**18**:404–13.

154. Rodin E. Decomposition and mapping of generalized spike-wave complexes. *Clin Neurophysiol.* 1999;**110**:1868–75.

155. Masterton RA, Carney PW, Abbott DF, Jackson GD. Absence epilepsy subnetworks revealed by event-related independent components analysis of functional magnetic resonance imaging. *Epilepsia.* 2013;**54**:801–8.

156. Luo C, Li Q, Lai Y, et al. Altered functional connectivity in default mode network in absence epilepsy: a resting-state fMRI study. *Hum Brain Mapp.* 2011;**32**:438–49.

157. McGill ML, Devinsky O, Kelly C, et al. Default mode network abnormalities in idiopathic generalized epilepsy. *Epilepsy Behav.* 2012;**23**:353–9.

158. Kay BP, DiFrancesco MW, Privitera MD, Gotman J, Holland SK, Szaflarski JP. Reduced default mode network connectivity in treatment-resistant idiopathic generalized epilepsy. *Epilepsia.* 2013;**54**:461–70.

159. Song M, Du H, Wu N, et al. Impaired resting-state functional integrations within default mode network of generalized tonic-clonic seizures epilepsy. *PLOS ONE.* 2011;**6**:e17294.

160. Masterton RA, Carney PW, Jackson GD. Cortical and thalamic resting-state functional connectivity is altered in childhood absence epilepsy. *Epilepsy Res.* 2012;**99**:327–34.

161. Bai X, Guo J, Killory B, et al. Resting functional connectivity between the hemispheres in childhood absence epilepsy. *Neurology.* 2011;**76**:1960–7.

162. Yang T, Ren J, Li Q, et al. Increased interhemispheric resting-state in idiopathic generalized epilepsy with generalized tonic-clonic seizures: a resting-state fMRI study. *Epilepsy Res.* 2014;**108**:1299–305.

163. Killory BD, Bai X, Negishi M, et al. Impaired attention and network connectivity in childhood absence epilepsy. *NeuroImage.* 2011;**56**:2209–17.

164. Wang Z, Zhang Z, Jiao Q, et al. Impairments of thalamic nuclei in idiopathic generalized epilepsy revealed by a study combining morphological and functional connectivity MRI. *PLOS ONE.* 2012;**7**:e39701.

165. Kay BP, Holland SK, Privitera MD, Szaflarski JP. Differences in paracingulate connectivity associated with epileptiform discharges and uncontrolled seizures in genetic generalized epilepsy. *Epilepsia.* 2014;**55**:256–63.

166. Kim JB, Suh SI, Seo WK, Oh K, Koh SB, Kim JH. Altered thalamocortical functional connectivity in

idiopathic generalized epilepsy. *Epilepsia.* 2014;55:592–600.

167. McGill ML, Devinsky O, Wang X, et al. Functional neuroimaging abnormalities in idiopathic generalized epilepsy. *NeuroImage Clin.* 2014;6: 455–62.

168. Ji GJ, Zhang Z, Xu Q, et al. Identifying corticothalamic network epicenters in patients with idiopathic generalized epilepsy. *AJNR Am J Neuroradiol.* 2015;36:1494–500.

169. Vollmar C, O'Muircheartaigh J, Barker GJ, et al. Motor system hyperconnectivity in juvenile myoclonic epilepsy: a cognitive functional magnetic resonance imaging study. *Brain.* 2011;134:1710–9.

170. Wandschneider B, Centeno M, Vollmar C, et al. Motor co-activation in siblings of patients with juvenile myoclonic epilepsy: an imaging endophenotype? *Brain.* 2014;137:2469–79.

171. Wang X, Jiao D, Zhang X, Lin X. Altered degree centrality in childhood absence epilepsy: a resting-state fMRI study. *J Neurol Sci.* 2017;373: 274–9.

172. Zhang Z, Liao W, Chen H, et al. Altered functional-structural coupling of large-scale brain networks in idiopathic generalized epilepsy. *Brain.* 2011;134:2912–28.

173. Li R, Liao W, Li Y, et al. Disrupted structural and functional rich club organization of the brain connectome in patients with generalized tonic-clonic seizure. *Hum Brain Mapp.* 2016;37:4487–99.

174. Morgan VL, Sonmezturk HH, Gore JC, Abou-Khalil B. Lateralization of temporal lobe epilepsy using resting functional magnetic resonance imaging connectivity of hippocampal networks. *Epilepsia.* 2012;53:1628–35.

175. Bettus G, Bartolomei F, Confort-Gouny S, et al. Role of resting state functional connectivity MRI in presurgical investigation of mesial temporal lobe epilepsy. *J Neurol Neurosurg Psychiatry.* 2010;81: 1147–54.

176. Zhang ZQ, Lu GM, Zhong Y, et al. MRI study of mesial temporal lobe epilepsy using amplitude of low-frequency fluctuation analysis. *Human Brain Mapp.* 2010;31:1851–61.

177. Weaver KE, Chaovalitwongse WA, Novotny EJ, Poliakov A, Grabowski TG, Ojemann JG. Local functional connectivity as a pre-surgical tool for seizure focus identification in non-lesion, focal epilepsy. *Front Neurol.* 2013;4:43.

178. Lee HW, Arora J, Papademetris X, et al. Altered functional connectivity in seizure onset zones revealed by fMRI intrinsic connectivity. *Neurology.* 2014;83:2269–77.

179. Stufflebeam SM, Liu H, Sepulcre J, Tanaka N, Buckner RL, Madsen JR. Localization of focal epileptic discharges using functional connectivity magnetic resonance imaging. *J Neurosurg.* 2011;114: 1693–7.

180. Chiang S, Levin HS, Haneef Z. Computer-automated focus lateralization of temporal lobe epilepsy using fMRI. *J Magn Reson Imaging.* 2015;41:1689–94.

181. Liao W, Ji GJ, Xu Q, et al. Functional connectome before and following temporal lobectomy in mesial temporal lobe epilepsy. *Sci Rep.* 2016;6:23153.

182. He X, Doucet GE, Pustina D, Sperling MR, Sharan AD, Tracy JI. Presurgical thalamic "hubness" predicts surgical outcome in temporal lobe epilepsy. *Neurology.* 2017;88:2285–93.

183. Negishi M, Martuzzi R, Novotny EJ, Spencer DD, Constable RT. Functional MRI connectivity as a predictor of the surgical outcome of epilepsy. *Epilepsia.* 2011;52:1733–40.

184. Xu Q, Zhang Z, Liao W, et al. Time-shift homotopic connectivity in mesial temporal lobe epilepsy. *AJNR Am J Neuroradiol.* 2014;35:1746–52.

185. Munsell BC, Wee CY, Keller SS, et al. Evaluation of machine learning algorithms for treatment outcome prediction in patients with epilepsy based on structural connectome data. *NeuroImage.* 2015;118: 219–30.

186. Bonilha L, Jensen JH, Baker N, et al. The brain connectome as a personalized biomarker of seizure outcomes after temporal lobectomy. *Neurology.* 2015;84:1846–53.

187. Doucet GE, Pustina D, Skidmore C, Sharan A, Sperling MR, Tracy JI. Resting-state functional connectivity predicts the strength of hemispheric lateralization for language processing in temporal lobe epilepsy and normals. *Hum Brain Mapp.* 2015;36:288–303.

188. Pravata E, Sestieri C, Mantini D, et al. Functional connectivity MR imaging of the language network in patients with drug-resistant epilepsy. *AJNR Am J Neuroradiol.* 2011;32:532–40.

189. McCormick C, Quraan M, Cohn M, Valiante TA, McAndrews MP. Default mode network connectivity indicates episodic memory capacity in mesial temporal lobe epilepsy. *Epilepsia.* 2013;54:809–18.

190. Voets NL, Zamboni G, Stokes MG, Carpenter K, Stacey R, Adcock JE. Aberrant functional connectivity in dissociable hippocampal networks is associated with deficits in memory. *J Neurosci.* 2014;34:4920–8.

191. Doucet G, Osipowicz K, Sharan A, Sperling MR, Tracy JI. Extratemporal functional connectivity impairments at rest are related to memory

performance in mesial temporal epilepsy. *Hum Brain Mapp*. 2013;34:2202–16.

192. Holmes M, Folley BS, Sonmezturk HH, et al. Resting state functional connectivity of the hippocampus associated with neurocognitive function in left temporal lobe epilepsy. *Hum Brain Mapp*. 2014;35: 735–44.

193. McDonald CR, Ahmadi ME, Hagler DJ, et al. Diffusion tensor imaging correlates of memory and language impairments in temporal lobe epilepsy. *Neurology*. 2008;71:1869–76.

194. Diehl B, Busch RM, Duncan JS, Piao Z, Tkach J, Luders HO. Abnormalities in diffusion tensor imaging of the uncinate fasciculus relate to reduced memory in temporal lobe epilepsy. *Epilepsia*. 2008;49: 1409–18.

195. Yogarajah M, Powell HW, Parker GJ, et al. Tractography of the parahippocampal gyrus and material specific memory impairment in unilateral temporal lobe epilepsy. *NeuroImage*. 2008;40: 1755–64.

196. McDonald CR, Leyden KM, Hagler DJ, et al. White matter microstructure complements morphometry for predicting verbal memory in epilepsy. *Cortex*. 2014;58:139–50.

197. Vaessen MJ, Jansen JF, Vlooswijk MC, et al. White matter network abnormalities are associated with cognitive decline in chronic epilepsy. *Cereb Cortex*. 2012;22:2139–47.

198. Osipowicz K, Sperling MR, Sharan AD, Tracy JI. Functional MRI, resting state fMRI, and DTI for predicting verbal fluency outcome following resective surgery for temporal lobe epilepsy. *J Neurosurg*. 2016;124:929–37.

199. Doucet GE, Rider R, Taylor N, et al. Presurgery resting-state local graph-theory measures predict neurocognitive outcomes after brain surgery in temporal lobe epilepsy. *Epilepsia*. 2015;56:517–26.

Chapter 9

Mapping Metabolism and Inflammation in Epilepsy

Csaba Juhász and Sandeep Mittal

9.1 Introduction

Abnormalities of brain energy metabolism and neuroinflammation are both key aspects of the pathophysiology of epilepsy.[1–3] Both metabolic and inflammatory changes in the epileptic brain can be studied by in vivo molecular imaging. Historically, mapping of cerebral glucose metabolism was the first clinical application of positron emission tomography (PET) in the evaluation of epilepsy. Initial studies with 2-deoxy-2[^{18}F]fluoro-D-glucose (FDG) PET demonstrated localized glucose metabolic changes in the form of interictal hypo and ictal hypermetabolism in patients with focal epilepsy.[4,5] Since then, FDG-PET became a mainstream imaging modality in the presurgical evaluation of patients with medically refractory epilepsy. Subsequently, interest in neuroimaging of epilepsy-associated neuroinflammation emerged following early studies showing increased peripheral benzodiazepine bindings sites, localized on activated microglia, in resected human epileptic tissue.[6,7] Interest in epilepsy-associated neuroinflammation has further increased during the last decade, after the widespread recognition of the pivotal role of neuroinflammatory processes in the pathophysiology of common forms of human epileptogenesis.[1,8,9]

Most PET studies of human epilepsies included patients with medically refractory seizures. These studies provided a great amount of information on chronic metabolic abnormalities in long-standing epilepsy. However, in this patient population, it is nearly impossible to differentiate between changes directly related to epileptogenesis vs. those caused by an underlying process (lesion) or secondary brain tissue damage. Alternative approaches include the study of human epileptogenesis before or shortly after the onset of clinical seizures and imaging studies of epileptogenesis in animal models using small-animal scanners.[10]

The term "epileptogenesis" refers to a number of molecular, cellular, and brain structural changes, triggered by a cerebral insult, which leads to the first spontaneous seizure(s). The underlying processes of epileptogenesis include, among others, neurodegeneration, gliosis, blood-brain barrier damage, and neuroinflammation.[11] These molecular and cellular changes may persist well beyond the initial clinical seizure. Therefore, neuroimaging studies in new- or recent-onset epilepsies may provide a valuable insight in the processes of epileptogenesis. The period of epileptogenesis also provides a critical therapeutic window to intervene with the intent to prevent (or delay) the development of clinical seizures and/or prevent progression of epilepsy and secondary brain damage. There is intense research to understand details of molecular mechanisms of pre- and postclinical stages of epileptogenesis. A critical component of this effort is to identify specific biomarkers to predict development of epilepsy after an insult, identify the severity and extent of brain tissue capable of generating spontaneous seizures, measure progression, and evaluate potential therapeutic interventions.[12] For potential epilepsy biomarkers, molecular imaging is a particularly attractive tool to study specific molecular processes in various brain regions in vivo.

In the next sections, we summarize some key molecular imaging studies aiming at mapping brain metabolic and neuroinflammatory processes in both animal seizure models and human epilepsies, focusing on key imaging findings in early and late stages of epilepsy.

9.2 Imaging Markers of Glucose Metabolism in Animal Models of Epileptogenesis

The increasing availability and use of small animal PET scanners in the last 10 years allowed the study of the temporospatial evolution of global and regional brain glucose metabolism in experimental models of epileptogenesis. The most commonly studied models employ pilocarpine or kainic acid to induce status epilepticus (SE). In these models, SE is followed by a latent period before development of spontaneous seizures in a subset of animals. In the early phase (within 1–2 days) after SE, longitudinal FDG-PET imaging can demonstrate a global decrease of cerebral glucose utilization, with maximum decreases in the limbic structures and less robust or minimal decrease in the cortex.[13,14] This is followed by a metabolic recovery in most structures except the hippocampus and thalamus, which remain hypometabolic in the chronic phase (Figure 9.1). Severe hypometabolism in the entorhinal cortex was followed by development of spontaneous seizures, suggesting a critical role of the entorhinal cortex in the early stages of temporal lobe epilepsy (TLE).[13] Similarly, widespread brain glucose hypometabolism was reported after kainic acid-induced SE.[15] In this model, the hippocampal and thalamic/hypothalamic regions showed a persistent reduction in glucose metabolism, while FDG uptake in the amygdala/entorhinal cortex and motor/somatosensory cortices transiently returned to control levels one week following SE and decreased again at later time points. While hippocampal volumes decrease in the silent and chronic period of epileptogenesis, no correlation was found between the degree of the hypometabolism on FDG-PET and the severity of MRI-detected hippocampal atrophy or CA1 pyramidal cell loss.[14,15]

A recent study has also provided data on early metabolic changes after traumatic brain injury in rats.[16] In this model, about half of the animals developed spontaneous recurring seizures or epileptic discharges during a 6-month follow-up period. Only a few subtle MRI and FDG-PET abnormalities were predictive of posttraumatic epilepsy. Specifically, a multivariate logistic regression model that incorporated serial PET parameters from the ipsilateral

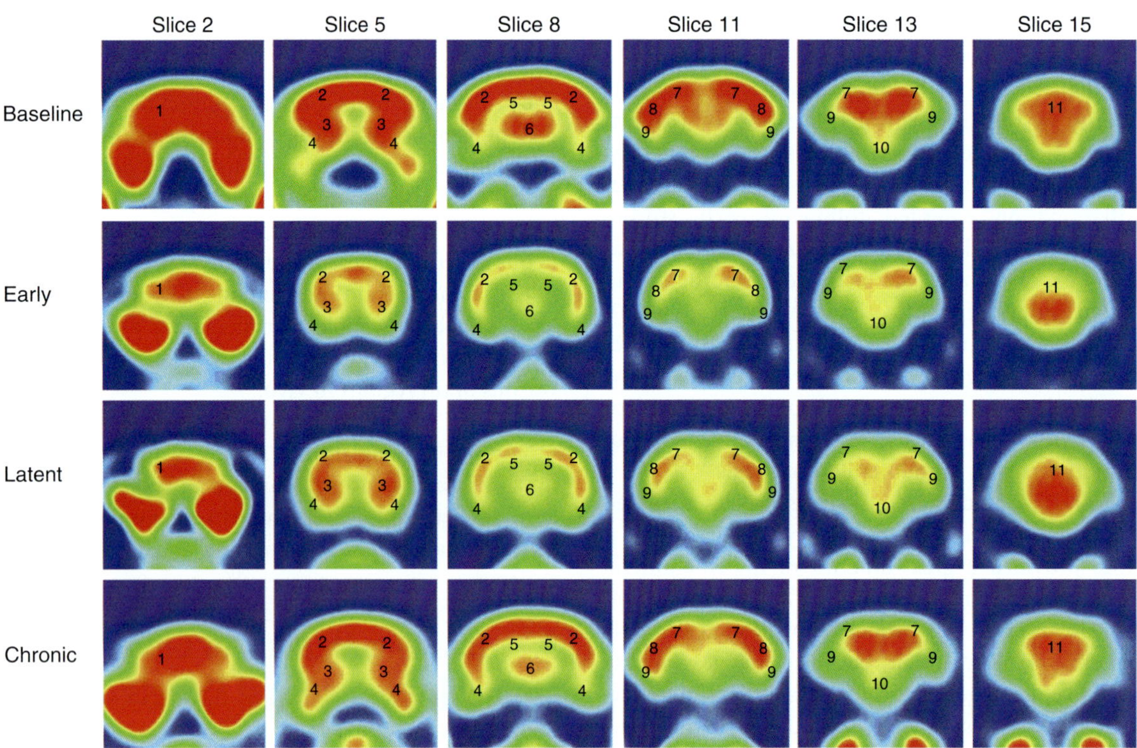

Figure 9.1. MRI coregistered micro-PET images (six selected coronal slices) in the different phases of epileptogenesis in rat brain. Status epilepticus was induced by pilocarpine. The glucose utilization in most brain structures, especially in limbic structures, decreased in the early and latent phases and returned to baseline levels in the chronic phase, with the exception of the hippocampus and thalamus. In the pons and the cerebellum, the glucose utilization remained at baseline levels in all phases. The numbered brain regions are as follows: (1) prefrontal cortex, (2) sensorimotor cortex, (3) striatum, (4) piriform cortex, (5) hippocampus, (6) thalamus, (7) visual cortex, (8) auditory cortex, (9) entorhinal cortex, (10) pons, and (11) cerebellum. Reproduced from Guo et al., *Neuroscience* 2009[13] with permission.

hippocampus predicted the outcome correctly. However, the relevance of these findings for human posttraumatic epilepsy remains to be demonstrated.

Altogether, the animal studies indicate that widespread cerebral hypometabolism is an early metabolic hallmark during SE-induced epileptogenesis, and persistent hippocampal hypometabolism is not simply due to pyramidal cell loss or structural atrophy. While increased metabolism is usually not observed, permanent hypometabolism develops in specific structures (hippocampus, entorhinal cortex, and thalamus) that commonly show focal hypometabolism (and also atrophy) in human TLE.[17,18] Whether limbic hypometabolism is also a common feature of the early stages of human epileptogenesis, where initial SE does not play a significant role in most cases, remains to be clarified.

9.3 Imaging Brain Metabolism in Human Epilepsy

9.3.1 Glucose Metabolic Changes in High-Risk Patients

Early stages of human epileptogenesis can be studied in patients with an epileptogenic brain insult before the onset of the first clinical seizure(s). The overall risk of epilepsy after common brain insults in adults is modest; for example, only up to 5% of stroke patients and around 10% of patients with traumatic brain injury go on to develop epilepsy.[19–21] Considering the relatively low epilepsy risk, these lesions warrant a widely available, easy-to-obtain biomarker to identify the subset of patients who are at highest risk for development of epilepsy. While MRI is routinely performed in these patient populations, widespread PET studies are not feasible. For application of PET imaging, a more realistic goal is to identify specific patient subgroups with a very high epilepsy risk, such as children with neurocutaneous disorders, including Sturge-Weber syndrome (SWS) and tuberous sclerosis complex (TSC). These disorders commonly manifest in infancy due to characteristic skin lesions, which typically precede development of clinical seizures. The overall risk of developing epilepsy is around 80% or higher in both diseases. In the vast majority of affected children, seizures start within the first two years of life. SWS is most commonly associated with partial seizures, while infantile spasms are more common in children with TSC. Prophylactic

antiepileptic/antiepileptogenic treatment would be an attractive approach to prevent development of seizures in affected children.

There had been two attempts to test prophylactic antiepileptic treatment in children with a neurocutaneous disorder. First, infants with SWS showed a 50% lower rate of epilepsy after prophylactic phenobarbitone administration.[22] Second, in a study of 45 infants with TSC, preventive vigabatrin treatment led to an increased rate of seizure freedom (93% vs. 35%), and lower incidence of drug-resistant epilepsy and mental retardation.[23] Novel antiepileptic drugs or emerging antiepileptogenic agents could be tested in these patient groups, especially if selected biomarkers could accurately identify those at highest risk for severe, drug-resistant seizures.

In an attempt to identify potential molecular imaging markers of epileptogenesis associated with SWS, we obtained FDG-PET scans in 60 children who were subjects of a prospective, longitudinal neuroimaging study.[24] Specifically, we analyzed the clinical correlates of *increased* cortical glucose metabolism, an interictal PET imaging feature previously observed in some children with SWS.[25] Altogether, 9 of the 60 children had increased metabolism on PET (Figure 9.2), while electroencephalography (EEG) during the FDG uptake period did not show seizures or frequent interictal spikes. These children were

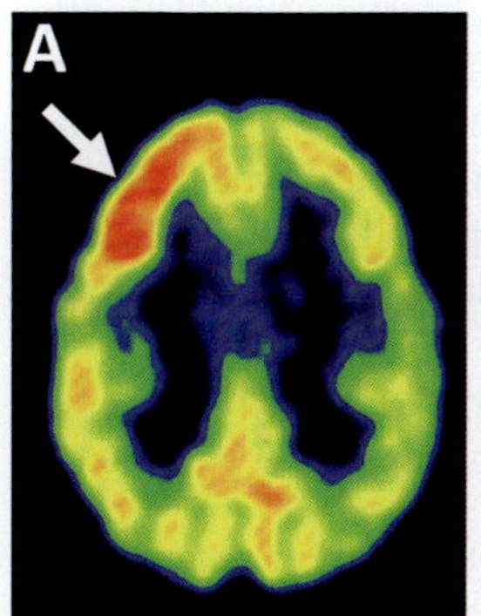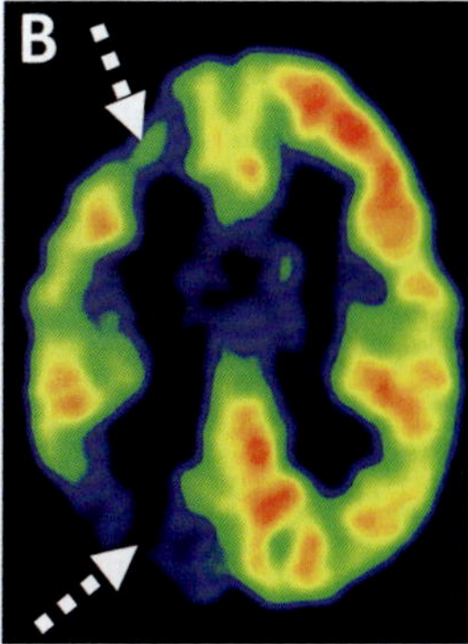

Figure 9.2. Glucose metabolic changes in Sturge-Weber syndrome (SWS) associated with epilepsy in an 11-month-old child. The leptomeningeal malformation affected the right hemisphere. (A) Interictal hypermetabolism in the right frontal cortex (solid arrow), already present before the onset of the first clinical seizure. Slightly increased metabolism is also seen in the right parietal region. (B) Follow-up PET scan at 3 years of age showed progression with marked hypometabolism in areas that were hypermetabolic on the initial scan (dotted arrows). By this age, the child had a history of seizures, which were controlled on antiepileptic medication.

younger than those who showed decreased glucose metabolism (1.9 vs. 4.5 years, respectively). Also, SWS children scanned before 1 year of age had a higher prevalence of hypermetabolism than older children (33% vs. 9%). Hypermetabolism typically occurred shortly before or after the onset of the first clinical seizure. In longitudinal studies, the hypermetabolic changes invariably switched to hypometabolism. These data suggested that interictal hypermetabolism in SWS is a transient metabolic phenomenon closely associated with imminent or recent onset of clinical seizures. Although these data were preliminary, they raised the intriguing possibility that interictal hypermetabolism in SWS may be an imaging biomarker for epileptogenesis.

The exact mechanisms of the interictal increase of glucose metabolism during the early stage of SWS need to be clarified. A plausible mechanism would be the metabolic effect of excess glutamate released as a result of chronic hypoxia due to the SWS-associated venous malformation. Because of the tight coupling between glucose metabolism and glutamate in human epileptic foci, ischemic injury of the cortex in SWS may lead to glutamate excitotoxicity in conjunction with glucose hypermetabolism. The role of excess glutamate in early SWS-induced seizures and hypermetabolism is also supported by preliminary MR spectroscopic findings demonstrating increased glutamate levels in the affected brain regions.[26] Similar increases in glutamate/glutamine levels have been reported in the cerebral cortex of children after traumatic brain injury.[27] These data suggest that pharmacologic interventions targeting glutamate may be beneficial to block the detrimental effects of excess glutamate and perhaps halt epileptogenesis in certain forms of human epilepsy.

9.3.2 Glucose Metabolist Changes in Recent-Onset Epilepsy

Studies of brain glucose metabolism in patients with new-onset seizures are scarce. In a study of 40 children who underwent FDG-PET scanning within one year after their third unprovoked seizure, focal hypometabolism was observed only in 20%, with abnormalities confined to the temporal lobe ipsilateral to their presumed epileptic focus.[28] This incidence was much lower than hypometabolism reported in patients with chronic "medically-controlled" epilepsy (48% abnormal PET)[29] and

those with intractable epilepsy where focal hypometabolism is present in the majority of cases. In a subsequent, longitudinal study of the same investigators, focal hypometabolism on the initial PET was associated with poor subsequent seizure control.[30] In addition, longer duration of epilepsy before the first PET scan was associated with subsequent development of hypometabolism on follow-up PET. These studies support the notion that early focal metabolic abnormalities, similar to early brain lesions visualized on structural imaging, indicate a risk for uncontrolled epilepsy. In this regard, early focal hypometabolic regions may be considered an imaging marker of subsequent intractability.

9.3.3 Glucose Metabolic Changes in Intractable Epilepsies

The vast majority of human glucose PET studies in epilepsy has been performed in populations with intractable seizures, often in surgical series. These subjects typically have a longstanding history of uncontrolled seizures, whose detrimental effects on brain function and structure may mask metabolic abnormalities specifically linked to epileptogenesis. In the majority of these patients, including those with TLE and extratemporal epilepsies, FDG-PET shows metabolic abnormalities in the form of interictal hypometabolism; the distribution of these metabolic changes largely depends on the type of epilepsy and the location of the epileptic focus.[31,32] Altogether, the majority of patients with medically intractable epilepsy show areas of hypometabolism in the epileptic hemisphere on interictal FDG-PET. Such hypometabolic areas, however, are not specific for epileptogenic regions (as defined by EEG), as they often extend to nonepileptic cortex and also subcortical structures such as the ipsilateral thalamus. Indeed, while the exact pathophysiology of impaired energy metabolism in epileptic brain regions remains unclear, hypometabolism can be present in any brain region showing neuronal loss or chronic synaptic deafferentation.[32] A recent study has also demonstrated mitochondrial dysfunction (reduced complex IV functioning) in resected brain tissue that showed PET hypometabolism before surgery in patients with intractable epilepsy and cortical dysplasia.[33]

Presence of focal hypometabolism is less frequent but not uncommon in nonselective epilepsy cohorts, including patients with medically controlled seizures. In a large, prospective series of 700 patients, Swartz et al.[34] found focal or regional FDG-PET abnormalities in 41–63% of cases depending on the lobar localization of the epileptic focus. In the nonsurgical subgroup, this rate varied between 28–37%. Altogether, these data suggested that patients with long-standing, drug-resistant epilepsy have more common (and more severe) brain metabolic abnormalities than those in an early stage of clinical epilepsy and epilepsies that can be controlled by antiepileptic drugs.

A small, longitudinal FDG-PET study, presented by our group, further supported the notion that cortical glucose hypometabolism in the epileptic brain can be progressive if seizures remain uncontrolled. In this study of 15 children with nonlesional epilepsy, the extent of cortical glucose hypometabolism measured in the epileptic hemisphere in 7–44 months intervals expanded in most patients who had persistent or increasing seizure frequency between the PET scans[35] (Figure 9.3). In contrast, decreased seizure frequency was accompanied by a decrease in the size of hypometabolic cortex. Similarly, dynamic glucose metabolic changes, associated with seizure frequency between scans, were observed in children with lesional neocortical epilepsy due to SWS.[36]

Hypometabolism in TLE is often not confined to the temporal lobe, but the exact significance of extratemporal metabolic abnormalities remains to be clarified. In a study of 47 patients with intractable unilateral mesial TLE, extratemporal hypometabolism, particularly contralateral to the epileptic focus, was a marker of poor surgical outcome, independent of other prognostic factors such as MRI, pathology, and EEG findings.[37] It is also notable that complete removal of the epileptic focus can reverse some of the hypometabolism observed in remote brain regions, as reported in several series. After temporal lobectomy or selective amygdalo-hippocampectomy, postsurgical PET could demonstrate normalization of glucose metabolism in some remote areas apparently connected to the primary (resected) epileptic focus.[38–40]

In summary, imaging studies in animal models and human epilepsy demonstrate early metabolic changes during the period of epileptogenesis and subsequent progression of hypometabolism if the seizures cannot be adequately controlled. Although the exact mechanisms of abnormal glucose metabolism in and beyond epileptic foci remain to be clarified, some studies suggest a potential role of increased glutamate and abnormal mitochondrial function. Longitudinal studies clearly suggest that some of the hypometabolic changes are the consequence of repeated seizures and are often multifocal and localized in an epileptic network, and some of them can be reversed by long-term seizure control by medication or surgery. Molecular imaging can provide invaluable data on metabolic changes during epileptogenesis and could also be useful to evaluate effects of antiepileptogenic interventions in the future.

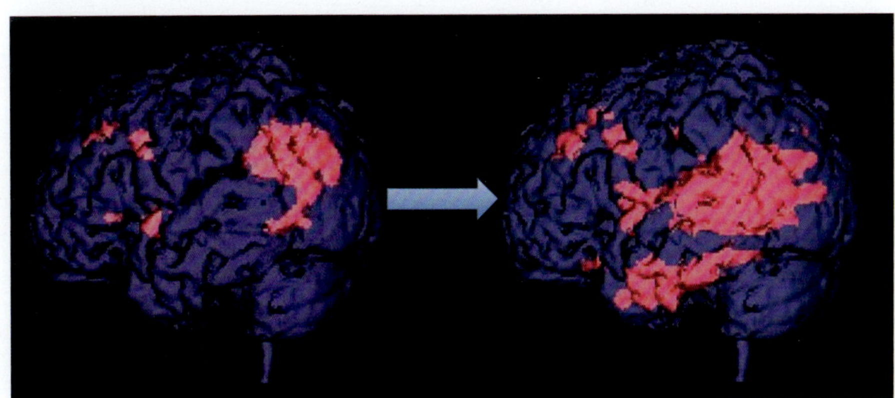

Figure 9.3. Progressive expansion of cortical hypometabolism in medically uncontrolled neocortical epilepsy. The initial FDG-PET scan (left image) showed a larger area of hypometabolism (projected on the 3D brain surface reconstructed from MRI) in the left parietal, posterior temporal cortex, and additional smaller hypometabolic areas in the frontal lobe. The child had one seizure per day at the time of the first scan, which increased to 10 per day in the next two years. Repeated PET scan 27 months later showed more extensive hypometabolism in the affected left hemisphere (right image). Reproduced from Benedek et al., *J Child Neurol.* 2006[35] with permission.

9.4 Molecular Imaging Correlates of Neuroinflammation

Activation of inflammatory pathways is increasingly implicated in the pathophysiology of human epilepsy.[8] Finding imaging markers to detect epilepsy-associated neuroinflammation would be important to identify epileptogenic lesions and foci amenable to anti-inflammatory treatment and also to monitor treatment effects in clinical trials. Glucose metabolism is generally decreased in epileptogenic brain regions showing low-level chronic inflammation. On clinical FDG-PET imaging, epileptogenic inflammatory lesions almost invariably display hypometabolism and cannot be reliably differentiated from most other epileptogenic etiologies.[41] The rare exception could be severe epilepsies associated with autoimmune encephalitis, where FDG-PET can be useful by showing areas of increased metabolism.[42] Still, in most common cases, imaging modalities specifically targeting various aspects of inflammatory processes are needed to capture neuroinflammatory components of epilepsy. Some of the promising imaging modalities already have been tested and proven to be potentially useful in various subgroups of human epilepsy to evaluate neuroinflammation associated with epileptogenesis (Table 9.1). Numerous other molecular imaging probes hold the promise for similar or additional benefit in studying epileptic foci (reviewed recently by Amhaoul et al.[9]), but most of these await human application. In the next section, we briefly review key results in both animal models of epilepsy and human applications while focusing on two molecular imaging approaches that showed promise in clinical applications: PET imaging of activated microglia and imaging of tryptophan metabolism via the kynurenine pathway.

9.4.1 Neuroinflammation in Animal Models of Epileptogenesis

PET imaging of activated microglia allows the study of brain inflammation during early epileptogenesis in experimental models. The most commonly used PET radiotracers bind to the translocator protein (TSPO, first described as the peripheral benzodiazepine receptor [PBR]), an 18 kDa protein localized mostly on the outer mitochondrial membrane of activated microglia and also other cells, such as reactive astrocytes. A rat study of the post-kainic-acid-induced SE model for TLE used [^{18}F]-PBR111 PET/CT for this purpose.[43] [^{18}F]-PBR111 is a PET radioligand with high specificity for the TSPO.[44] Animals with severe SE demonstrated much stronger expression of TSPO in the hippocampus and amygdala in vitro than those with mild symptoms. This TSPO overexpression was corroborated by increases in [^{18}F]-PBR111 volume of

Table 9.1 PET Radiotracers with Reports of Application in Detection of Neuroinflammation in Epileptic Foci or Lesions

Target/radiotracers	Human disease applications	Limitations
TSPO—activated microglia		
[^{11}C]-PK11195	Status epilepticus	Short half-life, high nonspecific binding
	Rasmussen's encephalitis	
	Cortical dysplasia	
[^{11}C]-PBR28	Temporal lobe epilepsy	Short half-life, affinity depends on TSPO gene polymorphism
	Neurocysticercosis	
[^{18}F]-PBR111	None (only rat TLE model)	Affinity depends on TSPO gene polymorphism
Tryptophan metabolism via the kynurenine pathway		
α[^{11}C]-methyl-L-tryptophan (AMT)	Tuberous sclerosis	Short half-life, limited sensitivity
	Cortical dysplasia	
	Epileptogenic tumors	

TSPO = 18-kDa translocator protein.

distribution in the same brain regions (Figure 9.4). This study provided proof-of-concept data that brain regions critical for seizure generation undergo substantial inflammatory changes during epileptogenesis in this seizure model.

In a more recent study, PET imaging was used to study the potential significance of abnormal TSPO binding in drug resistance in a chronic animal model of TLE.[45] The authors measured brain uptake of [11C]-PK11195 in rats with selection of phenobarbital responders vs. nonresponders. While responders showed uptake values similar to nonepileptic animals, nonresponders showed increased uptake of 26–39% in different brain regions, including hippocampus, putamen, occipital, and parietal cortices. Hippocampal uptake also showed a moderate correlation with seizure frequency. Interestingly, while hippocampi also showed signs of neurodegeneration and expression of activated microglia in both the responder and nonresponder epileptic groups, glucose uptake (measured by PET) did not differ between nonepileptic controls, drug responders, and nonresponders. This study provided evidence for chronic neuroinflammatory changes in the chronic TLE model and raised the possibility that a higher degree of inflammation is associated with drug resistance. However, the observed increases may also be secondary as a result of higher seizure frequency, and it remained unclear if imaging markers of neuroinflammation could predict drug resistance prospectively.

9.4.2 Imaging Activated Microglia in Humans

Initial human PET studies used the PET radioligand [11C]-PK11195 to detect inflammatory changes, in the form of increased binding, in patients with Rasmussen's encephalitis.[46] Increased TSPO binding was detected in the affected hemisphere in two patients. The study demonstrated the feasibility of using TSPO imaging to select potential biopsy site and perhaps assess efficacy of anti-inflammatory treatment. On the other hand, no increased binding was found in the epileptic hippocampus in the same study. In a subsequent study, increased [11C]-PK11195 binding was also reported in the left temporo-occipital cortex of a patient with severe, refractory seizures due to encephalitis of unknown origin.[47] In this case the focal PET

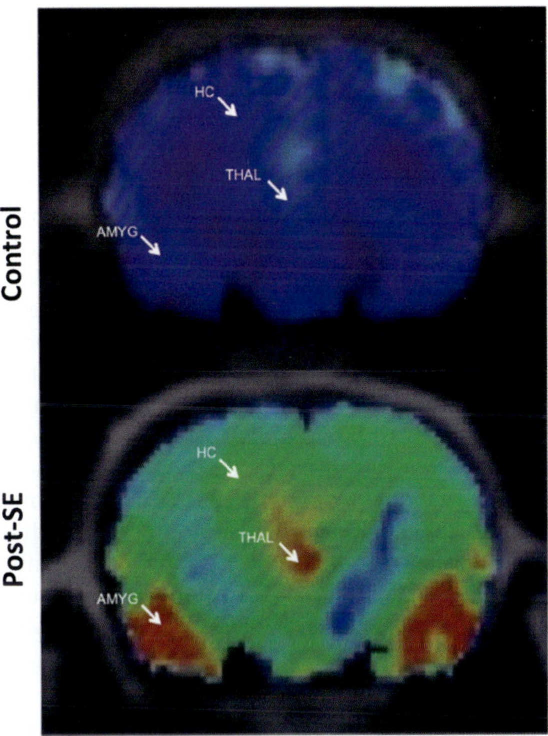

Figure 9.4. PET imaging of brain inflammation during early epileptogenesis in rat. High [18F]-PBR binding indicated activated microglia in the amygdala, thalamus, and hippocampus 7 days after post-kainic-acid-induced status epilepticus (SE). Reproduced from Dedeurwaerdere et al., *EJNMMI Res.* 2012[43] with permission.

findings helped in guiding the resection to control seizures.

Despite these promising studies in patients with severe epilepsy associated with encephalitis, molecular imaging studies targeting activated microglia in human epilepsy remained few and far between in the last decade, perhaps because of the relative insensitivity of [11C]-PK11195 to detect low-level brain inflammation as a result of low specific-to-nonspecific binding.[48] One case report showed images of increased PK binding in a frontal lobe focus associated with cortical dysplasia,[49] with histopathology confirming activated microglia; but this patient also had a history of severe uncontrolled seizures with partial SE. The relatively low sensitivity of [11C]-PK11195 may be overcome by the use of radioligands with better specific binding, such as [11C]-PBR28. Using this radioligand, a preliminary study evaluated TSPO binding in 16 patients with unilateral TLE compared to values of 30 healthy subjects.[50] Increased binding was detected in multiple regions ipsilateral to the seizure focus, including the hippocampus, parahippocampal gyrus, amygdala, fusiform gyrus. Interestingly, the asymmetry was more pronounced in patients with hippocampal sclerosis. A limitation of [^{11}C]-PBR28 is that its affinity depends on a single polymorphism in the TSPO gene.

Neurocysticercosis is one of the most common causes of focal epilepsy worldwide. Human neurocysticercosis and related animal models provide a unique opportunity to study mechanisms of epileptogenesis, including epilepsy-related neuroinflammation.[51] Calcified granulomas in neurocysticercosis are often associated with perilesional epilepsy and seizures. A PET study with [^{11}C]-PBR28 of nine patients with MRI-proven neurocysticercosis demonstrated increased binding by a mean of 13% in perilesional edema or degenerating cysts.[52] This increased binding persisted for 2–9 months in these patients, even after resolution of the edema. The study suggested that anti-inflammatory treatment may be useful to prevent or treat seizures in such patients.

9.4.3 Imaging Tryptophan Uptake in Humans

PET imaging of amino acid uptake is mostly used in clinical brain tumor imaging. Among the most commonly used amino acid PET tracers, O-(2-[^{18}F]fluoroethyl)-L-tyrosine (FET) shows high tumor-specific accumulation (related to upregulated amino acid transport in tumor tissue), while its uptake in inflammatory tissue is negligible.[53,54] [^{11}C]methionine, another frequently used clinical PET tracer for brain tumor imaging, was shown to accumulate in both tumor and inflammatory cells.[54] However, its potential use for detecting nontumoral epileptogenic lesions (such as cortical dysplasia) is not well established because of scarce reports with mixed results in such lesions.[55–57] In contrast, a wealth of data supports the clinical utility of alpha[^{11}C]methyl-L-tryptophan (AMT) PET for detecting a variety of epileptic foci and lesions in humans; these data are briefly summarized below.

The initial intent of developing a PET radioligand for imaging tryptophan metabolism was to track and quantify serotonin synthesis in the living brain, a valuable goal in neuropsychiatric disorders associated with abnormal serotonergic function. This has been accomplished with the radiosynthesis and subsequent validation of AMT.[58,59] AMT is transported across the blood-brain barrier by specific carriers (most notably the L-type amino acid transporter [LAT]), but it is not incorporated into proteins. Once in the cell, AMT is the substrate of the initial enzymatic steps of serotonin synthesis and is accumulated in serotonergic neurons in the form of alpha-methyl-L-serotonin. Thus, kinetic analysis of AMT uptake allows estimation of serotonin synthesis rate in the living human brain.[60]

The application of AMT-PET in human epilepsy was initially motivated by studies reporting increased serotonergic immunoreactivity in human epileptic brain tissue associated with cortical dysplasia.[61] However, subsequent studies of resected epileptic tissue, showing increased AMT uptake on PET, demonstrated that AMT accumulation in epileptic foci is more related to tryptophan metabolism via the inflammatory kynurenine pathway and may be an imaging marker of epilepsy-associated neuroinflammation.[62,63] Based on studies demonstrating increased levels of epileptogenic and neurotoxic metabolites of the kynurenine pathway (most notably, quinolinic acid, an NMDA receptor agonist), it has been long postulated that activation of this pathway plays a role in epileptogenesis.[64,65] Interest in this pathway has intensified in the last decade following reports showing that activation of this pathway, mostly due to induction of its key enzymes, such as indoleamine 2,3-dioxygenase (IDO) and tryptophan 2,3-dioxygenase (TDO), can lead to tumoral immune resistance.[66,67] Fueled by this interest, a number of potent enzyme inhibitors have been developed and are being tested clinically to block the activity of this pathway and reverse tumoral immune tolerance.[68] The results of these efforts may not only benefit tumor treatment but also provide novel therapeutic avenues in human epilepsy associated with inflammation.

The most extensively studied and validated clinical application of AMT-PET was the identification of epileptogenic lesions in patients with TSC. As reported initially[69] and confirmed by subsequent studies by multiple groups,[70–72] increased focal AMT uptake in the interictal state is a highly specific marker of epileptogenicity in TSC. In the largest study thus far (with 191 TSC patients included), excellent agreement was found between lateralizing ictal scalp EEG and AMT-PET; in 28 of 68 patients (41%) with lateralizing findings, AMT-PET was more localizing.[72] In addition, AMT-PET was localizing in 58% of the cases with nonlateralized ictal EEG. "Hot spots" on AMT uptake are not always confined to MRI-defined cortical tubers but can also extend to adjacent dysplastic epileptic cortex.[73] Importantly, resection of tubers (or dysplastic lesions) showing increased AMT uptake was associated with seizure-free surgical outcome.[73]

After the initial success of epileptic lesion localization in TSC, AMT-PET was successfully applied in other epilepsy populations with various etiologies. Most notably, studies in neocortical epilepsy demonstrated the ability of AMT-PET to detect epileptic cortical malformations, such as focal cortical dysplasia.[74,75] A subsequent, detailed comparison to histopathology suggested that, within various histologic types of cortical dysplasia, tryptophan accumulation on PET may be most specific to Type IIB cortical dysplasia.[76] Moderate increases in AMT uptake have also been reported in TLE with normal hippocampal volumes,[77] but the added clinical value of AMT-PET in presurgical localization of TLE remains unexplored.

While specificity of focal increased AMT uptake for epileptic cortex is very high, its sensitivity is suboptimal and depends on the underlying etiology of epilepsy. The highest sensitivity was observed in dysembryoplastic neuroepithelial tumors (DNTs) and TSC cases, where the epileptogenic lesions can be identified in up to 80% of the cases if the AMT scans are analyzed quantitatively together with other imaging modalities (MRI and FDG-PET).[78,79] The lowest sensitivity was detected in nonlesional neocortical epilepsies, where histopathology could not detect any specific lesion: in those cases, AMT-PET shows increased cortical uptake in less than 50% of the cases.[74,75] A preliminary study with direct comparison of AMT-PET vs. PET imaging of the GABA$_A$ receptors (using [11C]-flumazenil, performed in the same patients) also raised the possibility that these two molecular imaging tracers capture different aspects of epileptogenicity at different stages of clinical epilepsy:[80] while high AMT uptake was more common in younger patients with shorter epilepsy duration (Figure 9.5), decreased flumazenil uptake was often observed in older children with longer duration of intractable seizures. Altogether, the data thus far suggest that clinical utility of AMT-PET may depend on both underlying etiology and stage of chronic epilepsy. It is also clear that increased AMT uptake provides useful added information in delineation of cortical epileptic foci in important subgroups of patients with intractable epilepsy.

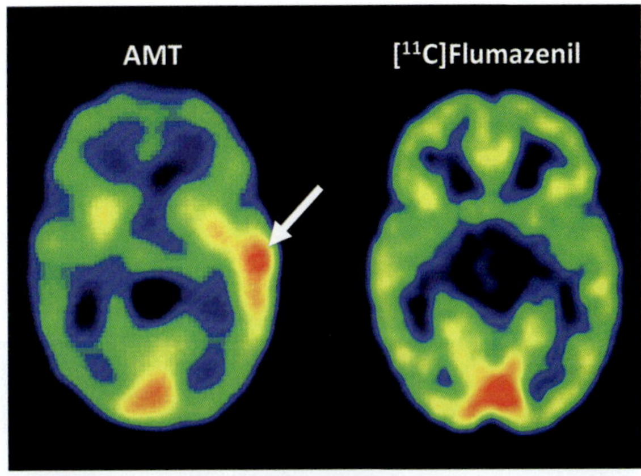

Figure 9.5. Focal increase of tryptophan uptake on alpha[11C]-methyl-L-tryptophan (AMT) PET in a young child with an epileptic focus in the temporal cortex (arrow). [11C]flumazenil PET in the same child showed no focal binding abnormalities.

Recent studies provide emerging evidence that increased AMT uptake in epileptogenic lesions is likely related to chronic inflammatory processes activated in epileptogenic lesions. Indeed, the three most common epileptogenic lesions showing tryptophan accumulation on PET (TSC, DNT, and Type II FCD) all express inflammatory markers such as interleukin-6 or interleukin-1β.[81–84] Proinflammatory cytokines, particularly interferon-γ but also IL-1β, can potentiate induction of IDO, thus leading to enhanced metabolism of tryptophan via the kynurenine pathway; this could lead to accumulation of both anti- and proepileptic kynurenine metabolites. To support this mechanism, increased expression of IDO, along with the LAT1 transporter, has been demonstrated in epileptogenic DNTs showing high AMT uptake on PET.[79]

In order to further establish the link between inflammation, activated kynurenine pathway, increased in vivo tryptophan uptake, and epileptogenicity, we performed an in-depth analysis of imaging, intracranial EEG, and immunohistochemical abnormalities of an inflammatory epileptogenic lesion associated with new-onset refractory SE.[85] In this patient, MRI performed after the onset of severe partial seizures showed a tumor-like lesion in the left temporal lobe, displaying T2/FLAIR hyperintensity without contrast enhancement (Figure 9.6). AMT-PET imaging showed high uptake in the lesion area extending behind it to the posterior temporal cortex. The extent of the epileptogenic region was mapped by chronic subdural EEG recording, which showed seizure onset in the left

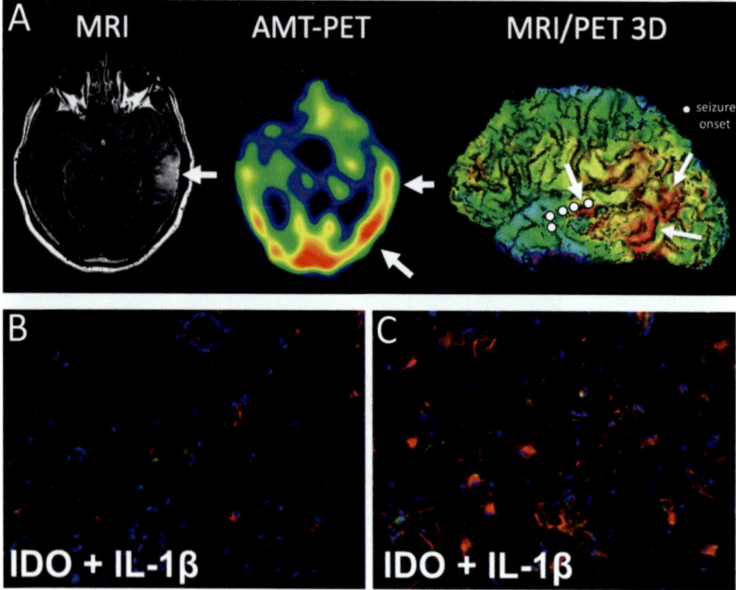

Figure 9.6. (A) Increased alpha[¹¹C]methyl-L-tryptophan (AMT) uptake (arrows on AMT-PET image and 3D MRI/PET surface) in the left temporal lobe in an adult patient with new-onset severe epilepsy. MRI showed increased FLAIR signal in the midtemporal region (without contrast enhancement; arrow). Intracranial EEG monitoring showed left temporal cortical seizure onset: affected electrodes are overlaid onto 3D PET/MRI surface, where cortex with high AMT uptake is in red. (B, C) Immunohistochemistry from image-guided biopsy specimens showed coexpression of interleukin-1β (IL-1β) and indoleamine 2,3-dioxygenase (IDO) in the AMT+ tissue only. Reproduced from Juhász et al., *Neurosurg Focus* 2013;[85] with permission.

temporal cortex overlapping with the AMT-positive region. Analysis of brain tissue sampled from this region demonstrated reactive astrocytosis, microglial activation and sparse lymphocytic inflammation, along with coexpression of IDO, IL-1β, and the IL-1β receptor (Figure 9.6). Sampling of tissue away from the AMT hot spot showed minimal expression of the same inflammatory markers. Extensive left temporal resection removed the inflammatory lesion along with the bulk of AMT-positive epileptic cortex. Postsurgical follow-up demonstrated long-term seizure freedom and no increased AMT uptake on PET. This study provided proof-of-principle evidence that high AMT uptake on PET can identify epileptogenic areas/lesions with inflammatory changes associated with an upregulated kynurenine pathway. Further studies with stereotactic, image-guided sampling of epileptic foci are required to establish if this close relation between molecular imaging findings and epilepsy-related inflammation is indeed present and common in focal epilepsies of various etiologies.

Acknowledgments

Some of the work presented here is the result of close collaborations with Harry T. Chugani, Diane C. Chugani, Otto Muzik, and other faculty and staff at the PET Center and Translational Imaging Laboratory at Wayne State University and the Children's Hospital of Michigan. Funding for these studies was provided by US National Institutes of Health, National Institute of Neurological Disorders and Stroke grants NS41922, NS34488, NS38324, and NS64989, and US National Institutes of Health, National Cancer Institute grant CA123451.

References

1. Vezzani A, Granata T. Brain inflammation in epilepsy: experimental and clinical evidence. *Epilepsia*. 2005;**46**: 1724–43.

2. Marchi N, Granata T, Janigro D. Inflammatory pathways of seizure disorders. *Trends Neurosci*. 2014;**37**:55–65.

3. Otáhal J, Folbergrová J, Kovacs R, et al. Epileptic focus and alteration of metabolism. *Int Rev Neurobiol*. 2014;**114**:209–43.

4. Kuhl DE, Engel J Jr, Phelps ME, et al. Epileptic patterns of local cerebral metabolism and perfusion in humans determined by emission computed tomography of ¹⁸FDG and ¹³NH₃. *Ann Neurol*. 1980;**8**:348–60.

5. Mazziotta JC, Engel J Jr. The use and impact of positron computed tomography scanning in epilepsy. *Epilepsia*. 1984;**25**(suppl 2):S86–104.

6. Johnson EW, de Lanerolle NC, Kim JH, et al. "Central" and "peripheral" benzodiazepine receptors: opposite changes in human epileptogenic tissue. *Neurology*. 1992;**42**:811–5.

7. Kumlien E, Hilton-Brown P, Spännare B, et al. In vitro quantitative autoradiography of [³H]-L-deprenyl and [³H]-PK11195 binding sites in human epileptic hippocampus. *Epilepsia*. 1992;**33**:610–7.

8. Vezzani A, French J, Bartfai T, et al. The role of inflammation in epilepsy. *Nat Rev Neurol.* 2011;7:31–40.

9. Amhaoul H, Staelens S, Dedeurwaerdere S. Imaging brain inflammation in epilepsy. *Neuroscience.* 2014;**279**:238–52.

10. O'Brien TJ, Jupp B. In-vivo imaging with small animal FDG-PET: a tool to unlock the secrets of epileptogenesis? *Exp Neurol.* 2009;**220**:1–4.

11. Pitkänen A, Lukasiuk K. Mechanisms of epileptogenesis and potential treatment targets. *Lancet Neurol.* 2011;**10**:173–86.

12. Engel J Jr, Pitkanen A, Loeb JA, et al. Epilepsy biomarkers. *Epilepsia.* 2013;**54**(suppl 4):61–9.

13. Guo Y, Gao F, Wang S, et al. In vivo mapping of temporospatial changes in glucose utilization in rat brain during epileptogenesis: an ^{18}F-fluorodeoxyglucose-small animal positron emission tomography study. *Neuroscience.* 2009;**162**:972–9.

14. Lee EM, Park GY, Im KC, et al. Changes in glucose metabolism and metabolites during the epileptogenic process in the lithium-pilocarpine model of epilepsy. *Epilepsia.* 2012;**53**:860–9.

15. Jupp B, Williams B, Binns D, et al. Hypometabolism precedes limbic atrophy and spontaneous recurrent seizures in a rat model of TLE. *Epilepsia.* 2012;**53**:1233–44.

16. Shultz SR, Cardamone L, Liu YR, et al. Can structural or functional changes following traumatic brain injury in the rat predict epileptic outcome? *Epilepsia.* 2013;**54**:1240–50.

17. Benedek K, Juhász C, Muzik O, et al. Metabolic changes of subcortical structures in intractable focal epilepsy. *Epilepsia.* 2004;**45**:1100–5.

18. Keller SS, O'Muircheartaigh J, Traynor C, Towgood K, Barker GJ, Richardson MP. Thalamotemporal impairment in temporal lobe epilepsy: a combined MRI analysis of structure, integrity, and connectivity. *Epilepsia.* 2014;**55**(2):306–15.

19. Camilo O, Goldstein LB. Seizures and epilepsy after ischemic stroke. *Stroke.* 2004;**35**:1769–75.

20. Lossius MI, Rønning OM, Slapø GD, et al. Poststroke epilepsy: occurrence and predictors—a long-term prospective controlled study (Akershus Stroke Study). *Epilepsia.* 2005;**46**:1246–51.

21. Wang H, Xin T, Sun X, et al. Post-traumatic seizures— a prospective, multicenter, large case study after head injury in China. *Epilepsy Res.* 2013;**107**:272–8.

22. Ville D, Enjolras O, Chiron C, et al. Prophylactic antiepileptic treatment in Sturge-Weber disease. *Seizure.* 2002;**11**:145–50.

23. Jóźwiak S, Kotulska K, Domanska-Pakiela D, et al. Antiepileptic treatment before the onset of seizures reduces epilepsy severity and risk of mental retardation in infants with tuberous sclerosis complex. *Eur J Paediatr Neurol.* 2011;**15**:424–31.

24. Alkonyi B, Chugani HT, Juhász C. Transient focal increase of interictal glucose metabolism in Sturge-Weber syndrome: implications for epileptogenesis. *Epilepsia.* 2011;**52**:1265–72.

25. Chugani HT, Mazziotta JC, Phelps ME. Sturge-Weber syndrome: a study of cerebral glucose utilization with positron emission tomography. *J Pediatr.* 1989; **114**: 244–53.

26. Juhász C, Hu J, Xuan Y, Chugani H. Imaging increased glutamate in children with Sturge-Weber syndrome: association with epilepsy severity. *Epilepsy Res.* 2016;**122**:66–72.

27. Ashwal S, Holshouser B, Tong K, et al. Proton MR spectroscopy detected glutamate/glutamine is increased in children with traumatic brain injury. *J Neurotrauma.* 2004;**21**:1539–52.

28. Gaillard WD, Kopylev L, Weinstein S, et al. Low incidence of abnormal (18)FDG-PET in children with new-onset partial epilepsy: a prospective study. *Neurology.* 2002;**58**:717–22.

29. Weitemeyer L, Kellinghaus C, Weckesser M, et al. The prognostic value of ^{18}F]FDG-PET in nonrefractory partial epilepsy. *Epilepsia.* 2005;**46**:1654–60.

30. Gaillard WD, Weinstein S, Conry J, et al. Prognosis of children with partial epilepsy: MRI and serial [18]FDG-PET. *Neurology.* 2007;**68**:655–9.

31. Henry TR. Positron emission tomography: glucose metabolism studies in temporal lobe epilepsy. In: Chugani HT, ed. *Neuroimaging in Epilepsy.* New York: Oxford University Press; 2011:122–40.

32. Juhász C, Chugani HT. Positron emission tomography: glucose metabolism in extratemporal lobe epilepsy. In: Chugani HT, ed. *Neuroimaging in Epilepsy.* New York: Oxford University Press; 2011:141–55.

33. Tenney JR, Rozhkov L, Horn P, et al. Cerebral glucose hypometabolism is associated with mitochondrial dysfunction in patients with intractable epilepsy and cortical dysplasia. *Epilepsia.* 2014;**55**:1415–22.

34. Swartz BE, Brown C, Mandelkern MA, et al. The use of 2-deoxy-2-^{18}F]fluoro-D-glucose (FDG-PET) positron emission tomography in the routine diagnosis of epilepsy. *Mol Imaging Biol.* 2002;**4**:245–52.

35. Benedek K, Juhász C, Chugani DC, et al. Longitudinal changes in cortical glucose hypometabolism in children with intractable epilepsy. *J Child Neurol.* 2006;**21**:26–31.

10 Interictal and Ictal Brain Network Changes in Focal Epilepsy

Mangor Pedersen, Amir Omidvarnia, and Graeme D. Jackson

10.1 Introduction

Focal epilepsy was in 2005 redefined as a disease where seizures occur as a result of abnormal brain network activity in confined areas of the brain.[1] This is a clear shift from more traditional definitions of focal seizures as arising from excessive activity or synchrony within a population of neurons (see Fisher et al.[2] for a good discussion on this topic dating back to Hughlings Jackson's work on epilepsy and seizures). In parallel with the increased interest in epilepsy and networks, our ability to analyze brain imaging data as complex networks has increased exponentially through *connectomics*.[3] This is particularly reflected in number of peer-reviewed articles in this research field over the last five years (Figure 10.1). Connectomics has now evolved into a research field with a clear final aim of systematically studying functional and structural connections of the brain by means of innovative imaging techniques.[4,5] The maps, called connectomes, can be constructed on a variety of spatial levels, including single neurons (microscale), populations of neurons (mesoscale), and the whole brain (macroscale). In addition to analyzing the structural wiring of the brain, connectomics allows functional connectivity patterns to be mapped, enabling inferences about information transfer within the brain.

The purpose of this chapter is to highlight the insight connectomics has offered into focal epilepsy as a disease that affects brain networks, and it is outside its scope to provide a comprehensive methodological overview of connectomics and its associated mathematical measures (including graph theory analyses). We refer the reader to Bullmore and Sporns[6] and Rubinov and Sporns[7] for thorough methodological overviews on connectomics.

10.2 Interictal Brain Network Findings in Focal Epilepsy

Several studies have in the last few years attempted to apply brain network metrics to patients with focal epilepsy using fMRI, diffusion MRI, and EEG, during the interictal state.[8–15] Van Diessen et al.[16] conducted a meta-analysis of brain network studies used in the focal epilepsies. They focused on two of the most commonly applied network measures, (1) clustering coefficient and (2) shortest path length. These two measures come from graph theory, which aims to quantify topological features of networks.[17] In brief, clustering coefficient counts the number of connected triangular nodes in a network, and is an estimate of network "cliquiness" (Figure 10.2, left). The shortest path length, on the other hand, counts the minimum number of links (i.e., connections) that connect two nodes, and is an index of efficient information transfer in the network (Figure 10.2, right). Brain networks have high clustering coefficient and low path length.

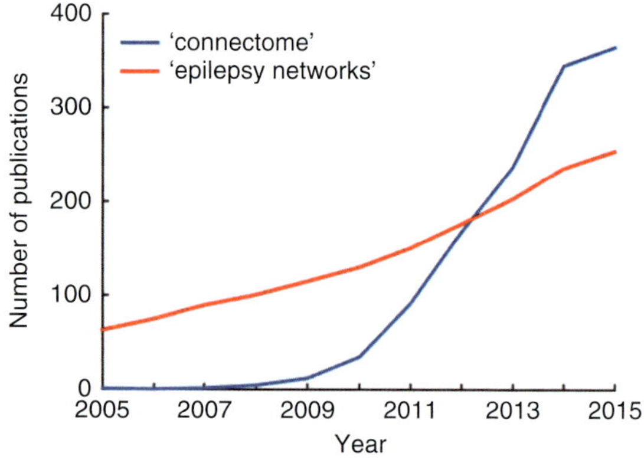

Figure 10.1. An overview of peer-reviewed publications with the keywords "connectome" and "epilepsy networks" between 2005 and 2015. Obtained from PubMed (http://www.ncbi.nml.hih.gov/pubmed).

This is associated with the well-known "small-world" network topology that is able to support both network segregation and integration (see Table 10.1 for a brief summary).[18] The result of the aforementioned meta-analysis demonstrated that the brains of focal epilepsy patients had significantly higher clustering coefficient and shortest path length, compared to controls. This finding accords with a regular network topology where brain nodes may isolate themselves from the wider networks, thus becoming less globally integrated. There is also evidence that this network configuration become more "extreme" during days with a high seizure load. Otte et al.[19] used task-free fMRI in rats with extratemporal focal epilepsy and reported that rats displayed more regularized brain network topology via increased clustering coefficient and shortest path length during days when they experienced an excessive amount of seizures. Thus, increased segregation and decreased integration of

networks may be related to severity as well as frequency of seizures.

These studies largely investigated whole-brain network function in focal epilepsy. A question that naturally arises at this point is whether network regularity is a brain-wide effect or is an exclusive feature of the seizure focus and/or surrounding epileptic brain regions. In an attempt to answer this, we calculated clustering coefficient and shortest path length using task-free fMRI data from within the lesional cortex in patients with polymicrogyria,[20] a malformation of cortical development resulting in widespread cortical lesions often associated with focal seizures.[21] Specifically, we compared clustering coefficient and shortest path length between the abnormal polymicrogyric cortex and the unaffected contralateral homologue. We observed that brain network regularity (increased clustering coefficient and shortest path length) was significantly higher within polymicrogyric lesions compared to the contralateral normal homologue. This finding suggests that the abnormal cortex itself may drive widespread network abnormalities and is in line with epilepsy syndromes where a complete resection of a small (and highly epileptogenic) lesion can reverse seizures and cognitive disturbances.[22,23] Further evidence for this is demonstrated by a series of electrophysiology studies investigating dynamic seizure networks.

10.3 Peri-ictal Brain Network Findings in Focal Epilepsy

Le Van Quyen et al.[24] outlined a (preconnectomics) model of how brain networks may change before,

Network analysis

High clustering coefficient Low shortest path length

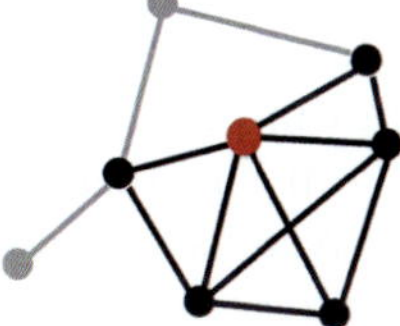 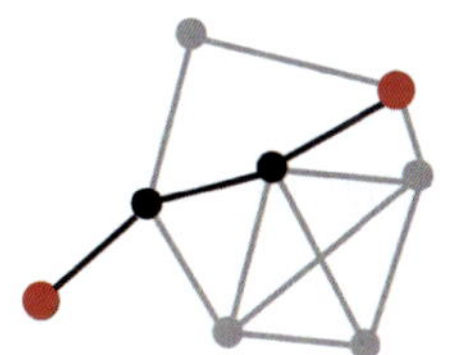

Figure 10.2. Two simple networks with high clustering coefficient (red node on the left) and low shortest path length (the links connecting the two red nodes on the right). These are features of an efficient network topology, often termed "small-worldness." See Watts and Strogatz.[18]

Table 10.1 Glossary of Brain Network Analysis

Connectomics	A new scientific field aiming to map the brain's functional and structural substrates
Graph theory	An established mathematical field on the study of network topology
Connectome	Representation of brain regions (nodes) and their interconnections (edges)
Clustering coefficient	Interconnectedness of (topologically) local brain nodes
Shortest path length	Number of shortest steps between two nodes—i.e., likely to represent efficiency in information transfer between nodes
Small-world network	A network with high segregation and integration among its nodes—high clustering coefficient and low shortest path length
Regular network	A network with high network segregation and low network integration—high clustering coefficient and high shortest path length
Random network	A network with low network segregation and high network integration—low clustering coefficient and low shortest path length

during, and after a focal seizure. In this model, the brain undergoes drastic topological changes during the preictal state where the local seizure network (nodes within and proximate to the seizure focus) isolates itself through inhibitory processes. As focal seizures progress, the local seizure network gradually becomes more integrated with wider brain networks allowing for seizure termination.

Since the proposal of this model, several electrophysiological studies have considered time-varying changes of seizure networks during epileptic events using the clustering coefficient (network segregation) and shortest path length (network integration). Both scalp-level and invasive electrophysiological recordings have been utilized to measure peri-ictal changes of brain networks. A common finding across these studies is that brain network properties undergo temporal changes during epileptic events (interictal discharges and/or seizure). During seizure initiation and propagation, there is an increase of clustering coefficient and shortest path length (Figure 10.3, black dotted line), resembling a regularized, or isolated, network topology (see Figure 10.3, upper left network).[25–28] In a network that is regularized, local clusters of neurons may isolate themselves from the rest of the network. Khambhati et al.[29] describe this topological seizure configuration as "network tightening," suggesting a mechanistic aim to contain seizure-related activity that has been brewing over a long period of time. Perhaps paradoxically, the studies presented above show a common pattern that upon termination of seizures, functional brain networks diverge their topology toward a more randomized configuration (Figure 10.3, lower left brain network). When a network enters a "random" mode, it consumes great effort and energy due to its emphasis on globally integrated (expensive) processes rather than locally segregated (cheap) processes[30] and may represent a seizure termination mechanism.

In sum, these findings broadly support Le Van Quyen's seizure network model and show that epileptogenic brain networks display characteristic abnormal behavior. Network regularity appears to be a promising brain configuration that may explain focal epilepsy phenotypes—whether these represent mechanisms to protect the brain against recurrent seizures, inhibit seizure spread or instigate seizures, remain uncertain.

10.4 "Common" Brain Networks in Focal Epilepsy?

The studies presented in the two previous sections raise an important question. Do patients with heterogeneous focal epilepsy (in terms of the seizure focus and symptomatology) have brain network abnormalities in common? In an attempt to answer this question, a few studies have used simultaneous EEG and fMRI. The first article to report "common" brain regions amongst focal epilepsy patients was Laufs et al.,[31] followed by studies from Fahoum et al.[32] and Flanagan et al.[33] These studies all found common brain regions on fMRI that were activated during epileptiform discharges, despite having variable seizure foci including both temporal lobe and neocortical epilepsy. One brain area found to be common to focal epilepsies was the ipsilateral piriform cortex. This cortex is a deep brain region involved in olfactory processing, and has been demonstrated to be highly epileptogenic in animal kindling models of epilepsy.[34] Other potential common brain regions of focal epilepsy include the thalamus, insula, cerebellum, and cingulate cortex.[32] We have also replicated these findings using fMRI connectivity measures in patients with extratemporal focal epilepsy.[35] It is therefore tempting to postulate that brain regions including the piriform cortex, cingulate cortex, cerebellum, and frontoinsular regions compose a common epilepsy network in focal epilepsy.

10.5 Future Directions I: Antiepileptic Drugs, Comorbidities, and Networks

A general limitation of studies summarized in this chapter is that patients are often on multiple antiepileptic drugs and have often endured seizures over a number of years before being considered for surgery. This raises a question of whether brain network abnormalities in focal epilepsy are related to commonly observed comorbidities and/or antiepileptic drug treatment.

Focal epilepsy patients often suffer from cognitive disturbances including attention problems, memory problems, depression, and anxiety.[36,37] A question regarding epilepsy and secondary behavioral problems is whether these debilitating cognitive and/or psychiatric problems exist prior to the onset of seizures, or does persistent seizure activity

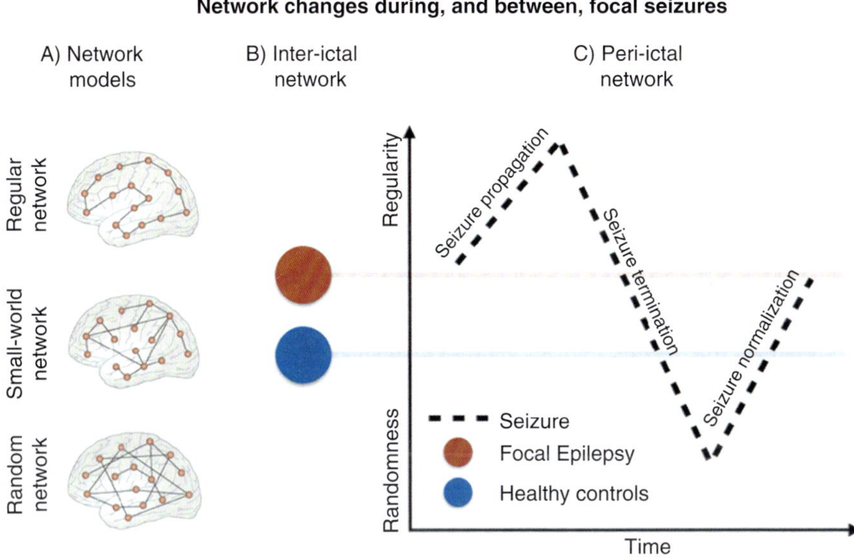

Figure 10.3. Network changes between and during focal seizures: (A) Three network topologies. Upper brain = regular network; middle brain = small-world network; lower brain = random network. (B) Summary of the interictal meta-analysis by van Diessen et al.[16] Patients with focal epilepsy (red circle) display a brain network topology skewed toward network regularity compared to healthy controls (blue circle). (C) Summary of dynamic network activity during focal seizures, based on electrophysiological data. Brains on the left-hand side of this figure are reused with permission from Bullmore and Sporns,[30] Nature Publishing Group.

cause behavioral problems? Taylor et al.[38] showed that patients with new-onset epilepsy display significantly more problems of memory and psychomotor speed than healthy control subjects. This finding suggests that cognitive problems are present before the onset of the first seizure. On the contrary, Thompson and Duncan[39] demonstrated that patients who experience more frequent secondary generalized tonic-clonic seizures scored worse between two sessions of cognitive tests spaced ten years apart than patients with infrequent secondary generalized tonic-clonic seizures. This finding suggests that recurrent seizures result in more severe cognitive problems. These two divergent results amplify the conundrum of whether cognitive comorbidities are present before or after the first overt seizure. Keezer et al.[40] have outlined several possible scenarios of comorbidities in epilepsy. A plausible theory is that seizures and comorbidities share a symbiotic relationship. That is, comorbidities may lead to worse seizures, and seizures may lead to worse comorbidities.

Antiepileptic drug treatment is also likely to impair brain network function. Jokeit et al.[41] found that the extent of fMRI activation in the temporal lobes was negatively correlated with levels of carbamazepine dosage. Valporate has also been found to be associated with reduced cortical thickness of the parietal lobes,[42] whereas Yasuda et al.[43] demonstrated that toprimate affects

cognitive brain networks (particularly the default mode network) using task-based fMRI data. Using task-free fMRI, Haneef and Chiang[10] found that epilepsy patients who were on carbamazepine displayed lower brain network activity compared to epilepsy patients who were not using carbamazepine. Combined, these studies suggest that antiepileptic drugs may affect brain network structure and function.

To overcome a potential bias due to effects of psychiatric/neuropsychological comorbidities and antiepileptic drugs, we need to conduct longitudinal studies where these factors can be monitored over time.

10.6 Future Directions II: Patient-Specific Brain Network Abnormalities

A machine learning approach such as multivariate pattern analysis has in recent years received widespread attention because it allows for classification of brain network data at the level of the individual.[44–47] First, multivariate pattern analysis searches for distinct brain patterns (e.g., clustering coefficient and shortest path length) between epilepsy and control groups ("training"). After training the data, we can make a single-subject inference by introducing a new subject into the analysis ("classification"). When classifying the new subject, the machine learning algorithm judges whether its brain patterns resemble the "epilepsy" or "control" group. The most common way to quantify the accuracy of such a classifier is via a leave-one-out cross-validation scheme. This cross-validation scheme excludes one subject and trains a model based on the remainder of the subjects. This is repeated, leaving out a different subject each time, until every subject has been excluded once. The final classifier accuracy is estimated based on the single-subject inferences achieved during the cross-validation procedure, i.e., percentage of subjects correctly classified (see Cristianini and Shawe-Taylor[48] and Rasmussen et al.[49] for methodological overviews). Some studies have already applied machine learning algorithms to epilepsy and neuro-imaging data (refer to Bernhardt et al.,[50] Pedersen et al.,[35,51] and Yang et al.[52] for examples). However, its full validity in predicting single-subject network characteristics in focal epilepsy remains uncertain at this stage. Further work with larger and homogenous "training" cohorts is needed to test the feasibility of machine learning as a reliable way of delineating patient-specific networks in focal epilepsy.

10.7 Conclusion

Both interictal and peri-ictal focal epilepsy networks are associated with "network regularity" expressed as increased clustering coefficient and shortest path length. Future brain network studies will benefit from controlling for "nonepileptic" variables such as common comorbidities and antiepileptic drug effects. Analyzing connectomic data at the level of the individual, enabled by machine learning approaches, is also likely to become a major contributor in enhancing our understanding of focal epilepsy as a brain network disease.

Acknowledgments

We acknowledge support by the National Health and Medical Research Council (NHMRC) of Australia (628952). The Florey Institute of Neuroscience and Mental Health acknowledges the strong support from the Victorian Government and in particular the funding from the Operational Infrastructure Support Grant. GDJ is supported by an NHMRC practitioner fellowship (1060312). MP was supported by the University of Melbourne international scholarships.

References

1. Berg AT, Berkovic SF, Brodie MJ, et al. Revised terminology and concepts for organization of seizures and epilepsies: report of the ILAE Commission on Classification and Terminology, 2005–2009. *Epilepsia.* 2010;**51**:676–85.

2. Fisher RS, van Emde Boas W, Blume W, et al. Epileptic seizures and epilepsy: definitions proposed by the International League Against Epilepsy (ILAE) and the International Bureau for Epilepsy (IBE). *Epilepsia.* 2005;**46**:470–2.

3. Sporns O, Tononi G, Kötter R. The human connectome: a structural description of the human brain. *PLOS Comput Biol.* 2005;**1**:e42.

4. Van Essen DC, Ugurbil K, Auerbach E, et al. The Human Connectome Project: a data acquisition perspective. *NeuroImage.* 2012;**62**:2222–31.

5. Van Essen DC, Ugurbil K. The future of the human connectome. *NeuroImage.* 2012;**62**:1299–310.

6. Bullmore E, Sporns O. Complex brain networks: graph theoretical analysis of structural and functional systems. *Nat Rev Neurosci.* 2009;**10**:186–98.

7. Rubinov M, Sporns O. Complex network measures of brain connectivity: uses and interpretations. *NeuroImage.* 2010;**52**:1059–69.

8. Bernhardt BC, Chen Z, He Y, Evans AC, Bernasconi N. Graph-theoretical analysis reveals disrupted small-world organization of cortical thickness correlation networks in temporal lobe epilepsy. *Cereb Cortex.* 2011;**21**:2147–57.

9. Doucet GE, Rider R, Taylor N, et al. Presurgery resting-state local graph-theory measures predict neurocognitive outcomes after brain surgery in temporal lobe epilepsy. *Epilepsia.* 2015;**56**:517–26.

10. Haneef Z, Chiang S. Clinical correlates of graph theory findings in temporal lobe epilepsy. *Seizure.* 2014;**23**:809–18.

11. Liao W, Zhang Z, Pan Z, et al. Altered functional connectivity and small-world in mesial temporal lobe epilepsy. *PLOS ONE.* 2014;**5**:e8525.

12. Pedersen M, Omidvarnia AH, Walz JM, Jackson GD. Increased segregation of brain networks in focal epilepsy: An fMRI graph theory finding. *NeuroImage Clin.* 2015;**8**:536–42.

13. Vaessen MJ, Braakman HMH, Heerink JS, et al. Abnormal modular organization of functional networks in cognitively impaired children with frontal lobe epilepsy. *Cereb Cortex.* 2013;**23**:1997–2006.

14. van Dellen E, Douw L, Baayen JC, et al. Long-term effects of temporal lobe epilepsy on local neural networks: a graph theoretical analysis of corticography recordings. *PLOS ONE.* 2009;**4**:e8081.

15. Vlooswijk MCG, Vaessen MJ, Jansen JFA, et al. Loss of network efficiency associated with cognitive decline in chronic epilepsy. *Neurology*. 2011;77:938–44.

16. van Diessen E, Zweiphenning WJEM, Jansen FE, et al. Brain network organization in focal epilepsy: a systematic review and meta-analysis. *PLOS ONE*. 2014;9:e114606.

17. Bondy A, Murty USR. *Graph Theory*. 1st corrected ed. New York: Springer; 2008.

18. Watts DJ, Strogatz SH. Collective dynamics of "small-world" networks. *Nature*. 1998;393:440–2.

19. Otte WM, Dijkhuizen RM, van Meer MPA, et al. Characterization of functional and structural integrity in experimental focal epilepsy: reduced network efficiency coincides with white matter changes. *PLOS ONE*. 2012;7:e39078.

20. Sethi M, Pedersen M, Jackson GD. Polymicrogyric cortex may predispose to seizures via abnormal network topology: an fMRI connectomics study. *Epilepsia*. 2016;57:e64–8.

21. Guerrini R. Polymicrogyria and epilepsy. *Epilepsia*. 2010;51:10–2.

22. Archer JS, Warren AEL, Stagnitti MR, et al. Lennox-Gastaut syndrome and phenotype: secondary network epilepsies. *Epilepsia*. 2014;55:1245–54.

23. Berkovic SF, Arzimanoglou A, Kuzniecky R, et al. Hypothalamic hamartoma and seizures: a treatable epileptic encephalopathy. *Epilepsia*. 2003;44:969–73.

24. Le Van Quyen M, Navarro V, Martinerie J, Baulac M, Varela FJ. Toward a neurodynamical understanding of ictogenesis. *Epilepsia*. 2003;44(suppl 12):30–43.

25. Ibrahim GM, Cassel D, Morgan BR, et al. Resilience of developing brain networks to interictal epileptiform discharges is associated with cognitive outcome. *Brain J Neurol*. 2014;137:2690–702.

26. Kramer MA, Eden UT, Kolaczyk ED, et al. Coalescence and fragmentation of cortical networks during focal seizures. *J Neurosci*. 2010;30:10076–85.

27. Schindler KA, Bialonski S, Horstmann MT, Elger CE, Lehnertz K. Evolving functional network properties and synchronizability during human epileptic seizures. *Chaos*. 2008;18:033119.

28. Warren CP, Hu S, Stead M, et al. Synchrony in normal and focal epileptic brain: the seizure onset zone is functionally disconnected. *J Neurophysiol*. 2010;104: 3530–9.

29. Khambhati A, Litt B, Bassett DS. Dynamic network drivers of seizure generation, propagation and termination in human epilepsy. *PLOS Comput Biol*. 2015;1004608.

30. Bullmore E, Sporns O. The economy of brain network organization. *Nat Rev Neurosci*. 2012;13:336–49.

31. Laufs H, Richardson MP, Salek-Haddadi A, et al. Converging PET and fMRI evidence for a common area involved in human focal epilepsies. *Neurology*. 2011;77:904–10.

32. Fahoum F, Lopes R, Pittau F, Dubeau F, Gotman J. Widespread epileptic networks in focal epilepsies: EEG-fMRI study. *Epilepsia*. 2012;53:1618–27.

33. Flanagan D, Badawy RAB, Jackson GD. EEG-fMRI in focal epilepsy: local activation and regional networks. *Clin Neurophysiol*. 2014;125:21–31.

34. Piredda G. A crucial epileptogenic site in the deep prepiriform cortex. *Nature*. 1985;317:623–5.

35. Pedersen M, Curwood EK, Vaughan DN, Omidvarnia AH, Jackson GD. Abnormal brain areas common to the focal epilepsies: multivariate pattern analysis of fMRI. *Brain Connect*. 2016;6: 208–15.

36. Berg AT. Epilepsy, cognition, and behavior: the clinical picture. *Epilepsia*. 2011;52:7–12.

37. Wilson SJ, Baxendale S. The new approach to classification: rethinking cognition and behavior in epilepsy. *Epilepsy Behav*. 2014;41:307–10.

38. Taylor J, Kolamunnage-Dona R, Marson AG, et al. Patients with epilepsy: cognitively compromised before the start of antiepileptic drug treatment? *Epilepsia*. 2010;51:48–56.

39. Thompson PJ, Duncan JS. Cognitive decline in severe intractable epilepsy. *Epilepsia*. 2005;46:1780–7.

40. Keezer MR, Sisodiya SM, Sander JW. Comorbidities of epilepsy: current concepts and future perspectives. *Lancet Neurol*. 2016;15:106–15.

41. Jokeit H, Okujava M, Woermann FG. Memory fMRI lateralizes temporal lobe epilepsy. *Neurology*. 2001;57: 1786–93.

42. Pardoe HR, Berg AT, Jackson GD. Sodium valproate use is associated with reduced parietal lobe thickness and brain volume. *Neurology*. 2013;80:1895–900.

43. Yasuda CL, Centeno M, Vollmar C, et al. The effect of topiramate on cognitive fMRI. *Epilepsy Res*. 2013;105: 250–5.

44. Orrù G, Pettersson-Yeo W, Marquand AF, Sartori G, Mechelli A. Using support vector machine to identify imaging biomarkers of neurological and psychiatric disease: a critical review. *Neurosci Biobehav Rev*. 2012;36:1140–52.

45. Pereira F, Mitchell T, Botvinick M. Machine learning classifiers and fMRI: a tutorial overview. *NeuroImage*. 2009;45:S199–209.

46. Richiardi J, Achard S, Bunke H, Van De Ville D. Machine learning with brain graphs: predictive modeling approaches for functional

imaging in systems neuroscience. *IEEE Signal Process Mag.* 2013;30:58–70.

47. Schrouff J, Rosa MJ, Rondina JM, et al. PRoNTo: pattern recognition for neuroimaging toolbox. *Neuroinformatics.* 2013;11:319–33.

48. Cristianini N, Shawe-Taylor J. *An Introduction to Support Vector Machines and Other Kernel-Based Learning Methods.* Cambridge: Cambridge University Press; 2000.

49. Rasmussen PM, Hansen LK, Madsen KH, Churchill NW, Strother SC. Model sparsity and brain pattern interpretation of classification models in neuroimaging. *Pattern Recognit Brain Decoding.* 2012;45:2085–100.

50. Bernhardt BC, Hong SJ, Bernasconi A, Bernasconi N. Magnetic resonance imaging pattern learning in temporal lobe epilepsy: classification and prognostics. *Ann Neurol.* 2015;77:436–46.

51. Pedersen M, Curwood EK, Archer JS, Abbott DF, Jackson GD. Brain regions with abnormal network properties in severe epilepsy of Lennox-Gastaut phenotype: multivariate analysis of task-free fMRI. *Epilepsia.* 2015a;56:1767–73.

52. Yang Z, Choupan J, Reutens D, Hocking J. Lateralization of temporal lobe epilepsy based on resting-state functional magnetic resonance imaging and machine learning. *Front Neurol.* 2015;6:184.

Chapter 11

Ictal Events Imaged through SPECT

Elson L. So, Vlastimil Sulc, Gregory Worrell, and Benjamin H. Brinkmann

11.1 Introduction

Brain perfusion alterations could conceivably serve as a promising "biomarker" of epileptic seizures. It has been known for some time that focal seizure activity is associated with increased perfusion at the region of the seizure activity (ictal hyperperfusion). It is also known that the focus of seizure activity could have reduced perfusion during the baseline or interictal state (interictal hypoperfusion). Using the appropriate radioligands, ictal hyperperfusion and interictal hypoperfusion could be detected by perfusion SPECT studies. Radioligands such as ^{99m}Tc-Hexamethylpropylene Amine Oxime (Tc-HMPAO) and ^{99m}Tc-Bicisate (Tc-ECD) have high first-pass extraction in the brain. Therefore, the temporal profile of cerebral uptake and binding of these radioligands corresponds closely to the duration of most focal seizure activity, permitting the pattern of cerebral distribution of the radioligand to map the focus of seizure activity.

A distinction would have to be made between SPECT being a biomarker of epileptic seizure versus a biomarker of epilepsy. The reason is that strictly speaking, ictal SPECT findings are applicable only to the seizure when the radioligand was injected. The SPECT findings could not be extrapolated to the individual's epilepsy, if the individual has more than one epileptogenic zone or more than one type of epilepsy. Also, the utility and utilization of SPECT findings as a biomarker of epileptic seizure or epilepsy are heavily dependent on a number of parameters, including how soon the radioligand was injected, and how extensive the seizure had spread at the time of injection. Incidentally, ictal and immediate postictal SPECT findings occasionally reveal images of streak patterns or multifocal patterns that suggest pathways of seizure spread or potential epileptic seizure networks.

The objective of this chapter is to critically appraise the current state of knowledge and usefulness of SPECT abnormalities as a biomarker of seizure activity, and potentially as a biomarker of epilepsy. Emphasis will also be given to both the requirements and the constraints that dictate optimization of the techniques for SPECT radioligand injection, ictal and interictal data acquisition, postacquisition data and image processing, and interpretation with correlation with other modalities such as EEG.

11.2 Ictal SPECT Image Acquisition Techniques

11.2.1 Peri-ictal SPECT Dynamics

Ictal SPECT involves injection of Tc99 m ethyl cysteinate diethylester (ECD) or Tc-99 m hexamethyl propylene amine oxime (HMPAO) cerebral blood flow tracer during the early phase of the seizure as blood flow to the seizure onset zone increases. The tracer is delivered to cerebral tissue in proportion to blood flow, crosses the blood-brain barrier in lipophilic form, and is converted to a hydrophylic form intracellularly. In this form the tracer's distribution is stable for multiple hours with minimal back diffusion and represents a snapshot of the average cerebral blood flow pattern 60–90 seconds after bolus injection of the tracer.[1,2] Tracer injection early in the seizure produces an increased isotope concentration in the seizure onset zone, representing the increased blood flow associated with initiation of the seizure. Late injection of the isotope may capture a reduced cerebral perfusion state and commensurate decreased isotope concentration in the seizure onset zone, though the timing of the change to decreased perfusion is highly variable, and its magnitude is smaller making localization of the seizure onset zone more difficult.[3,4] Following isotope injection and cessation of seizure activity the patient is scanned on a gamma camera system using a high-resolution fan beam collimator. Tomographic images are produced by a filtered back projection or an iterative reconstruction algorithm,[5] and algorithms to compensate for photon attenuation in brain tissue are applied[6] to produce high-quality images of the isotope's distribution. Separate acquisition of an interictal or baseline SPECT image provides a patient-specific reference study to aid comparison and analysis of the ictal SPECT. The interictal tracer injection is ideally performed after a minimum of 24 hours of freedom from seizures or auras to exclude any seizure-related effects, although this may not be possible in patients with frequent seizures or in status epilepticus.

11.2.2 Subtraction SPECT Coregistered to MRI (SISCOM)

While visual review of ictal and interictal SPECT studies can be effective, subtraction methods provide a more effective means of quantifying the spatial distribution of differences in isotope concentration.[7] Ictal and interictal SPECT volumes are spatially aligned using a voxel intensity based coregistration algorithm.[8] Because isotope uptake is highly variable between studies, it is necessary to normalize the pixel intensities of SPECT images prior to analysis. Normalization is performed to equalize the mean intracerebral pixel intensities of the ictal and interictal scans. The spatially aligned image volumes are subtracted, producing a volumetric map of the intracerebral differences in isotope concentration. Spatially, the majority of these differences are low in magnitude and represent normal variation in blood flow patterns, secondary perfusion changes outside the seizure onset zone, fluctuations in the counting statistics of isotope decay, or image reconstruction artifacts, but the perfusion changes in the seizure onset zone are expected to be of higher magnitude than these effects. The standard deviation of difference values in the subtraction volume is calculated, and pixels above or below two standard deviations are reviewed as perfusion abnormalities and potentially delineating the seizure onset zone. Evaluation of the perfusion differences in context with the underlying brain anatomy is facilitated by spatial alignment of the ictal and interictal SPECT and subtraction images with a high resolution MRI.[7]

11.2.3 SPM-Based Subtraction SPECT

Thresholding the subtraction image results in a map of the largest magnitude perfusion changes between the interictal and ictal images. This map includes normal physiological variations in cerebral blood flow in addition to perfusion changes related to seizure initiation. In order to statistically characterize perfusion changes relative to normal, a map of normal variation in cerebral perfusion can be used to test the significance of ictal-interictal differences.[9] Paired resting SPECT images of neurologically normal healthy volunteers acquired using either Tc99 m ECD[10] or HMPAO[11] are nonlinearly registered, or spatially normalized, into a common standardized template space[12] for intersubject comparison. The spatially normalized images constitute a spatially varying statistical distribution characterizing normal variation between resting SPECT scans of an individual. A patient's ictal, interictal, and subtraction SPECT images can be nonlinearly transformed into the template space, and a statistical test is performed to identify areas where ictal-interictal perfusion changes are statistically significantly different from normal variation. This process is typically accomplished using the open-source statistical parametric mapping (SPM) software,[12] running under Matlab (MathWorks, Natick, MA). Spatial smoothing is applied to reduce noise in the images, a cluster size threshold is applied to exclude voxel clusters smaller than the native resolution of the gamma camera system, and Bonferroni or false discovery rate corrections are applied to account for multiple spatial comparisons. Following statistical evaluation the significant perfusion difference map is transformed back into the patient's image space, and aligned to a high resolution MRI to show the relationship with cerebral neuroanatomy.

11.3 SPECT as a Marker of the Seizure-Onset Zone

11.3.1 Interictal SPECT

The characteristic interictal SPECT perfusion pattern at the seizure-onset focus is hypoperfusion relative to other parts of the brain. Most series that reported the accuracy of interictal SPECT were based on comparison with surface EEG seizure localization, but not with the epileptogenic zone. On this count alone, interictal SPECT can be said to have no proven value to serve as an imaging biomarker of epilepsy or seizures. Furthermore, interictal SPECT has low sensitivity for correctly localizing the seizure focus—correctly localizing in only 28% of extratemporal and 48% of temporal seizure patients.[13] A meta-analysis of the literature found that when using surgical outcome as a reference, interictal SPECT has 4.4% false positive rate.[14] When the reference is the localization by other established diagnostic tests, the false positive rate is 7.4%. Interictal hypoperfusion changes tend to be more extensive than expected. For example in temporal lobe epilepsy, the hypoperfusion pattern may involve both temporal and extratemporal areas.[15] Moreover, in the presence of an MRI epileptogenic lesion such as mesial temporal sclerosis, interictal SPECT is unlikely to give additional localizing information.

Epileptogenic foci have been observed to sometimes be associated with relative reduction of central benzodiazepine receptors. 123I-iomazenil SPECT maps the distribution of central benzodiazepine receptors. SPM analysis of interictal 123I-iomazenil SPECT has more recently been reported in mesial temporal lobe epilepsy to be more accurate than 99 mTc-ECD perfusion SPECT.[16] Nonetheless, only 10% of the 123I-iomazenil SPECT localized to the operated medial temporal lobe.

11.3.2 Ictal SPECT by Visual Inspection

The early method of interpreting ictal SPECT studies consist of visually comparing perfusion images derived from ictal SPECT injection with those derived from interictal injection. The result of the comparison is often indeterminate or equivocal. One reason for this is that the amount of radioligand injected and the latency to image acquisition may vary between the two studies, resulting in varying degrees of image intensities. Also, the images obtained may not be at the same anatomical level of the brain; and without coregistration of the SPECT images on MRI images, there is no brain structural detail to aid in the localization of the perfusion changes. Therefore, agreement by visual comparison among reviewers had been low at 41%, compared with SISCOM at 84%. Sensitivity of visual comparison is also low at 39%, compared with 88% with SISCOM. Digital subtraction SPECT such as SISCOM and SPM analyzed SPECT should now be the standard for the use of SPECT in epilepsy.

11.3.3 SISCOM

Computer-aided digital subtraction SPECT has been convincingly demonstrated to be superior to interpretation by visual comparison of the ictal and the interictal images.

Most published studies involving digital subtraction SPECT are those with the SISCOM method. It is unclear whether there is a digital subtraction method that is superior to others, and reports of a number of different subtraction methods have not been validated in a sufficiently large sample of patients.

A much-needed application of ictal SPECT is when the MRI is negative.[17] Unfortunately, the number of studies that have addressed this situation is limited. Nonetheless, subtraction ictal SPECT with MRI coregistration is regularly used in some centers for localizing peri-ictal perfusion abnormalities, which can then serve as a target for intracranial electrode implantation to map the seizure-onset zone.[18] In a group of 26 MRI-negative refractory temporal or extratemporal epilepsy patients, 92% had a focus of abnormal perfusion detected by reviewers who were blinded to other clinical data.[7]

When the MRI was negative in refractory temporal lobe epilepsy, SISCOM was still localizing in 82%.[19] The resection of the localized abnormality was associated with higher probability of seizure freedom. Seizure-free outcome was also better when subtle MRI structural abnormalities existed at the resection site, or when there were no extratemporal or contralateral spikes or sharp waves.

As for MRI-negative extratemporal epilepsy, the yield of SISCOM was 77% in a small series of 17 patients.[20] Of the patients, 55% had excellent postsurgical outcome when there was a SISCOM focus and it was surgically resected. On the other hand, none had excellent outcome when the SISCOM was negative, or when a SISCOM focus was not included in the resection. The difference was statistically significant.

Another study involving a larger sample of patients showed that 68% of 85 MRI-negative extratemporal epilepsy patients had localizing SISCOM abnormalities.[21] SISCOM result contributed to the decision to proceed with intracranial EEG recording, but it did not by itself determine which patients proceeded to intracranial electrode implantation. Only 24 patients eventually had resective surgery, and this small number of patients prevented the ability of SISCOM result to distinguish those who had excellent outcome from those who did not.

In a larger group of 130 patients with either temporal or extratemporal epilepsy and with or without MRI lesion, SISCOM was positive in 61%.[17] A hypothesis for further epilepsy surgery evaluation was generated in 70% of the SISCOM positive patients. Factors that contributed to the likelihood for SISCOM to generate a hypothesis were shorter duration of seizures, focal seizures, and lesional MRI result. However, only 38% of these patients eventually had resective surgery. The SISCOM abnormality was concordant with the site of surgery in 82% of the patients, and the concordance was associated with 60% rate of seizure freedom at two years' postsurgical follow-up. In contrast, seizure freedom was observed in only 17% of the 6 patients who had nonlocalizing SISCOM findings. Statistical analysis for the difference was not provided.

11.3.4 SPM-Based Subtraction SPECT

This section discusses the yield of comparing patients' ictal/interictal subtraction data with normative data by using SPM (SPM subtraction-SPECT). Experience in small series of patients showed that the technique could identify focal or regional perfusion abnormalities in 60% to 93% of patients.[11,22,23] One particular study disappointingly found that the SPM was not superior to visual comparison of ictal and interictal images.[23] In contrast, McNally and colleagues reported that their method of ictal-interictal SPECT analyzed by SPM (ISAS) was 93% sensitive and 87% specific in temporal lobe seizures, and 77% sensitive and 93% specific in neocortical seizures.[11] However, these favorable results were achieved with radioligand injection during seizures, because postictal injections are associated with poorly localizing findings, despite being lateralizing in 80% of the injections. SPM-subtraction SPECT has been further applied to analysis of subregions of the mesial temporal lobe in patients with mesial temporal sclerosis. Analysis of the hippocampal subregion correctly lateralized the seizure onset in 91% of patients, and in 87% with amygdala subregion analysis. It should be noted that the foregoing reports of favorable yield of SPM-subtraction SPECT involved predominantly temporal lobe epilepsy and lesional epilepsy. Moreover, some of the studies consist mostly of patients who had responded to epilepsy surgery.[11] Therefore, the reported results would not be generalizable to all patients who present for epilepsy surgery evaluation, especially those whose seizure focus is not as obvious, and in whom localizing tests such as SPECT are most needed.

The technique of SPM-based subtraction SPECT analysis that we had developed was statistical ictal SPECT coregistered to MRI (STATISCOM). We conducted an experiment to compare STATISCOM with SISCOM in 87 temporal lobe epilepsy surgery patients.[10] Patients were consecutively included in the study. The following statistically significant results were found. Kappa score for agreement between blinded reviewers was good for STATISCOM at .81 and poorer for SISCOM at .36. The sensitivity of STATISCOM in revealing a focus of increased perfusion was 84% vs. 66% with SISCOM. STATISCOM was superior to SISCOM in distinguishing between mesial and neocortical temporal seizures (68% vs. 24% of the patients). The superiority of STATISCOM in this regard was also true for nonlesional patients, with 80% sensitivity vs. 47% for SISCOM. The ability of STATISCOM to distinguish between mesial and lateral neocortical temporal lobe seizures had prognostic significance. Postsurgical seizure-free rate was 81% when the subregional localization was present and 53% when it was absent. In contrast, subregional localization by SISCOM did not have prognostic implication.

One way of assessing the potential for SPECT perfusion abnormality to serve as a focal epileptogenic biomarker independent from the presence of a structural lesion is in MRI-negative epilepsy. A three-way comparison study was performed to evaluate STATISCOM, ISAS, and SISCOM in 21 temporal epilepsy surgery and 28 extratemporal epilepsy surgery patients.[24] The range of age at epilepsy onset in the 49 patients was from the first year in life to 34 years. Agreement among the reviewers was equally high for STATISCOM and ISAS, and both were superior to SISCOM in either temporal or extratemporal epilepsy. In temporal epilepsy, kappa score was .55 for STATISCOM, .67 for ISAS, and .32 for SISCOM. In extratemporal epilepsy, kappa score was .56 for STATISCOM, .46 for ISAS, and .40 for SISCOM. The temporal lobe hyperperfusion colocalized with the resected site in 71% of the patients with STATISCOM, 67% with ISAS, and 38% with SISCOM. The extratemporal lobe hyperperfusion colocalized with the resected site in 57% of the patients with STATISCOM, 53% with ISAS, and 36% with SISCOM. In either temporal or extratemporal epilepsy, the rate with either STATISCOM or ISAS was significantly better than SISCOM, but there was no statistically significant difference between STATISCOM and ISAS. In temporal lobe surgery, there was significant association between STATISCOM or ISAS colocalization and excellent postsurgical seizure control, but not for SISCOM colocalization. For extratemporal surgery, there was

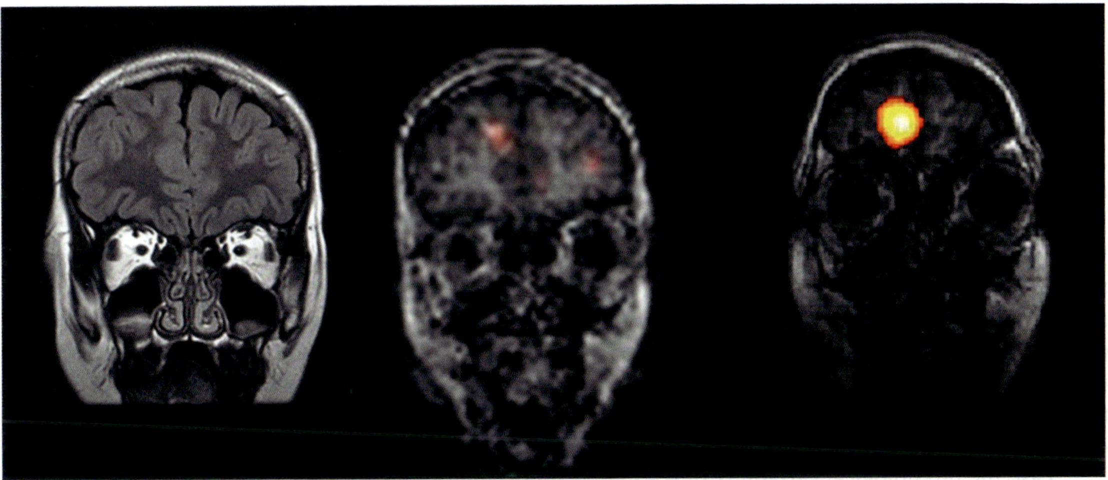

Figure 11.1. This 11-year-old boy was previously healthy and without risk factors for epilepsy. He developed unprovoked spells at age 6.5 years. Spells consisted of a vague aura, staring, flushed complexion, giggling, then restless movement of the legs, but no impairment of awareness. The spells became multiple and daily in occurrence, and uncontrolled by multiple antiepileptic medication trials. Interictal EEGs were unremarkable, but MRI showed subtle signal abnormality at the right mesial frontal white matter and blurring of the nearby cortical margin. Video-scalp EEG monitoring recorded seizures that were mostly bilateral and indeterminate in onset, but with some seizures that were most prominent on the right. SISCOM was nonlocalizing, showing multiple foci (center image), whereas STATISCOM showed hyperperfusion at the right frontal region (right image). Intracranial EEG recording delineated the seizure onset zone at the location of the STATISCOM focus, which was then resected. The patient became seizure-free without antiepileptic medications.

no association between any of the SPECT method result and postsurgical outcome.

11.3.5 Combined Results from SPECT and Other Functional Imaging Studies

There is the potential for localization accuracy to be improved when SPECT results are combined with those of other functional imaging studies. Perissinotti and colleagues reported in a cohort of children with MRI lesional and nonlesional epilepsy that SISCOM alone agreed with the presumed epileptogenic zone in 67%, and FDG-PET alone in 57%. When the two were combined, the rate improved to 76%.[25] However, the presumed epileptogenic zone was determined by video-scalp EEG monitoring and clinical data, not by seizure freedom following focal resective surgery. Only 18 patients underwent surgery, 14 of whom were found to have cortical dysplasia. MRI identified the presumed epileptogenic zone in 72% of these 18 patients, but the rate for the combination of SISCOM and FDG-PET was 100%. The small number of patients did not permit the determination of whether the rates are statistically different between those who were postsurgically seizure-free and those who were not. Such a comparison

would be necessary to determine the biomarker value of SPECT by itself or in combination with other functional studies.

The chance of excellent outcome following extratemporal lobe surgery is enhanced when both digital subtraction ictal SPECT and FDF-PET findings colocalize with both temporal lobe MRI lesion and the video-EEG localization. Concordance of both ictal SPECT and PET with MRI and video-EEG was associated with 65% rate of Class I Engel outcome, whereas the rate was significantly lower at only 33% for nonconcordance.[26]

The value of concordance among diagnostic modalities was also demonstrated in a small cohort of MRI-negative temporal or extratemporal patients.[27] All patients who had concordance of SISCOM with MSI and intracranial EEG were seizure-free after resective surgery.

11.4 Role of SPECT in Evaluating Epileptic Networks

As discussed above, SPECT provides information on regional cerebral blood flow (CBF) changes that occur during seizures (ictal) that can be directly compared to interictal (between seizures) brain states. In focal

patients with MRI-negative but histologically proven cortical dysplasia.[44] Currently available SPM-SPECT analysis could potentially confirm the reliability of SPECT as a biomarker for epilepsy due to cortical dysplasia. There is also the possibility that SPECT may become a component of a "biomarker panel" that consists of results from a number of different tests, which when taken together very highly indicate epilepsy or the epileptogenic zone.

References

1. Léveillé J, Demonceau G, De Roo M, et al. Characterization of technetium-99 m-l,l-ECD for brain perfusion imaging, part 2: biodistribution and brain imaging in humans. *J Nucl Med.* 1989;**30**(11):1902–10.

2. Neirinckx R, Canning LR, Piper IM, et al. Technetium-99 m d,1-HMPAO: a new radiopharmaceutical for SPECT imaging of regional cerebral blood perfusion. *J Nucl Med.* 1987;**28**:191–202.

3. Avery RA, Spencer SS, Spanaki MV, et al. Effect of injection time on postictal SPET perfusion changes in medically refractory epilepsy. *Eur J Nucl Med.* 1999;**26**(8):830–6.

4. Newton M, Berkovic SF, Austin MC, et al. Postictal switch in blood flow distribution and temporal lobe seizures. *J Neurol Neurosurg Psychiatr.* 1992;**55**:891–4.

5. Yokei T, Shinohara H, Onishi H. Performance evaluation of OSEM reconstruction algorithm incorporating three-dimensional distance-dependent resolution compensation for brain SPECT: a simulation study. *Ann Nucl Med.* 2002;**16**(1):11–8.

6. Chang L. A method for attenuation correction in radionuclide computed tomography. *IEEE Trans Nucl Sci.* 1978;**25**(1):638–43.

7. O'Brien T, So EL, Mullan BP, et al. Subtraction ictal SPECT co-registered to MRI improves clinical usefulness of SPECT In localizing the surgical seizure focus. *Neurology.* 1998;**50**(2):445–54.

8. Brinkmann B, O'Brien TJ, Aharon S, et al. Quantitative and clinical analysis of SPECT image registration for epilepsy studies. *J Nucl Med.* 1999;**40**:1098–105.

9. Brinkmann B, O'Brien TJ, Mullan BP, et al. Subtraction ictal SPECT coregistered to MRI for seizure focus localization in partial epilepsy. *Mayo Clin Proc.* 2000;**75**:615–24.

10. Kazemi NJ, Worrell GA, Stead SM, et al. Ictal SPECT statistical parametric mapping in temporal lobe epilepsy surgery. *Neurology.* 2010;**74**(1):70–6.

11. McNally KA, Paige AL, Varghese G, et al. Localizing value of ictal-interictal SPECT analyzed by SPM (ISAS). *Epilepsia.* 2005;**46**(9):1450–64.

12. Friston K, Penny W, Ashburner J, et al. *Statistical Parametric Mapping: The Analysis of Functional Brain Images.* New York: Academic Press; 2007.

13. Newton M, Berkovic SF, Austin MC, et al. SPECT in the localisation of extratemporal and temporal seizure foci. *J Neurol Neurosurg Psychiatr.* 1995;**59**:26–30.

14. Devous MD Sr, Thisted RA, Morgan GF, et al. SPECT brain imaging in epilepsy: a meta-analysis. *J Nucl Med.* 1998;**39**(2):285–93.

15. Harvey A, Bowe JM, Hopkins IJ, et al. Ictal 99Tc-HMPAO single photon emission computed tomography in children with temporal lobe epilepsy. *Epilepsia.* 1993;**34**:869–77.

16. Fujitani S, Matsuda K, Nakamura F, et al. Statistical parametric mapping of interictal 123I-iomazenil SPECT in temporal lobe epilepsy surgery. *Epilepsy Res.* 2013;**106**(1–2):173–80.

17. von Oertzen TJ, Mormann F, Urbach H, et al. Prospective use of subtraction ictal SPECT coregistered to MRI (SISCOM) in presurgical evaluation of epilepsy. *Epilepsia.* 2011;**52**(12):2239–48.

18. O'Brien T, So EL, Mullan BP, et al. Subtraction SPECT co-registered to MRI improves postictal SPECT localization of seizure foci. *Neurology.* 1999;**52**:137–46.

19. Bell M, Rao S, So EL, et al. Epilepsy surgery outcomes in temporal lobe epilepsy with a normal MRI. *Epilepsia.* 2009;**50**(9):2053–60.

20. O'Brien T, So EL, Mullan BP, et al. Subtraction peri-ictal SPECT is predictive of extratemporal epilepsy surgery outcome. *Neurology.* 2000;**55**:1668–77.

21. Noe K, Sulc V, Wong-Kisiel L, et al. Long-term outcomes after nonlesional extratemporal lobe epilepsy surgery. *JAMA Neurol.* 2013;**70**(8):1003–8.

22. Lee JD, Kim HJ, Lee BI, et al. Evaluation of ictal brain SPECT using statistical parametric mapping in temporal lobe epilepsy. *Eur J Nucl Med.* 2000;**27**(11):1658–65.

23. Amorim BJ, Ramos CD, dos Santos AO, et al. Brain SPECT in mesial temporal lobe epilepsy: comparison between visual analysis and SPM. *Arquivos De Neuro-Psiquiatria.* 2010;**68**(2):153–60.

24. Sulc V, Stykel S, Hanson DP, et al. Statistical SPECT processing in MRI-negative epilepsy surgery. *Neurology.* 2014;**82**:932–9.

25. Perissinotti A, Setoain X, Aparicio J, et al. Clinical role of subtraction ictal SPECT coregistered to MR imaging and F-18-FDG PET in pediatric epilepsy. *J Nucl Med.* 2014;**55**(7):1099–105.

26. Chandra PS, Vaghania G, Bal CS, et al. Role of concordance between ictal-subtracted SPECT and PET in predicting long-term outcomes after epilepsy surgery. *Epilepsy Res.* 2014;**108**(10):1782–9.

27. Schneider F, Wang IZ, Alexopoulos AV, et al. Magnetic source imaging and ictal SPECT in MRI-negative neocortical epilepsies: additional value and comparison with intracranial EEG. *Epilepsia*. 2013;**54**(2):359–69.

28. Chassagnon S, Andre V, Koning E, Ferrandon A, Nehlig A. Optimal window for ictal blood flow mapping: insight from the study of discrete temporo-limbic seizures in rats. *Epilepsy Res*. 2006;**69**:100–18.

29. Weinand M, Carter LP, Patton DD, et al. Long-term surface cortical cerebral blood flow monitoring in temporal lobe epilepsy. *Neurosurgery*. 1994;**35**:657–64.

30. Baumgartner C, Serles W, Leutmezer F, et al. Preictal SPECT in temporal lobe epilepsy: regional cerebral blood flow is increased prior to electroencephalography-seizure onset. *J Nucl Med*. 1998;**39**(6):978–82.

31. Brinkmann B, Jones DT, Stead M, et al. Statistical parametric mapping demonstrates asymmetric uptake with Tc-99 m ECD and Tc-99 m HMPAO SPECT in normal brain. *J Cereb Blood Flow Metab*. 2012;**32**:190–8.

32. Spencer S. Neural networks in human epilepsy: evidence of and implications for treatment. *Epilepsia*. 2002;**43**:219–27.

33. Van Paesschen W, Dupont P, Sunaert S, Goffin K, Van Laere K. The use of SPECT and PET in routine clinical practice in epilepsy. *Curr Opin Neurol*. 2007;**20**(2):194–202.

34. Van Paesschen W, Dupont P, Van Driel G, Van Billoen H, Maes A. SPECT perfusion changes during complex partial seizures in patients with hippocampal sclerosis. *Brain*. 2003;**126**(pt 5):1103–11.

35. Bohnen NI, O'Brien TJ, Mullan BP, So EL. Cerebellar changes in partial seizures: clinical correlations of quantitative SPECT and MRI analysis. *Epilepsia*. 1998;**39**:640–50.

36. Shin WC, Hong SB, Tae WS, Seo DW, Kim SE. Ictal hyperperfusion of cerebellum and basal ganglia in temporal lobe epilepsy: SPECT subtraction with MRI coregistration. *J Nucl Med*. 2001;**42**(6):853–8.

37. Raichle M, MacLeod AM, Snyder AZ, et al. A default mode of brain function. *Proc Natl Acad Sci USA*. 2001;**98**:676–82.

38. Blumenfeld H, McNally KA, Vanderhill SD, et al. Positive and negative network correlations in temporal lobe epilepsy. *Cereb Cortex*. 2004;**14**:892–902.

39. Blumenfeld H, Varghese GI, Purcaro MJ, et al. Cortical and subcortical networks in human secondarily generalized tonic-clonic seizures. *Brain*. 2009;**132**:999–1012.

40. Cleeren E, Casteels C, Goffin K, Janssen P, Van Paesschen W. Ictal perfusion changes associated with seizure progression in the amygdala kindling model in the rhesus monkey. *Epilepsia*. 2015;**56**(9):1366–75.

41. Elwan SA, Wu G, Huang SS, Najm IM, So NK. Ictal single photon emission computed tomography in epileptic auras. *Epilepsia*. 2014;**55**(1):133–6.

42. Barba C, Barbati G, Di Giuda D, et al. Diagnostic yield and predictive value of provoked ictal SPECT in drug-resistant epilepsies. *J Neurol*. 2012;**259**(8):1613–22.

43. Wieser H, Bancaud J, Talairach J, et al. Comparative value of spontaneous and chemically and electrically induced seizures in establishing the lateralization of temporal lobe seizures. *Epilepsia*. 1979;**20**:47–59.

44. Kudr M, Krsek P, Marusic P, et al. SISCOM and FDG-PET in patients with non-lesional extratemporal epilepsy: correlation with intracranial EEG, histology, and seizure outcome. *Epileptic Disord*. 2013;**15**(1):3–13.

Chapter

12

Imaging Cortical and Subcortical Circuitry in Generalized Epilepsies

Fenglai Xiao, Lorenzo Caciagli, Britta Wandschneider, and Matthias Koepp

12.1 Introduction

Imaging, in particular group analysis of structural MRI in homogenous patient populations, has provided further insight into disease pathophysiology and refined the cortico-thalamic network theory of genetic generalized epilepsy (GGE).[1–4] The identification of key regions within networks provides putative biomarkers that can be targeted by clinical treatment strategies. Connectivity analysis provides a versatile tool to assess pathological interactions of different regions in GGE.

12.2 Thalamus

The thalamus as well as thalamocortical networks are critically important for the generation of generalized seizures. Simultaneous EEG-fMRI studies found that hemodynamic changes associated with generalized spike-wave discharges (GSWDs) could be consistently observed in the thalamus as well as within areas belonging to the default mode network (DMN).[5,6] Hence, thalamus and DMN have been the focus of investigations aiming to elucidate the pathophysiology of GGE syndromes.

There is compelling evidence that dysfunction in the cortico-thalamic circuitry contributes to the pathogenesis of generalized seizures.[2,3,7–9] Cortico-thalamic interactions in GGE have been characterized by morphometric covariance,[3] functional,[7,8] and anatomic connectivity.[9,10]

In a recent study on GGE patients, Wang and colleagues utilized bilateral thalamic medial dorsal nuclei and pulvinar, identified in a structural analysis as seed regions for functional connectivity (FC) analysis, and observed reduced connectivity between thalamic nuclei and bilateral orbital frontal cortex, caudate nucleus, putamen, and amygdala,[9] demonstrating the different involvement of the thalamocortical networks in GGE. Using similar methods, Kim

et al. found decreased thalamocortical FC between anteromedial thalamus and medial prefrontal cortex and precuneus/PCC in a mixed cohort of GGE patients.[7] Moreover, greater reduction in FC between anteromedial thalamus and medial prefrontal in relation to increasing disease duration in a mixed GGE group suggested a cumulative influence of the disease-underlying factors on thalamocortical connectivity.

A recent resting-state functional MRI study examined 97 GGE patients and found increased FC in thalamocortical networks.[8] The thalamus was functionally divided into subregions analogous to histological parcellations,[8] which is in accord with distinct cortical lobes relating to five parallel corticothalamic networks.[11] Patients showed increased FC strength in four corticothalamic networks compared with controls, involving (1) the prefrontal-thalamic, (2) motor/premotor-thalamic, (3) parietal/occipital-thalamic, and (4) temporal-thalamic networks but no changes in sensory-thalamic network

Abnormalities in the motor-/premotor-thalamic system may be expected in view of the sustained muscle rigidity as well as rapid muscle contraction/relaxation experienced by patients during seizures. The premotor cortex and specific nuclei (pulvinar and ventrolateral nucleus) showed increased FC within the motor-thalamic but not the sensory-thalamic network. This may be attributed to the disruption of functional integrity in the sensorimotor network of the neocortex[12] and may reflect the higher excitability of the motor system of patients with GGE compared with controls.[13] Altered connectivity of the posterior aspect of the middle temporal gyrus may relate to the abnormal semantic processing observed in many patients with GGE.[14] Similarly, altered prefrontal-thalamic connectivity in different subtypes suggests that these network changes may be a common feature of GGE.

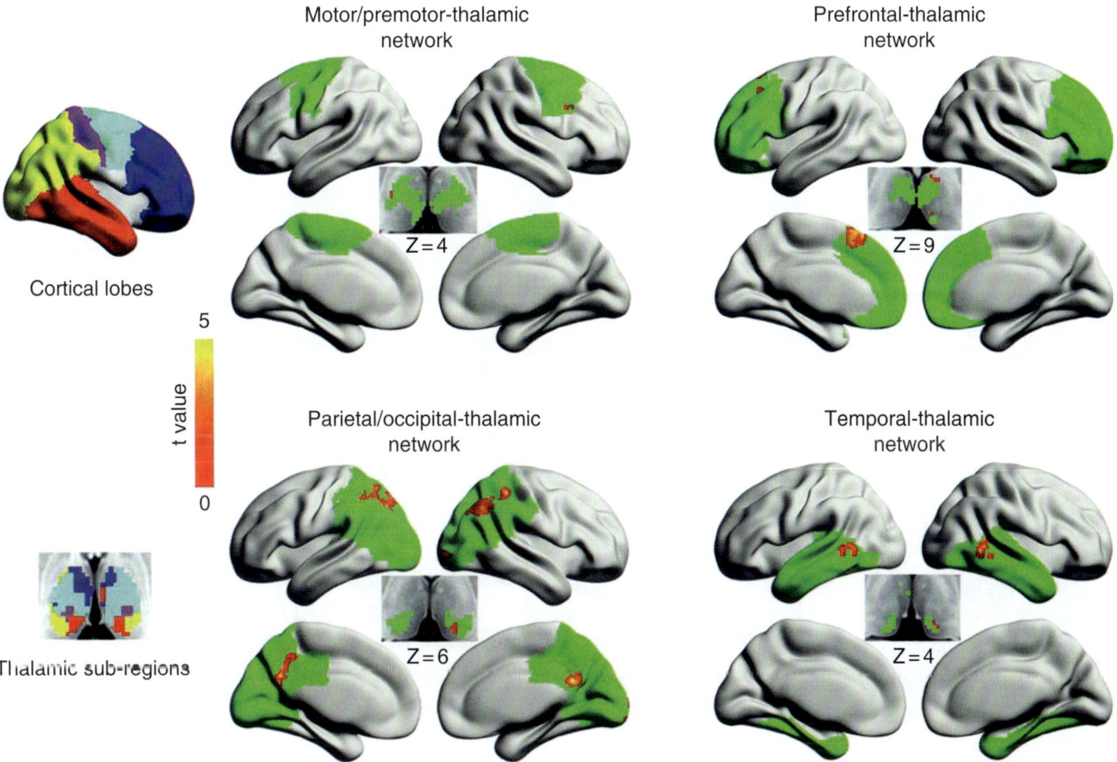

Figure 12.1. Comparison of voxel-wise thalamocortical functional connectivity between patients with generalized epilepsy and healthy controls. The left column shows cortical lobes and corresponding thalamic subregions obtained by using the winner-take-all approach. The right column shows the between-group difference in functional connectivity strength. Abnormal voxels are labeled by warm colors and rendered on cortical surface models and thalamic sections (green). The color scale represents *t* values, as determined with 2-sample *t* tests (*p* < .05, corrected). The *z* plane coordinates refer to the Montreal Neurological Institute space. Reproduced from Ji et al.[8] with permission.

12.3 Default Mode Network

The DMN is composed of precuneus/posterior cingulate cortex, medial prefrontal cortex, lateral parietal cortex, and inferior temporal cortex.[15,16] The nodes of this network were shown to exhibit coherent spontaneous fluctuations of their resting-state BOLD fMRI time courses,[15] and constitute a system engaged during internally focused activities, including evaluating salience of internal and external cues, remembering the past, and planning the future.[15] Of note, cortical DMN regions are known to be functionally and structurally connected to thalamus,[11] and deactivation of the default mode areas along with thalamic activation during GSWDs was consistently found by EEG-fMRI studies.[5,6] Dynamic connectivity maps exploring thalamic and DMN interactions during the preictal, ictal, and postictal states of absence seizures show an evolving anticorrelation between the thalamus and the DMN, which is consistent with an inhibitory effect of seizures on the default mode of brain function, gradually fading out after seizure onset.[17] Two recent resting-state studies found that DMN connectivity is lower in patients with GGE than in healthy controls. Impaired DMN connectivity found in studies probing the integrity of structural and functional studies might relate to impaired consciousness observed in GGE patients, frequently reported in other mental diseases.[18] Another resting-state study on a large sample of patients with GGE showed that connectivity is further reduced in the subgroup with treatment-resistant seizures.[19] Furthermore, in these studies, the strength of DMN connectivity was negatively correlated with disease duration.[20,21] It is notable that DMN connectivity declines with duration of epilepsy[20,21] and that this effect is significantly more pronounced in treatment-resistant patients.[19] Although longitudinal studies would be needed to confirm that seizure duration correlates with changes in DMN FC over time, these

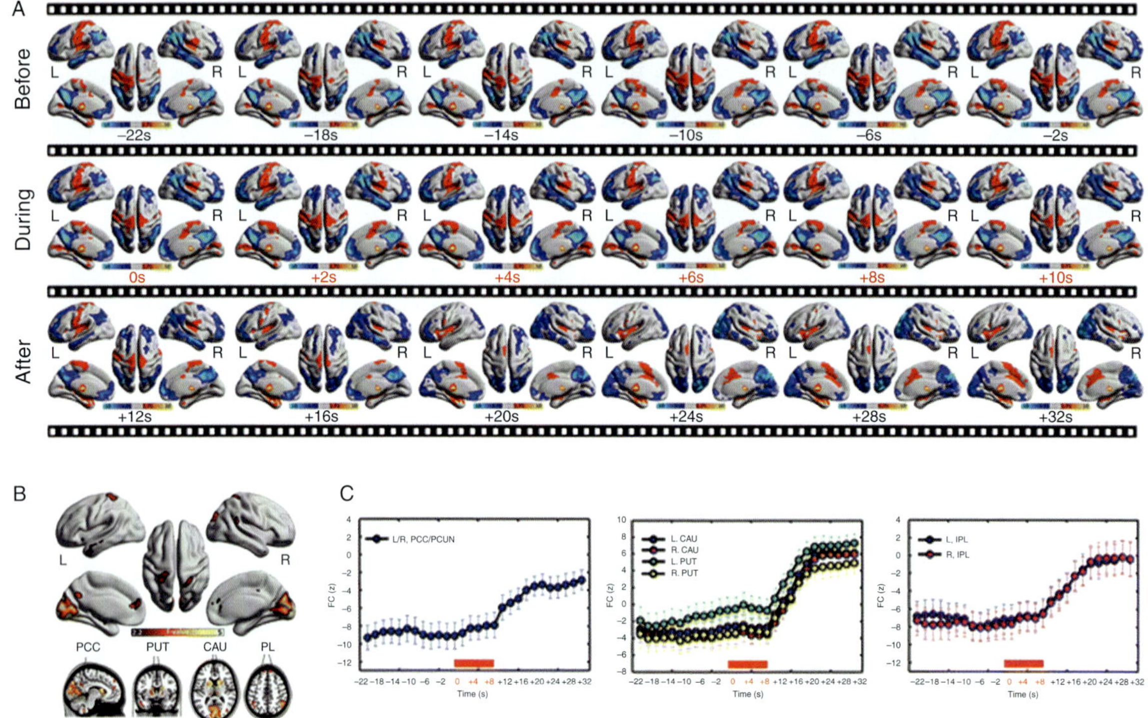

Figure 12.2. Dynamic changes of the default-mode network during absence seizures. (A) Significant positive (warm color) and negative (cool color) correlations were derived from one-sample t tests (uncorrected heighted threshold $p < .05$, and extend threshold $k = 5$ voxels). The results are presented on inflated surface maps for preictal (upper), ictal (middle), and postictal (lower) periods from -22 to $+32$ s, relative to seizure onset. (B) Significant change of the default mode network functional connectivity map across the preictal, ictal, and postictal intervals (one-way within-subject ANOVA analysis, FDR-corrected height threshold $p < .05$, and extent threshold $k = 5$ voxels). Significant network changes (warm color) are presented on inflated surface maps (upper) and axial maps (lower). (C) Mean functional connectivity strength (z value) time courses in five anatomical clusters (thalamus, caudate, and inferior parietal lobule). Error bars refer to the standard error of the mean across all seizures. The mean seizure duration is shown in the red box. Reproduced from Liao et al.[17] with permission.

findings might indicate a detrimental effect of epilepsy-related factors on the brain, supporting a clinical decision to treat aggressively.

12.4 Frontal Lobes

GGE is typically associated with normal intelligence, but many patients exhibit specific frontal-lobe cognitive deficits.[22] The presence of specific hypo- and hyperconnectivity of frontal lobe areas indicated that GGE may not homogeneously affect the entire brain, and that the frontal lobe might be the epicenter of the effects of GGE.[23–25]

Vollmar et al. demonstrated aberrant patterns of FC between the supplementary motor area (SMA) and the motor cortex during a working memory task, which was paralleled by abnormal structural connectivity metrics.[23,24] Precipitation of myoclonic

jerks by cognitive tasks is a known trait of some JME patients, known as praxis induction.[26,27] Therefore, the abnormal connectivity of SMA, prefrontal cognitive areas, and motor system was interpreted as a potential substrate underlying seizures triggered by cognitive effort. Regions of cortical hyperexcitability may overlap with areas physiologically activated during cognitive or motor activities. Hence, a complex task involving several functional cortical systems may summon a "critical mass of activated cortex," which leads to seizure precipitation. The abnormal motor cortex coactivation during a working memory task may represent the functional correlate of this mechanism.

In an EEG-fMRI study, Szaflarski et al. found greater spike-related activation of paracingulate cortex in valproate-unresponsive GGE patients compared with valproate-responsive subjects.[28] A further

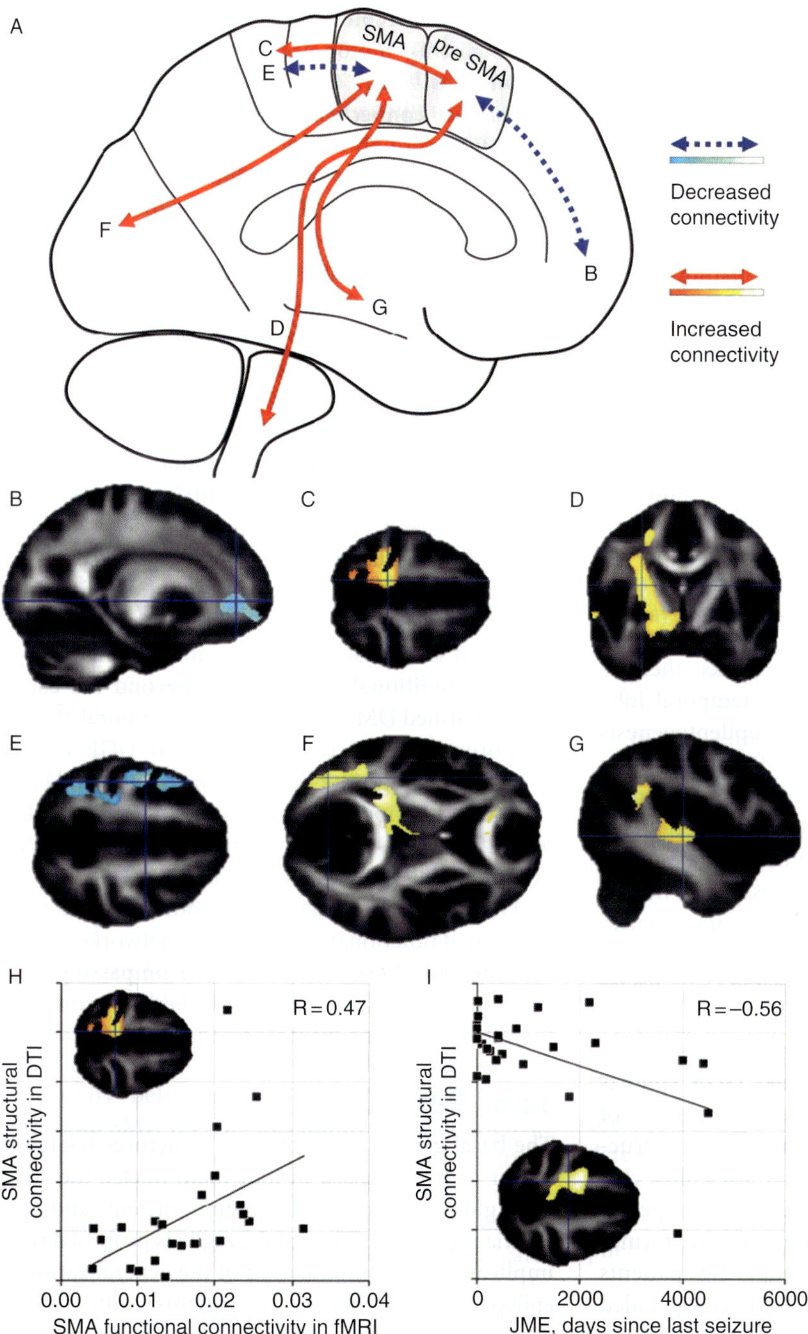

Figure 12.3. Alterations of structural connectivity in juvenile myoclonic epilepsy (JME). (A) Summary of findings in JME patients. The anterior pre-supplementary motor area (SMA) cluster showed reduced connectivity to anterior frontopolar regions (B) and increased connectivity to the medial central region (C) as well as to descending motor pathways (D). The posterior SMA cluster showed reduced connectivity to the central region (E) but increased connectivity to the temporal neocortex (F) and occipital cortex (G). (H) Functional connectivity between the prefrontal cortex and SMA, as found with fMRI, correlated significantly with the structural connectivity shown with diffusion tensor imaging (DTI). (I) Connectivity of the SMA also correlated with disease activity. Reproduced from Vollmar et al.[23] with permission.

study detected a positive correlation between GSWD frequency and FC between paracingulate and frontal regions related to attention and execution functions as well as precentral gyrus. This might indicate an increased crosstalk between motor, default mode, and executive networks in patients with frequent GSWDs,[29] suggesting that differences in paracingulate connectivity may be associated with frequent GSWDs and uncontrolled seizures in GGE. Recent investigations combining resting-state fMRI and

15. Raichle ME, MacLeod AM, Snyder AZ, Powers WJ, Gusnard DA, Shulman GL. A default mode of brain function. *Proc Natl Acad Sci USA*. 2001;**98**(2):676–82.

16. Buckner RL, Andrews Hanna JR, Schacter DL. The brain's default network. *Ann N Y Acad Sci*. 2008;**1124**(1):1–38.

17. Liao W, Zhang Z, Mantini D, et al. Dynamical intrinsic functional architecture of the brain during absence seizures. *Brain Struct Funct*. 2014;**219**(6):2001–15.

18. Whitfield-Gabrieli S, Ford JM. Default mode network activity and connectivity in psychopathology. *Annu Rev Clin Psychol*. 2012;**8**(1):49–76.

19. Kay BP, DiFrancesco MW, Privitera MD, Gotman J, Holland SK, Szaflarski JP. Reduced default mode network connectivity in treatment-resistant idiopathic generalized epilepsy. *Epilepsia*. 2013;**54**(3):461–70.

20. Luo C, Li Q, Lai Y, et al. Altered functional connectivity in default mode network in absence epilepsy: a resting-state fMRI study. *Hum Brain Mapp*. 2011;**32**(3):438–49.

21. McGill ML, Devinsky O, Kelly C, et al. Default mode network abnormalities in idiopathic generalized epilepsy. *Epilepsy Behav*. 2012;**23**(3):353–9.

22. Wandschneider B, Thompson PJ, Vollmar C, Koepp MJ. Frontal lobe function and structure in juvenile myoclonic epilepsy: a comprehensive review of neuropsychological and imaging data. *Epilepsia*. 2012;**53**(12):2091–8.

23. Vollmar C, O'Muircheartaigh J, Barker GJ, et al. Motor system hyperconnectivity in juvenile myoclonic epilepsy: a cognitive functional magnetic resonance imaging study. *Brain*. 2011;**134**(6):1710–9.

24. Vollmar C, O'Muircheartaigh J, Symms MR, et al. Altered microstructural connectivity in juvenile myoclonic epilepsy: the missing link. *Neurology*. 2012;**78**(20):1555–9.

25. Ji G-J, Zhang Z, Xu Q, Zang Y-F, Liao W, Lu G. Generalized tonic-clonic seizures: aberrant interhemispheric functional and anatomical connectivity. *Radiology*. 2014;**271**(3):839–47.

26. Koepp MJ, Caciagli L, Pressler RM, Lehnertz K, Beniczky S. Reflex seizures, traits, and epilepsies: from physiology to pathology. *Lancet Neurol*. 2016;**15**(1):92–105.

27. Yacubian EM, Wolf P. Praxis induction. Definition, relation to epilepsy syndromes, nosological and prognostic significance. A focused review. *Seizure*. 2014;**23**(4):247–51.

28. Szaflarski JP, Kay B, Gotman J, Privitera MD, Holland SK. The relationship between the localization of the generalized spike and wave discharge generators and the response to valproate. *Epilepsia*. 2013;**54**(3):471–80.

29. Kay BP, Holland SK, Privitera MD, Szaflarski JP. Differences in paracingulate connectivity associated with epileptiform discharges and uncontrolled seizures in genetic generalized epilepsy. *Epilepsia*. 2014;**55**(2):256–63.

30. Yang T, Ren J, Li Q, et al. Increased interhemispheric resting-state in idiopathic generalized epilepsy with generalized tonic-clonic seizures: a resting-state fMRI study. *Epilepsy Res*. 2014;**108**:1299–305.

31. Bai X, Guo J, Killory B, et al. Resting functional connectivity between the hemispheres in childhood absence epilepsy. *Neurology*. 2011;**76**(23):1960–7.

32. Tae WS, Hong SB, Joo EY, et al. Structural brain abnormalities in juvenile myoclonic epilepsy patients: volumetry and voxel-based morphometry. *Korean J Radiol*. 2006;**7**(3):162–72.

33. Lin K, de Araújo Filho GM, Pascalicchio TF, et al. Hippocampal atrophy and memory dysfunction in patients with juvenile myoclonic epilepsy. *Epilepsy Behav*. 2013;**29**(1):247–51.

34. Ronan L, Alhusaini S, Scanlon C, Doherty CP, Delanty N, Fitzsimons M. Widespread cortical morphologic changes in juvenile myoclonic epilepsy: evidence from structural MRI. *Epilepsia*. 2012;**53**(4):651–8.

35. Lin JJ, Dabbs K, Riley JD, et al. Neurodevelopment in new-onset juvenile myoclonic epilepsy over the first 2 years. *Ann Neurol*. 2014;**76**(5):660–8.

36. Caplan R, Levitt J, Siddarth P, et al. Frontal and temporal volumes in childhood absence epilepsy. *Epilepsia*. 2009;**50**(11):2466–72.

37. Chan CHP, Briellmann RS, Pell GS, Scheffer IE, Abbott DF, Jackson GD. Thalamic atrophy in childhood absence epilepsy. *Epilepsia*. 2006;**47**(2):399–405.

38. Tondelli M, Vaudano AE, Ruggieri A, Meletti S. Cortical and subcortical brain alterations in juvenile absence epilepsy. *YNICL*. 2016;**12**:306–11.

39. Lee CY, Tabesh A, Spampinato MV, Helpern JA, Jensen JH, Bonilha L. Diffusional kurtosis imaging reveals a distinctive pattern of microstructural alternations in idiopathic generalized epilepsy. *Acta Neurol Scand*. 2014;**130**(3):148–55.

40. Focke NK, Diederich C, Helms G, Nitsche MA, Lerche H, Paulus W. Idiopathic-generalized epilepsy shows profound white matter diffusion-tensor imaging alterations. *Hum Brain Mapp*. 2014;**35**(7):3332–42.

41. Liu M, Concha L, Beaulieu C, Gross DW. Distinct white matter abnormalities in different idiopathic generalized epilepsy syndromes. *Epilepsia*. 2011;**52**(12):2267–75.

42. Caeyenberghs K, Powell H, Thomas RH, et al. Hyperconnectivity in juvenile myoclonic epilepsy: a network analysis. *YNICL*. 2015;**7**:98–104.

43. Xue K, Luo C, Zhang D, et al. Diffusion tensor tractography reveals disrupted structural connectivity in childhood absence epilepsy. *Epilepsy Res.* 2014;**108**: 125–38.

44. Bernhardt BC, Bernasconi N, Concha L, Bernasconi A. Cortical thickness analysis in temporal lobe epilepsy: Reproducibility and relation to outcome. *Neurology.* 2010;**74**(22):1776–84.

45. Carney PW, Masterton RAJ, Harvey AS, Scheffer IE, Berkovic SF, Jackson GD. The core network in absence epilepsy. Differences in cortical and thalamic BOLD response. *Neurology.* 2010;**75** (10):904–11.

46. Hamandi K, Salek-Haddadi A, Laufs H, et al. EEG-fMRI of idiopathic and secondarily generalized epilepsies. *NeuroImage.* 2006;**31**(4):1700–10.

47. Dong L, Luo C, Zhu Y, et al. Complex discharge-affecting networks in juvenile myoclonic epilepsy: a simultaneous EEG-fMRI study. *Hum Brain Mapp.* 2016;**37**:3515–29.

48. Norden AD, Blumenfeld H. The role of subcortical structures in human epilepsy. *Epilepsy Behav.* 2002;**3** (3):219 31.

49. Li Q, Luo C, Yang T, et al. EEG-fMRI study on the interictal and ictal generalized spike-wave discharges in patients with childhood absence epilepsy. *Epilepsy Res.* 2009;**87**(2–3):160–8.

50. Blumenfeld H, Varghese GI, Purcaro MJ, et al. Cortical and subcortical networks in human secondarily generalized tonic-clonic seizures. *Brain.* 2008;**132**(4):999–1012.

51. Luo C, Li Q, Xia Y, et al. Resting state basal ganglia network in idiopathic generalized epilepsy. *Hum Brain Mapp.* 2011;**33**(6):1279–94.

52. Luo C, Xia Y, Li Q, et al. Diffusion and volumetry abnormalities in subcortical nuclei of patients with absence seizures. *Epilepsia.* 2011;**52**(6):1092–9.

53. Li Y, Du H, Xie B, et al. Cerebellum abnormalities in idiopathic generalized epilepsy with generalized tonic-clonic seizures revealed by diffusion tensor imaging. *PLOS ONE.* 2010;**5**(12):e15219.

54. Gottesman II, Gould TD. The endophenotype concept in psychiatry: etymology and strategic intentions. *Am J Psychiatry.* 2003;**160**(4):636–45.

55. Delgado-Escueta AV, Koeleman BP, Bailey JN, Medina MT, Durón RM. The quest for juvenile myoclonic epilepsy genes. *Epilepsy Behav.* 2013;**28**: S52–7.

56. Zifkin B, Andermann E, Andermann F. Mechanisms, genetics, and pathogenesis of juvenile myoclonic epilepsy: review. *Curr Opin Neurol.* 2005;**18**(2): 147–53.

57. Nair RR, Thomas SV. Genetic liability to epilepsy in Kerala State, India. *Epilepsy Res.* 2004;**62**(2):163–70.

58. Marini C, Scheffer IE, Crossland KM, et al. Genetic architecture of idiopathic generalized epilepsy: clinical genetic analysis of 55 multiplex families. *Epilepsia.* 2004;**45**(5):467–78.

59. Winawer MR, Marini C, Grinton BE, et al. Familial clustering of seizure types within the idiopathic generalized epilepsies. *Neurology.* 2005;**65**(4): 523–8.

60. Vijai J, Cherian PJ, Sylaja PN, Anand A, Radhakrishnan K. Clinical characteristics of a South Indian cohort of juvenile myoclonic epilepsy probands. *Seizure.* 2003;**12**(7):490–6.

61. Koepp MJ, Thomas RH, Wandschneider B, Berkovic SF, Schmidt D. Concepts and controversies of juvenile myoclonic epilepsy: still an enigmatic epilepsy. *Expert Rev Neurother.* 2014;**14**(7):819–31.

62. Bigos KL, Weinberger DR. Imaging genetics—days of future past. *NeuroImage.* 2010;**53**(3):804–9.

63. Siniatchkin M, Koepp M. Neuroimaging and neurogenetics of epilepsy in humans. *Neuroscience.* 2009;**164**(1):164–73.

64. Alhusaini S, Whelan CD, Sisodiya SM, Thompson PM. Quantitative magnetic resonance imaging traits as endophenotypes for genetic mapping in epilepsy. *YNICL.* 2016;**12**:526–34.

65. Blokland GA, McMahon KL, Thompson PM, Martin NG, de Zubicaray GI, Wright MJ. Heritability of working memory brain activation. *J Neurosci.* 2011;**31**(30):10882–90.

66. Spencer MD, Holt RJ, Chura LR, et al. Atypical activation during the Embedded Figures Task as a functional magnetic resonance imaging endophenotype of autism. *Brain.* 2012;**135**(11): 3469–80.

67. Callicott JH, Egan MF, Mattay VS, et al. Abnormal fMRI response of the dorsolateral prefrontal cortex in cognitively intact siblings of patients with schizophrenia. *Am J Psychiatry.* 2003;**160**(4): 709–19.

68. Wandschneider B, Centeno M, Vollmar C, et al. Motor co-activation in siblings of patients with juvenile myoclonic epilepsy: an imaging endophenotype? *Brain.* 2014;**137**:2469–79.

69. Levav M, Mirsky AF, Herault J, Xiong L, Amir N, Andermann E. Familial association of neuropsychological traits in patients with generalized and partial seizure disorders. *J Clin Exper Neuropsychol.* 2002;**24**(3):311–26.

70. Iqbal N, Caswell H, Muir R, et al. Neuropsychological profiles of patients with juvenile myoclonic epilepsy

and their siblings: an extended study. *Epilepsia*. 2015;56(8):1301–8.

71. Wandschneider B, Kopp UA, Kliegel M, et al. Prospective memory in patients with juvenile myoclonic epilepsy and their healthy siblings. *Neurology*. 2010;75(24):2161–7.

72. Helmstaedter C, Aldenkamp AP, Baker GA, Mazarati A, Ryvlin P, Sankar R. Disentangling the relationship between epilepsy and its behavioral comorbidities—the need for prospective studies in new-onset epilepsies. *Epilepsy Behav*. 2014;31: 43–7.

13 Prevention of Epileptogenesis in Animal Models

Chapter

Asht Mangal Mishra and Hal Blumenfeld

13.1 Introduction

Epileptogenesis is the process of development of epilepsy after exposure of the brain to initial precipitating injury in combination with genetic factors.[1] There is typically a time gap between the exposure of the brain to these influences and development of spontaneous recurrent seizures.[2] Epileptogenesis is difficult to study because patients are diagnosed after they have already developed spontaneous seizures and associated comorbidities. Based on brain regions involved, epilepsy is broadly classified into one of the following two categories: (1) generalized epilepsy, involving widespread brain regions in both hemispheres, or (2) focal epilepsy, where only select brain regions are involved. Approximately 50 million people worldwide are living with epilepsy.[3]

Because patients are usually diagnosed after full manifestation of epilepsy, it is difficult at the present time to prevent this disorder in humans. For this reason, we have tried treating epilepsy before it develops in rodent models as a proof-of-principle approach intended for eventual clinical translation. We succeeded in demonstrating primary prevention of a genetic form of epilepsy in an animal model. Primary prevention of epilepsy in an animal model is an important focus of this chapter, which we place in the broader context of neuroimaging epilepsy biomarkers.

Neuroimaging has a crucial role in the measurement of long-term structural and functional changes in the brains of both partial and generalized epilepsy patients noninvasively. Conventional neuroimaging methods comprise magnetic resonance imaging (MRI) morphometric analysis, proton magnetic resonance spectroscopy (^{1}H-NMR), and positron emission tomography (PET). To measure subtle structural and functional network changes during the process of epileptogenesis, additional advanced imaging techniques, such as resting functional connectivity in functional MRI (fMRI) and diffusion tensor MRI (DTI), are being used in animal models.

To cure epilepsy, we first need to understand epileptogenesis in animal models and in patients, and then we need therapies that can prevent or reverse epileptogenesis. Antiepileptic medicines have shown promise in suppressing primary generalized but not in secondary epilepsies.[4]

In this chapter we review neuroimaging research strategies to identify and prevent epileptogenesis in focal epilepsy in the first section and in generalized epilepsy in the second section. We review studies from well-validated animal models of both acquired and genetic epilepsies using various neuroimaging modalities. Focal seizures discussed in this chapter include febrile seizures, models of traumatic brain injury (TBI) and posttraumatic epilepsy (PTE), kindling of temporal brain regions, such as amygdala, for TLE model and proconvulsant-induced SE model, such as pilocarpine or kainic acid, and consequent spontaneous seizures modeling temporal lobe epilepsy (TLE). Our discussion of generalized epilepsy will focus on childhood absence epilepsy. Once we understand the underlying structural and functional abnormalities in epileptogenesis in animal models,[5,6] it may be possible to implement primary prevention with treatment drugs and to track underlying structural and functional changes in human patients, as we have done successfully in a generalized epilepsy animal model.[7,8]

13.2 Focal Epilepsy

13.2.1 Magnetic Resonance Imaging

MRI has been the diagnostic tool of epilepsy for many decades.[9,10] It can detect structural as well as functional changes in many type of seizures. Therefore, abnormalities on MRI have been used as a biomarker of epileptogenesis in other animal models of focal epilepsy. Febrile seizures (FSs) are a common type of seizures in children that are associated with a sudden rise in body temperature.[11] The hyperthermia-induced prolonged FSs model in rat shows that T2 signals from serial MRIs before and after seizure onset were increased in the dorsal hippocampus, piriform cortex, and amygdala by 75% and 87.5% of these rats one and eight days after seizures, respectively.[12] In another rat study, hippocampal T2 signal increased at 1 month[13] after the induction of FSs. These abnormal T2 signals have an increased likelihood of developing spontaneous seizures,[12,13] but itself was not found to be a predictor of epileptogenesis[13] (Figure 13.1). Febrile status epilepticus (FSE) shows reduced T2 relaxation times in amygdala on high-magnetic-field MRI hours after FSE and predicted experimental TLE.[14] These reduced T2 signals in temporal lobe structures, like amygdala, signify injury-sensor high-mobility group box 1, a trigger of inflammatory cascades, and indicate underlying changes in neuronal integrity that promote epileptogenesis and provide early markers for subsequent TLE development.[14]

PTE is another type of partial epilepsy that results from an insult to the brain from an external mechanical force, leading to TBI with permanent damage to brain cells. Periods of increased abnormal plasticity occur in the process of posttraumatic epileptogenesis. This change mimics the critical period of development.[15,16] Quantitative T2 image signal evaluation has been used to analyze the severity of cortical damage and neural and motor impairment in a lateral fluid percussion injury (LFPI) rat model of TBI, and T2 signal intensities are correlated with motor deficits and histological lesion volume and also differentiate moderate from severe TBI.[17] Conventional and advanced quantitative MRI changes, such as T2, T1ρ, and diffusion tensor metrics acquired after trauma in the LFPI rat model, predict functional and pathological changes several months after trauma.[18] These results show that early changes in the hippocampus and perifocal brain regions may alert us to long-term brain changes after trauma.

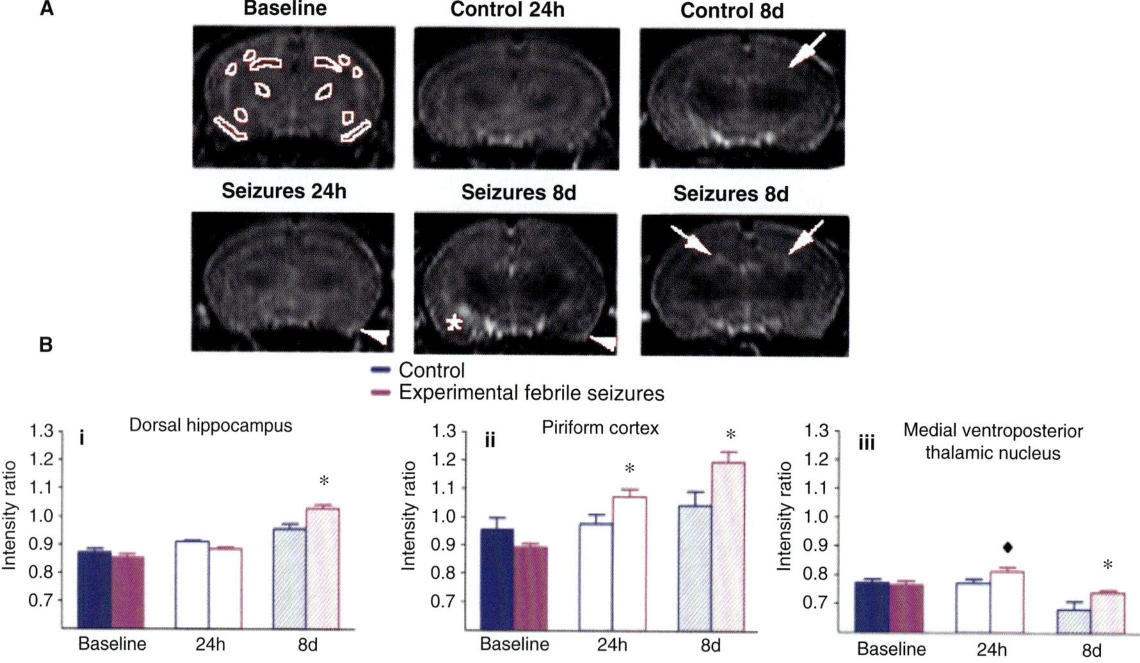

Figure 13.1. T2-weighted MR imaging signal intensities after prolonged experimental febrile seizures (FSs) in rat. (A) The control and the seizure animal groups were imaged before the seizures induction on postnatal day 10, P10, to provide baseline intensities, and on P12 (24 hours after seizure induction, seizure group) as well as 8 days after the seizures. Regions used for analysis are delineated in white in baseline image. Increased signal intensity can be appreciated in the dorsal hippocampus (white arrows), amygdala (asterisk), and piriform cortex (white arrowheads). (B) T2-weighted signal intensity modulation by prolonged experimental FSs: (i) a significant increase of T2-weighted signal occurred in dorsal hippocampus, by 8 days after the seizures, (ii) in piriform cortex, signal intensity increased significantly at 24 hours and 8 days after seizure, (iii) in the medial ventroposterior thalamic nucleus, signal intensity increased significantly at 8 days and there was a strong trend for increased signal at 24 hours (diamond) after seizure. Reproduced from Dubé et al.[12] with permission.

In lithium-pilocarpine-induced status epilepticus (SE), rats without MRI abnormalities or altered T2 relaxation times did not end up developing epilepsy, whereas those with increases in T2 relaxation times and abnormalities in the T2-weighted images, specifically at the level of piriform and entorhinal cortices, subsequently developed epilepsy at 4 months after induction.[19] Apart from these findings, other studies have also documented more widespread T2 signal decreases in the acute stages after seizure induction and associated with spontaneous seizures in a later stage.[20] Acute MRI changes in the parietal cortex, hippocampus, piriform cortex, and thalamus after lithium-pilocarpine-induced seizure have been shown to be predictive of later hippocampal damage.[21] These findings suggest temporal intricacies in T2 signal changes post-SE partial epilepsy models likely predict the evolving processes of epileptogenesis.[20,22] Similar to chemical-induced SE models, SE induced by electrical stimulation of the rat amygdala resulted in increased T2 signal intensity in the amygdala beginning 2 days after SE and pathologic T2 was observed in the piriform cortex, midline thalamus, and hippocampus. These initial MRI changes showed a low predictive value for the subsequent severity of epilepsy,[23] and acutely increased T2 signal intensity in the hippocampus, results in acquired epileptogenesis in limbic system.[24] These MRI signal changes have the potential to become a clinically useful biomarker after careful further study.

13.2.2 MRI Morphometric Analysis

Conventional MRI can delineate different brain structures on the basis of gray and white matter structures within the brain. MRI morphometry and volumetry take advantage of this property of MRI for image analysis to compare the size and shape of various brain structures. Voxel-based morphometry (VBM) is performed by spatially normalizing MRI images across all the subjects in the study into the same stereotactic space, segmenting the gray matter from the spatially normalized images and smoothing the gray matter segments, and then performing voxel-wise statistical comparison of the smoothed gray matter images from the two groups to produce a parametric map of structural regions.[25] Initial studies using VBM analysis included a study of hippocampi of London taxi drivers showing that the posterior hippocampi of taxi drivers were larger than those of control participants.[26] VBM analysis has been used to study epilepsy and other brain disorders. Volumetry is another forms of analysis where the volumes of regions of interest are calculated for comparison. VBM and volumetric changes in the limbic structures like CA1 and dentate gyrus in the hippocampus have been reported in SE[21,24,27] as well as in the LFPI model of PTE.[28–31] These techniques may play a role in assessing structural changes in the cortical and subcortical structures compromised in epileptogenesis.

13.2.3 Proton Magnetic Resonance Spectroscopy

^{1}H NMR is a technique for identification and quantitation of the metabolites in healthy and brain with various diseases.[32,33] Major metabolites detected by this technique are N-acetylaspartate (NAA), creatine (Cr), phospholipids with choline (Cho), lactate (Lac), myo-inositol (mIns), and glutathione (GSH). NAA is a marker of neuronal and axonal change; Cr and/or phosphocreatine show brain energy metabolism; Cho shows membrane turnover, inflammation, or demyelination; Lac is a marker of anaerobic glycolysis;[33] mIns and GSH are considered important markers of inflammation.[34] Imbalance of excitatory and inhibitory neurotransmitters like glutamate, glutamine, and gamma-aminobutyric acid (GABA) likely show the process of epileptogenesis.[35,36] Acute and chronic reductions in NAA[35,37] and reductions in excitatory and inhibitory neurotransmitters[38] have been shown in the post-SE rat model of TLE. Increases in mIns and GSH have been shown after seizure induction in rats.[35] The above findings suggest capability of PMRS in identifying biomarkers associated with structural and/or functional changes occurring in the epileptogenic process.

13.2.4 Positron Emission Tomography

PET measures emissions from a large number of metabolically active radiotracers. PET visualizes distribution of the tracer by visualizing receptor density in the brain and identifies pathology with altered metabolism in epilepsy. For the same reason, PET was applied to epilepsy research several years after it was first developed.[39] Recent studies in the field show many radiotracers are helpful in identifying epileptogenesis in rodent models of epilepsies.[40–44] The rat model of TLE shows severe glucose hypometabolism associated with epileptogenesis.[45]

13.2.5 Functional Magnetic Resonance Imaging

Functional MRI uses paramagnetic properties of deoxyhemoglobin in the blood to generate MRI signal. Because the level of deoxyhemoglobin changes with brain function, this property of hemoglobin was used to develop a new MRI image contrast named blood-oxygen-level-dependent (BOLD) fMRI by Ogawa and colleagues.[46] Since EEG detects location and magnitude of brain activity, EEG-correlated fMRI has been used to investigate epileptogenic networks in focal epilepsy.[47,48] Epilepsy is known to involve abnormal neural networks,[47] and fMRI can detect neuronal networks involved during seizures. The chemical-induced seizure model shows robust fMRI changes correlated with seizure.[49] The above-

mentioned studies paved the way to use fMRI as a biomarker to track disease progression and treatment efficacy in focal epilepsy.

13.2.6 fMRI with Resting Functional Connectivity

Functional connectivity is an fMRI technique where temporal correlation of a neurophysiological signal is measured from remotely located brain areas.[50] Biswal and colleagues demonstrated correlation of low frequency (<0.1 Hz) changes in the primary sensory cortex induced by hand movement both within and between hemispheres.[51] In our recent study of the LFPI rat model of PTE, we reported decreased connectivity between the ipsilateral and contralateral parietal cortex and between the parietal cortex and hippocampus on the side of injury as compared to sham-operated animals (Figure 13.2)[6] four months after induction of injury. These injured animals have increased seizure susceptibility

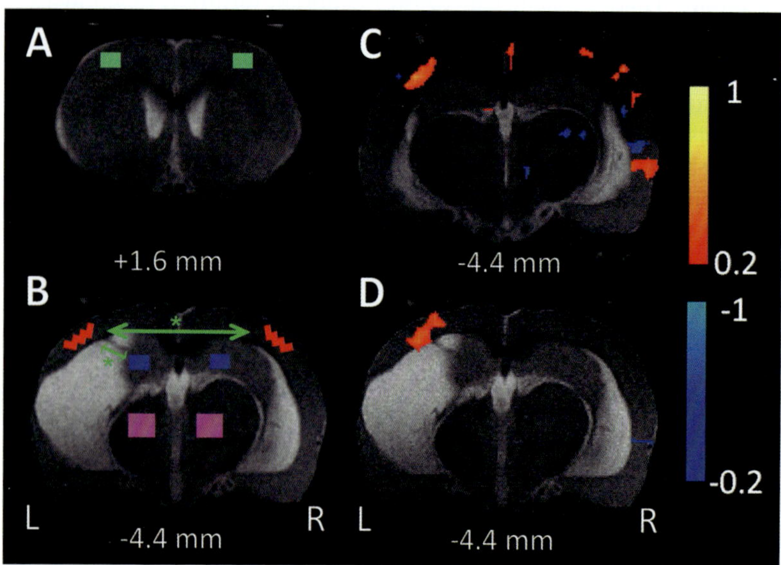

Figure 13.2. Decreased resting functional connectivity in TBI rats vs. sham-operation control rats. Two coronal MRI slices (A, B) from a rat with lateral fluid percussion injury (FPI) (note the tissue loss in the left hemisphere) demonstrating the brain regions of interest (ROIs) that were used on coronal BOLD-fMRI images for resting BOLD-fMRI signal correlation analysis. Four ROIs were made in each hemisphere (8 ROIs total): frontal cortex, parietal cortex, hippocampus, and thalamus (L, left; R, right). Panel B shows schematically using arrows the significant differences in connectivity (1) between the ipsilateral and contralateral parietal cortices and (2) between the ipsilateral parietal cortex and hippocampus in the rats with lateral FPI and control rats. Statistical significance: *p < .05. (C) Example of resting functional connectivity in a control rat. Pearson correlation values are shown using the left parietal cortex ROI (B) correlated to the whole brain. Some positive correlation is seen with the contralateral parietal cortex and ipsilateral hippocampus. (D) Example of the same analysis for the left parietal cortex ROI in a TBI animal shows reduced connectivity. Warm colors represent increases and cool colors decreases in resting-state connectivity; scale bars are for Pearson correlation with display threshold = 0.2. Voxels lying outside the brain or in CSF are not shown. Reproduced from Mishra et al.[6] with permission.

evaluated by the pentylenetetrazole test. Injured animals also show abnormal negative connectivity between the ipsilateral and contralateral parietal cortex. This is the first evidence on decreased functional connectivity after experimental TBI.[6] These abnormal connectivity changes can serve as a useful biomarker for identifying individuals with a risk of developing PTE and may be useful for monitoring prevention of epileptogenesis by early treatment.

13.2.7 Diffusion Imaging

Diffusion-weighted imaging (DWI) measures the magnitude of diffusivity of water protons, whereas DTI measures both the magnitude and the directionality.[52] Water diffusion is hindered by extracellular as well as intracellular microstructures, and therefore these techniques can detect subtle changes in the pathological brain[5,53,54] where magnitude and directionality of water diffusion is altered. Apparent diffusion coefficient (ADC) is an average measure of magnitude of water diffusion in biological tissue. One of the several metrics of DTI is fractional anisotropy (FA), which measures the degree of anisotropy of water diffusion in a tissue under observation and reflects alignment of cellular structures within the tissue like white matter fiber tracts[5,54] and other cellular structures that simulate white matter fibers.[55] It has a value of 0 for least anisotropic tissue and 1 for most anisotropic.[52,56] In an animal model of PTE, diffusion values have been predictive of increased seizure susceptibility.[29] The FPI-induced rat model of moderate TBI shows that hippocampal DWI correlate with EEG parameters and with mossy fiber sprouting density in the ipsilateral hippocampus at early as well as chronic time points after TBI.[29] ADC changes are also correlated with seizure susceptibility in FPI model of PTE.[29,57]

White matter fiber tractography is another DTI method of generating a map of connection of white matter fibers traveling from one to another direction in the brain by using diffusion images acquired in multiple directions. This technique uses known white matter fiber bundle as a starting point, which allows assessment of white matter integrity neighboring to that start point called seed region. Decreased ADC and increased FA have been reported to be associated with microstructural alterations in the hippocampus of rats after FSs.[58] The rat model of status epilepticus induced by pilocarpine shows that DWI is a sensitive technique for the early identification of seizure-induced neuronal cell damage.[59] A decrease

in FA was seen in multiple brain areas and was associated with altered tissue microstructure in the kainic-acid-induced rat model of epilepsy.[60]

All the above studies show that DWI and DTI measurements are sensitive in detecting epilepsy-related changes during epileptogenesis.[61]

13.3 Generalized Epilepsy

In this section we discuss imaging findings in generalized epilepsy focusing on childhood absence epilepsy (CAE). Widespread cortical and subcortical network impairment leads to transient altered consciousness in absence seizures. Chronic resting-state dysfunction in the cortical and subcortical network may lead to interictal psychosocial comorbidities in absence epilepsy.

Early and effective seizure treatment in rodent models and possibly human patients can have long-term beneficial effects on both seizures and epilepsy comorbidities. Early and sustained treatment of seizures with ethosuximide (ESX) in a rodent absence epilepsy model before seizure onset led to markedly reduced seizures, which continued for at least 3 months after termination of treatment[68] (Figure 13.3). Subsequent work examined developmental changes in network connectivity related to epileptogenesis[5,6,69] as well as activity-dependent changes in ion channel expression during the course of epileptogenesis.[68,70,71] We found that early treatment in at least two rodent absence models and with antiseizure medication led to beneficial effects not only on seizures, but also on epileptogenesis-related structural and molecular changes as well as behavioral comorbidities.[7,8,72]

Effective treatment with ESX may also have a beneficial disease-modifying effect in human patients with absence epilepsy, based on improved long-term seizure remission off medications.[73] With better genetic understanding of epilepsy it will hopefully soon be possible to initiate treatment prior to disease onset with the goal of improving long-term outcome.

Above results are suggestive of a "critical period" early in development during which treatment can block both molecular and electrophysiological symptoms of epileptogenesis. With the help of noninvasive biomarkers we can monitor progression of epileptogenesis severity and efficacy of therapy after a particular treatment.

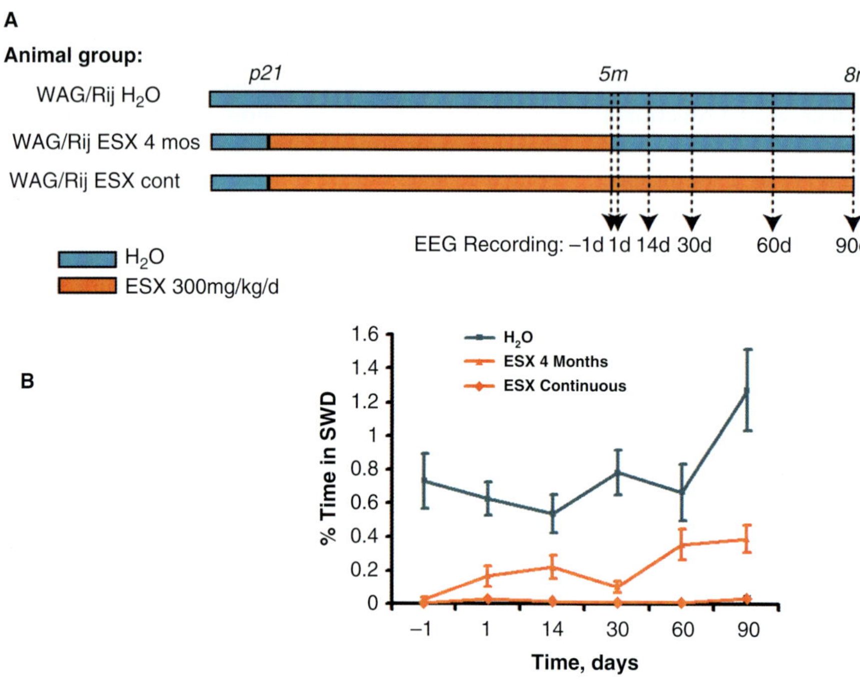

Figure 13.3. Early ethosuximide (ESX) treatment persistently suppressed the development of spike-wave discharges (SWDs) in WAG/Rij rats, even after stopping treatment. (A) Epileptic WAG/Rij rats were given either normal drinking water (H_2O group), ESX from age p21 through age 5 months and then normal drinking water from age 5 to 8 months (ESX 4-month group), or ESX continuously from age p21 through 8 months (ESX continuous group). EEG was recorded 1 day before stopping ESX and 1, 14, 30, 60, and 90 days after stopping ESX. (B) Quantification of effects of early ESX treatment on percentage time in SWD. Even after stopping ESX, percentage time in SWD remained markedly reduced in the treated rats (ESX 4-month group) when comparing all time points for days 1 through 90 to rats on normal H_2O. Reproduced from Blumenfeld et al.[8] with permission.

13.3.1 MRI Morphometric Analysis

VBM can reveal subtle structural changes, when conventional MRI scans reveal normal anatomy. VBM show gray matter abnormalities in the thalamus and frontal cortex, supporting the role of the thalamocortical network in the mechanisms of generalized seizures.[74] VBM should be checked for consistency in user-specific parameters like type and level of statistical correction, modulation, smoothing kernel, adjustment for brain size, subgroup analysis and software version before it can be used clinically as a biomarker.[75,76]

13.3.2 Functional MRI

fMRI studies in rat models of genetic absence epilepsy report BOLD fMRI signal abnormalities during spontaneous spike-wake discharges.[47,77,78] Increases were most prominent in focal regions of somatosensory cortex, motor cortex and thalamus, while the occipital region was spared.[78–80] These regions were shown to have both increased neuronal firing and increased CBF during SWD.[78,80] There are some inconsistencies in findings like BOLD fMRI increase versus decrease[81] and anesthesia among some

limitations.[47] In spite of these variations, fMRI has been proved to be useful in identifying functional abnormalities in epileptogenesis.

fMRI studies of bicuculline-induced generalized tonic-clonic seizures show that BOLD fMRI increases were still largest in somatosensory cortex, while decreases were seen in the hippocampus.[77,82,83]

Blumenfeld and colleagues have performed simultaneous EEG, fMRI, and behavioral testing during typical childhood absence seizures to demonstrate that seizures with impaired consciousness exhibit widespread bilateral changes in cortical and thalamic networks.[62,63] In the interictal period, children with absence epilepsy exhibit heightened features of attention impairment, anxiety, and depression,[64,65] which may be related to long-term alterations in resting network functional connectivity.[66,67] Studies of CAE patients show fMRI increases in the bilateral thalamus and occipital cortex, while decreased fMRI signals were observed in the bilateral lateral parietal cortex, precuneus, cingulate gyrus, and basal ganglia.[62,66,84–86] The fMRI time course signal changes of childhood absence seizure show small early fMRI increases in the medial/lateral parietal cortex and orbital/medial frontal >5 seconds before seizure onset, followed by intense fMRI decreases continuing up to 20 s after seizure offset.[63] This dynamic pattern of fMRI signal changes shows the temporal complexity of seizure-related fMRI changes. These changes were not detected by hemodynamic response function modeling. These studies show that identification of a seizure network via these fMRI changes provides potential targets for therapy, and for monitoring of therapeutic efficacy.

13.3.3 Functional MRI with Resting Functional Connectivity

fMRI-based resting functional connectivity is a noninvasive method for assessing brain network connectivity.[87,88] This technique can easily be performed and translated from animal studies to use in human epilepsy patients. Resting functional connectivity is useful in studying long-range interactions. Since generalized epilepsy such as absence seizure involves the thalamocortical network, it may be a robust method for measuring long-term changes, even when seizures are not occurring. It will also be useful to assess network changes during treatment period when seizures are blocked by medication.

In WAG/Rij rats, we observed increased cortical-cortical correlations at rest, when no seizures were present.[69] These effects were not present in nonepileptic controls (Figure 13.4). We see strongest connectivity between regions most intensely involved in seizures, predominantly in the bilateral somatosensory and adjacent cortices. Group statistics revealed higher resting interhemispheric cortical-cortical correlations in WAG/Rij rats compared to nonepileptic controls (Figure 13.4). Our findings suggest that activity-dependent plasticity may lead to long-term changes in epileptic networks that can be detected even at rest. With the help of further studies, we should be able to determine whether preventing epileptogenesis, with antiepileptic drugs, will also prevent the above-mentioned abnormal functional connectivity in epileptic networks in this rat model.[8] A study of human absence epilepsy children showed significantly increased interhemispheric correlation between orbitofrontal cortex regions.[66] The above-mentioned studies showed that abnormalities were observed in both animal model and human CAE patients during resting state. Therefore, these findings have potential to

serve as an interictal biomarker of seizure severity in CAE and possibly in other generalized epilepsies.

13.3.4 Diffusion Tensor Imaging

DTI measures anisotropy of water diffusion in biological systems. Water diffusion is highly anisotropic in white matter structures[56] in the nervous system because of its hindrance in the perpendicular direction by the axonal membrane and myelin sheath, which jointly modulate the degree of anisotropy. White matter microstructures are altered in epilepsy; for instance, infantile spasm, a form of pediatric epilepsy, has been associated with abnormal myelination.[89,90] Therefore, DTI detects subtle white matter changes in epilepsy and other disorders where white matter is affected.[91–93]

We studied beneficial effects of chronic and early pharmacological treatment with ESX on epileptogenesis in two genetic absence epilepsy models. In our study of WAG/Rij rat model of absence seizures, we showed reduced FA and increased perpendicular diffusivity in the anterior corpus callosum, which was absent in young rats before seizure manifestation or in nonepileptic controls[5] (Figure 13.5A). Genetic Absence Epilepsy Rats from Strasbourg (GAERS) show more pronounced reduced FA and increased perpendicular diffusivity in anterior corpus callosum and internal capsule.[5] The white matter structural changes we see can be due to a reduction in myelin[94, 95] and/or decreased axon fiber density.[96,97] They may also be due to changes in the density and orientation of crossing fibers in pathways connecting regions of seizure activity.

In our recent study we showed antiepileptogenesis in WAG/Rij rats. These rats received either ESX for 2 months (postnatal months 2–3 or 4–5) or ESX for 4 months[2–5] in their drinking water. Control rats drank plain water. We measured EEG during treatment, and 6 days and 2 months posttreatment. A behavioral test was performed 6 days posttreatment. Ex vivo DTI was performed after treatment. SWD were suppressed during treatment, 6 days, and 2 months posttreatment in the 4-month group. Duration of afterdischarges elicited by cortical electrical stimulation 6 days posttreatment was also suppressed. Increased DTI metrics FA in corpus callosum and internal capsule were found[7] (Figure 13.5B–C). We didn't see large effects on any parameters after shorter treatments with ESX. Finally, we showed widespread effects of chronic ESX treatment not only within but also outside the circuitry in which SWD are initiated and generated,

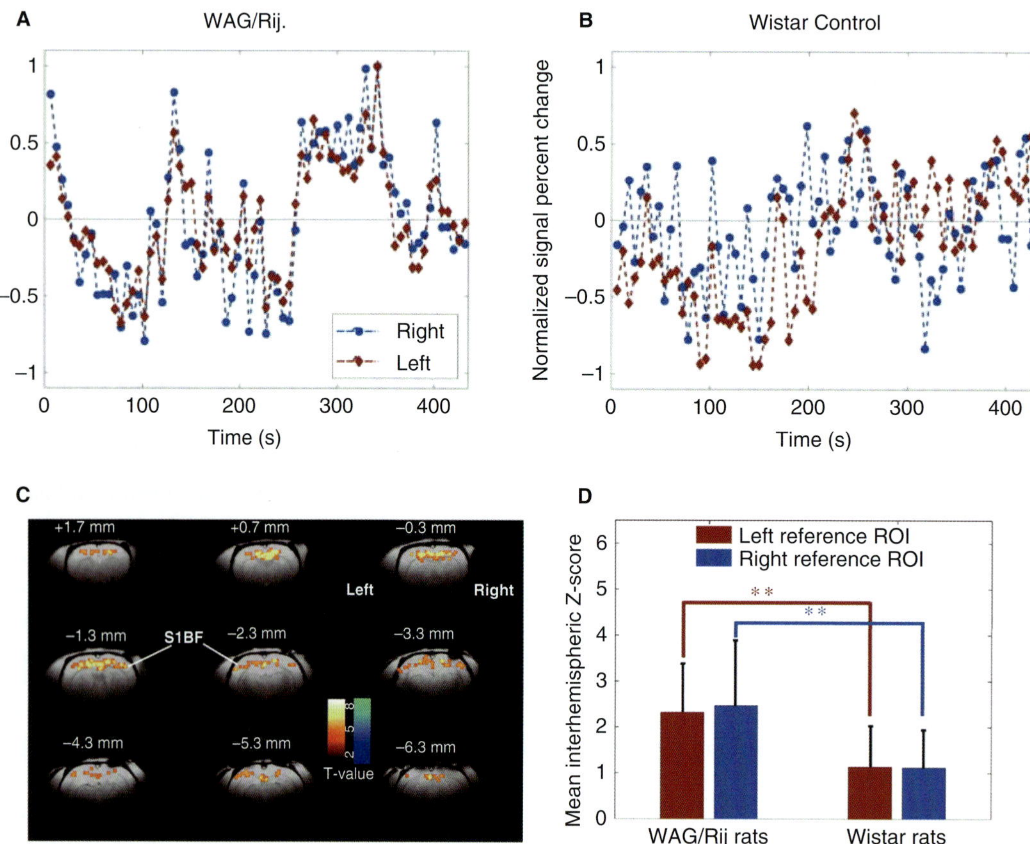

Figure 13.4. Resting functional connectivity differences in epileptic WAG/Rij rats versus Wistar control rats. (A) Examples of left and right mean fMRI signal time courses in reference regions at rest when SWDs are not occurring. WAG/Rij rat (A) shows high correlation between slow (<0.1 Hz) changes in left and right reference regions. Nonepileptic Wistar control rat (B) shows less interhemispheric correlation in left and right reference regions. (C) WAG/Rij rats show increased connectivity in contralateral somatosensory and adjacent cortices. Left cortex reference region of interest versus whole brain. Increases are in warm colors and decreases are in cold colors. Very similar results were obtained using the right cortex ROI as reference. Wistar control rats show only localized connectivity in the ipsilateral (left) somatosensory cortex (data not shown). (D) Interhemispheric resting-state cortical connectivity. Increased interhemispheric resting cortical connectivity in WAG/Rij rats (mean ± standard deviation of z scores, from left and right brain regions) compared to nonepileptic Wistar controls. **$p = .01$, one-way ANOVA followed by Tukey's HSD method for post hoc pairwise comparisons. Results were very similar using either the left ROI time course as reference (red) and calculating mean z score for all voxels in the right ROI, or using the right ROI time course as reference (blue). Reproduced from Mishra et al.[69] with permission.

including preventing epileptogenesis and reducing depressive-like symptoms. Our findings suggest that chronic ESX treatment of seizures in the cortex in absence epilepsy may lead to recovery of microstructural changes in white matter pathways during repeated seizure discharges.

The animal models will enable histological studies to determine microstructural mechanisms for changes observed on DTI. In addition, DTI can also be a potential imaging biomarker for monitoring treatment by showing structural changes in affected brain regions in epilepsy.

13.4 Conclusions

The prevention of epileptogenesis has been shown in a few animal models of generalized epilepsy, whereas the process of epileptogenesis has been studied in numerous additional animal models. Advanced neuroimaging methods enable the detection of subtle changes in brain networks during epileptogenesis by using network-based methods such as DTI, fMRI, and fMRI with resting functional connectivity, which are being further developed and applied to understand clinical cases in human epilepsy. Use of these imaging techniques as

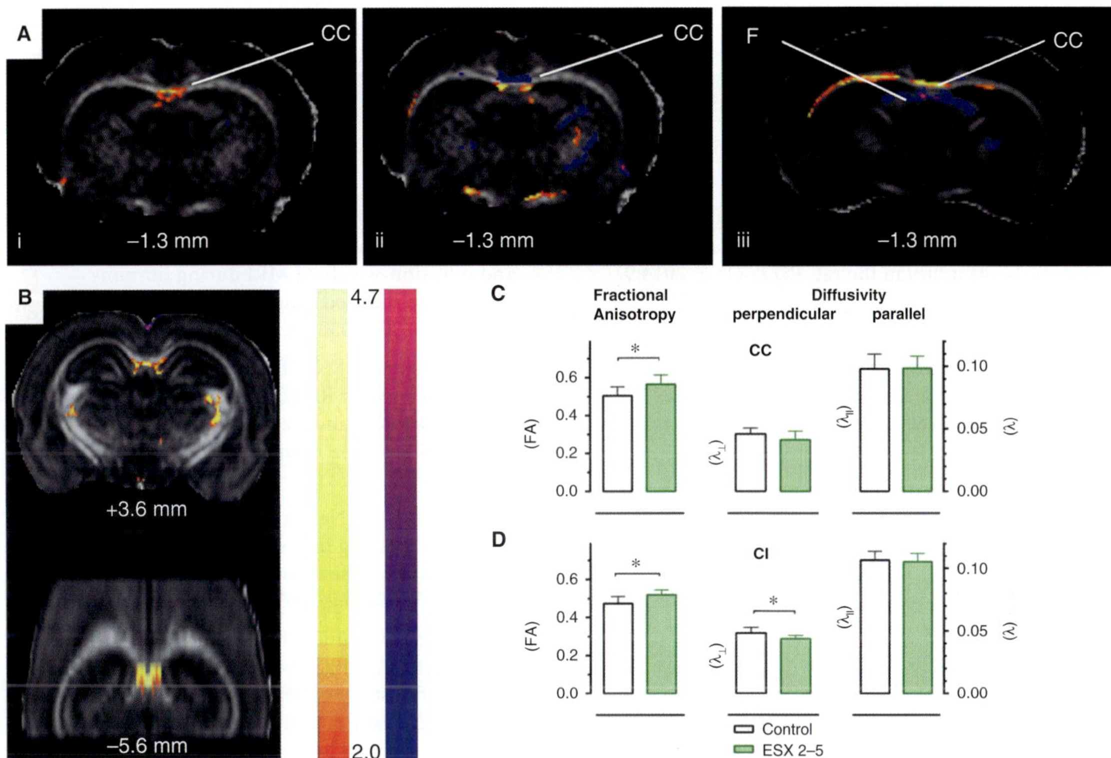

Figure 13.5. Prevention of epileptogenesis in WAG/Rij rats. (A) Epileptic adult WAG/Rij (8 months) and GAERS (1.7 months) rats, but not young WAG/Rij rats (1.7 months), have decreased fractional anisotropy (FA) in anterior corpus callosum compared to controls. T-maps at −1.3 mm from bregma are shown, with warm colors representing decreased FA when compared with nonepileptic controls, and cool colors represent increased FA. (i) In adult epileptic WAG/Rij rats, decreased FA was observed in the anterior corpus callosum (CC). (ii) In young WAG/Rij rats, anterior corpus callosum does not show decreased FA when compared with controls. (iii) In adult epileptic GAERS, extensive decreased FA was observed in the anterior corpus callosum. Unlike the WAG/Rij rats, the fornix showed increased FA in GAERS when compared with controls, which may represent a strain difference not directly related to seizures. t value threshold = 2.00, extent threshold = 50 voxels, and FA threshold = 0.30. (B) Chronic ESX treatment improves white matter FA based on midline CC and in the internal capsule (IC) compared to age-matched. Image is horizontal slice extending from AP +3.6 mm to −5.6 mm from bregma. (C) Midline corpus callosum shows significantly increased FA, and a trend in decreased perpendicular diffusivity (λ) in ESX treated WAG/Rij rats compared to untreated control WAG/Rij rats, with no change in parallel diffusivity ($\lambda\|$). (D) Internal capsule also exhibited significantly increased FA, along with significantly decreased λ in ESX treated WAG/Rij rats compared to untreated controls, with no change in $\lambda\|$. Reproduced from Chahboune et al.[5] and from Van Luijtelaar et al.[7] with permission.

biomarkers of epileptogenesis may ultimately help in studying the prevention of epileptogenesis in people with epilepsy, leading to a marked improvement in clinical outcome and quality of life.

Acknowledgments

This work was supported by grant NIH R01 NS049307, by the Epilepsy Foundation Award ID 123505 (AMM), and by the Betsy and Jonathan Blattmachr family.

References

1. Shorvon SD. The etiologic classification of epilepsy. *Epilepsia.* 2011;52:1052–7.

2. Giblin KA, Blumenfeld H. Is epilepsy a preventable disorder? New evidence from animal models. *Neuroscience.* 2010;16(3):253–75.

3. Banerjee PN, Filippi D, Allen Hauser W. The descriptive epidemiology of epilepsy—a review. *Epilepsy Res.* 2009;85:31–45.

4. Blumenfeld H. New strategies for preventing epileptogenesis: perspective and overview. *Neurosci Lett.* 2011;**497**(3):153–4.

5. Chahboune H, Mishra AM, DeSalvo MN, et al. DTI abnormalities in anterior corpus callosum of rats with spike-wave epilepsy. *NeuroImage.* 2009;**47**(2): 459–66.

6. Mishra AM, Bai X, Sanganahalli BG, Waxman SG. Decreased resting functional connectivity after traumatic brain injury in the rat. *PLOS ONE.* 2014;**9** (4):e95280.

7. Van Luijtelaar G, Mishra AM, Edelbroek P, et al. Anti-epileptogenesis: electrophysiology, diffusion tensor imaging and behavior in a genetic absence model. *Neurobiol Dis.* 2013;**60**:126–38.

8. Blumenfeld H, Klein JP, Schridde U, et al. Early treatment suppresses the development of spike-wave epilepsy in a rat model. *Epilepsia.* 2008;**49**(3):400–9.

9. Shorvon SD. A history of neuroimaging in epilepsy 1909–2009. *Epilepsia.* 2009;**50**:39–49.

10. Filler AG. The history, development and impact of computed imaging in neurological diagnosis and neurosurgery: CT, MRI, and DTI. *Nat Proc.* 2009. Available at:http://precedings.nature.com/documents/3267/version/5.

11. Patterson JL, Carapetian SA, Hageman JR, Kelley KR. Febrile seizures. *Pediatr Ann.* 2013;**42**:249–54.

12. Dubé C, Yu H, Nalcioglu O, Baram TZ. Serial MRI after experimental febrile seizures: altered T2 signal without neuronal death. *Ann Neurol.* 2004;**56**(3): 709–14.

13. Dube M, Ravizza T, Hamamura M, et al. Epileptogenesis provoked by prolonged experimental febrile seizures: mechanisms and biomarkers. *J Neurosci.* 2010;**30**(22):7484–94.

14. Choy M, Dubé CM, Patterson K, et al. Neurobiology of disease: a novel, noninvasive, predictive epilepsy biomarker with clinical potential. *J Neurosci.* 2014;**34** (26):8672–84.

15. Pagni CA, Zenga F. Prevention and treatment of post-traumatic epilepsy. *Expert Rev Neurother.* 2006;**6** (8):1223–33.

16. Pitkänen A, McIntosh T. Animal models of post-traumatic epilepsy. *J Neurotrauma.* 2006;**23**(2): 241–61.

17. Kharatishvili I, Sierra A, Immonen RJ, Gröhn OHJ, Pitkänen A. Quantitative T2 mapping as a potential marker for the initial assessment of the severity of damage after traumatic brain injury in rat. *Exp Neurol.* 2009;**217**(1):154–64.

18. Immonen RJ, Kharatishvili I, Gröhn H, Pitkänen A, Gröhn OHJ. Quantitative MRI predicts long-term structural and functional outcome after experimental traumatic brain injury. *NeuroImage.* 2009;**45**(1):1–9.

19. Roch C, Leroy C, Nehlig A, Namer IJ. Predictive value of cortical injury for the development of temporal lobe epilepsy in 21-day-old rats: an MRI approach using the lithium-pilocarpine model. *Epilepsia.* 2002;**43**(10): 1129–36.

20. van Eijsden P, Notenboom RGE, Wu O, et al. In vivo 1H magnetic resonance spectroscopy, T2-weighted and diffusion-weighted MRI during lithium-pilocarpine-induced status epilepticus in the rat. *Brain Res.* 2004;**1030**(1):11–8.

21. Choy M, Cheung KK, Thomas DL, Gadian DG, Lythgoe MF, Scott RC. Quantitative MRI predicts status epilepticus-induced hippocampal injury in the lithium-pilocarpine rat model. *Epilepsy Res.* 2010;**88** (2–3):221–30.

22. Roch C, Leroy C, Nehlig A, Namer IJ. Magnetic resonance imaging in the study of the lithium-pilocarpine model of temporal lobe epilepsy in adult rats. *Epilepsia.* 2002;**43**(4):325–35.

23. Nairismägi J, Gröhn OHJ, Kettunen MI, Nissinen J, Kauppinen RA, Pitkänen A. Progression of brain damage after status epilepticus and its association with epileptogenesis: a quantitative MRI study in a rat model of temporal lobe epilepsy. *Epilepsia.* 2004;**45**(9): 1024–34.

24. Jupp B, Williams JP, Tesiram YA, Vosmansky M, O'Brien TJ. Hippocampal T2 signal change during amygdala kindling epileptogenesis. *Epilepsia.* 2006;**47** (1):41–6.

25. Ashburner J, Friston KJ. Voxel-based morphometry—the methods. *NeuroImage.* 2000;**11**(6 pt 1):805–21.

26. Maguire EA, Gadian DG, Johnsrude IS, et al. Navigation-related structural change in the hippocampi of taxi drivers. *Proc Natl Acad Sci USA.* 2000;**97**(8):4398–403.

27. Wolf OT, Dyakin V, Patel A, et al. Volumetric structural magnetic resonance imaging (MRI) of the rat hippocampus following kainic acid (KA) treatment. *Brain Res.* 2002;**934**(2):87–96.

28. Shultz SR, Cardamone L, Liu YR, et al. Can structural or functional changes following traumatic brain injury in the rat predict epileptic outcome? *Epilepsia.* 2013;**54** (7):1240–50.

29. Kharatishvili I, Immonen R, Gro O, Pitka A, Gröhn O, Pitkänen A. Quantitative diffusion MRI of hippocampus as a surrogate marker for post-traumatic epileptogenesis. *Brain.* 2007;**130**(130):3155–68.

30. Liu YR, Cardamone L, Hogan RE, et al. Progressive metabolic and structural cerebral perturbations after traumatic brain injury: an in vivo imaging study in the rat. *J Nucl Med.* 2010;**51**(11):1788–95.

31. Immonen R, Kharatishvili I, Gröhn O, Pitkänen A, Pitkanen A. MRI biomarkers for post-traumatic epileptogenesis. *J Neurotrauma*. 2013;**30**:1305–9.

32. Gupta RK, Cloughesy TF, Sinha U, et al. Relationships between choline magnetic resonance spectroscopy, apparent diffusion coefficient and quantitative histopathology in human glioma. *J Neurooncol*. 2000;**50**(3):215–26.

33. Mishra AM, Gupta RK, Jaggi RS, et al. Role of diffusion-weighted imaging and in vivo proton magnetic resonance spectroscopy in the differential diagnosis of ring-enhancing intracranial cystic mass lesions. *J Comput Assist Tomogr*. 2004;**28**(4):540–7.

34. Nagarajan R, Sarma MK, Thames AD, Castellon SA, Hinkin CH, Thomas MA. 2D MR spectroscopy combined with prior-knowledge fitting is sensitive to HCV-associated cerebral metabolic abnormalities. *Int J Hepatol*. 2012;**2012**:179365.

35. Filibian M, Frasca A, Maggioni D, Micotti E, Vezzani A, Ravizza T. In vivo imaging of glia activation using 1H-magnetic resonance spectroscopy to detect putative biomarkers of tissue epileptogenicity. *Epilepsia*. 2012;**53**(11):1907–16.

36. Duncan J. Magnetic resonance spectroscopy. *Epilepsia*. 1996;**37**(7):598–605.

37. Lee EM, Park GY, Im KC, et al. Changes in glucose metabolism and metabolites during the epileptogenic process in the lithium-pilocarpine model of epilepsy. *Epilepsia*. 2012;**53**(5):860–9.

38. Alvestad S, Hammer J, Qu H, Håberg A, Ottersen OP, Sonnewald U. Reduced astrocytic contribution to the turnover of glutamate, glutamine, and GABA characterizes the latent phase in the kainate model of temporal lobe epilepsy. *J Cereb Blood Flow Metab*. 2011;**31**(8):1675–86.

39. Kuhl DE, Engel J Jr, Phelps ME KA. Epileptic patterns of local computed, cerebral metabolism and perfusion in man: investigation by emission tomography of 18F-fluorodeoxyglucose and 13N-ammonia. *Trans Am Neurol Assoc*. 1978;**103**:52–3.

40. Dedeurwaerdere S, Callaghan PD, Pham T, et al. PET imaging of brain inflammation during early epileptogenesis in a rat model of temporal lobe epilepsy. *EJNMMI Res*. 2012;**2**(1):60.

41. Jones NC, Nguyen T, Corcoran NM, et al. Targeting hyperphosphorylated tau with sodium selenate suppresses seizures in rodent models. *Neurobiol Dis*. 2012;**45**(3):897–901.

42. Virdee K, Cumming P, Caprioli D, et al. Applications of positron emission tomography in animal models of neurological and neuropsychiatric disorders. *Neurosci Biobehav Rev*. 2012;**36**(4):1188–216.

43. Jupp B, Williams J, Binns D, et al. Hypometabolism precedes limbic atrophy and spontaneous recurrent seizures in a rat model of TLE. *Epilepsia*. 2012;**53**(7):1233–44.

44. Goffin K, Paesschen W Van, Dupont P, Laere K Van. Longitudinal microPET imaging of brain glucose metabolism in rat lithium-pilocarpine model of epilepsy. *Exp Neurol*. 2009;**217**(1):205–9.

45. Guo Y, Gao F, Wang S, et al. In vivo mapping of temporospatial changes in glucose utilization in rat brain during epileptogenesis: an 18F-fluorodeoxyglucose-small animal positron emission tomography study. *Neuroscience*. 2009;**162**(4):972–9.

46. Ogawa S, Menon RS, Tank DW, et al. Functional brain mapping by blood oxygenation level-dependent contrast magnetic resonance imaging. A comparison of signal characteristics with a biophysical model. *Biophys J*. 1993;**64**(3):803–12.

47. Blumenfeld H. Functional MRI studies of animal models in epilepsy. *Epilepsia*. 2007;**48**:18–26.

48. Hyder F. Dynamic imaging of brain function. *Methods Mol Biol*. 2009;**489**:3–21.

49. Ogawa SLT. Blood oxygen level dependent MRI of the brain: effects of seizure induced by kainic acid in the rat. *Proc Soc Magn Reson Med*. 1992;**1**:501.

50. Friston KJ, Frith CD, Liddle PF, Frackowiak RS. Functional connectivity: the principal-component analysis of large (PET) data sets. *J Cereb Blood Flow Metab*. 1993;**13**(1):5–14.

51. Biswal B, Yetkin FZ, Haughton VM, Hyde JS. Functional connectivity in the motor cortex of resting human brain using echo-planar MRI. *Magn Reson Med*. 1995;**34**(4):537–41.

52. Basser PJ, Pierpaoli C. Microstructural and physiological features of tissues elucidated by quantitative-diffusion-tensor MRI. *J Magn Reson B*. 1996;**111**(3):209–19.

53. Ranjan P, Mishra AM, Kale R, Saraswat VA, Gupta RK. Cytotoxic edema is responsible for raised intracranial pressure in fulminant hepatic failure: in vivo demonstration using diffusion-weighted MRI in human subjects. *Metab Brain Dis*. 2005;**20**(3):181–92.

54. Kale RA, Gupta RK, Saraswat VA, et al. Demonstration of interstitial cerebral edema with diffusion tensor MR imaging in type C hepatic encephalopathy. *Hepatology*. 2006;**43**(4):698–706.

55. Gupta RK, Hasan KM, Mishra AM, et al. High fractional anisotropy in brain abscesses versus other cystic intracranial lesions. *AJNR Am J Neuroradiol*. 2005;**26**(5):1107–14.

56. Beaulieu C. The basis of anisotropic water diffusion in the nervous system—a technical review. *NMR Biomed*. 2002;**15**(7–8):435–55.

57. Frey L, Lepkin A, Schickedanz A, Huber K, Brown MS, Serkova N. ADC mapping and T1-weighted signal changes on post-injury MRI predict seizure susceptibility after experimental traumatic brain injury. *Neurol Res.* 2014;**36**(1):26–37.

58. Jansen JFA, Lemmens EMP, Strijkers GJ, et al. Short- and long-term limbic abnormalities after experimental febrile seizures. *Neurobiol Dis.* 2008;**32**(2):293–301.

59. Wall CJ, Kendall EJ, Obenaus A. Rapid alterations in diffusion-weighted images with anatomic correlates in a rodent model of status epilepticus. *Am J Neuroradiol.* 2000;**21**(10):1841–52.

60. Sierra A, Laitinen T, Lehtimäki K, Rieppo L, Pitkänen A, Gröhn O. Diffusion tensor MRI with tract-based spatial statistics and histology reveals undiscovered lesioned areas in kainate model of epilepsy in rat. *Brain Struct Funct.* 2011;**216**(2):123–35.

61. Nehlig A. Hippocampal MRI and other structural biomarkers: experimental approach to epileptogenesis. *Biomark Med.* 2011;**5**(5):585–97.

62. Guo JN, Kim R, Chen Y, et al. Impaired consciousness in patients with absence seizures investigated by functional MRI, EEG, and behavioural measures: a cross-sectional study. *Lancet Neurol.* 2016;**15**:1336–45.

63. Bai X, Vestal M, Berman R, et al. Dynamic time course of typical childhood absence seizures: EEG, behavior, and functional magnetic resonance imaging. *J Neurosci.* 2010;**30**(17):5884–93.

64. Vega C, Vestal M, DeSalvo M, et al. Differentiation of attention-related problems in childhood absence epilepsy. *Epilepsy Behav.* 2010;**19**(1):82–5.

65. Vega C, Guo J, Killory B, et al. Symptoms of anxiety and depression in childhood absence epilepsy. *Epilepsia.* 2011;**52**(8):e70–4.

66. Bai X, Guo J, Killory B, et al. Resting functional connectivity between the hemispheres in childhood absence epilepsy. *Neurology.* 2011;**76**(23):1960–7.

67. Killory BD, Bai X, Negishi M, et al. Impaired attention and network connectivity in childhood absence epilepsy. *NeuroImage.* 2011;**56**:2209–17.

68. Blumenfeld H, Klein JP, Schridde U, et al. Early treatment suppresses the development of spike-wave epilepsy in a rat. *Epilepsia.* 2008;**49**(3):400–9.

69. Mishra AM, Bai X, Motelow JE, et al. Increased resting functional connectivity in spike-wave epilepsy in WAG/Rij rats. *Epilepsia.* 2013;**54**(7):1214–22.

70. Klein JP, Khera DS, Nersesyan H, Kimchi EY, Waxman SG, Blumenfeld H. Dysregulation of sodium channel expression in cortical neurons in a rodent model of absence epilepsy. *Brain Res.* 2004;**1000**(1–2):102–9.

71. Blumenfeld H, Lampert A, Klein JP, et al. Role of hippocampal sodium channel Nav1.6 in kindling epileptogenesis. *Epilepsia.* 2009;**50**(1):44–55.

72. Dezsi G, Ozturk E, Stanic D, et al. Ethosuximide reduces epileptogenesis and behavioral comorbidity in the GAERS model of genetic generalized epilepsy. *Epilepsia.* 2013;**54**(4):635–43.

73. Berg AT, Levy SR, Testa FM, Blumenfeld H. Long-term seizure remission in childhood absence epilepsy: might initial treatment matter? *Epilepsia.* 2014;**55**(4):551–7.

74. Yasuda CL, Betting LE, Cendes F. Voxel-based morphometry and epilepsy. *Expert Rev Neurother.* 2010;**10**(6):975–84.

75. Bruggemann JM, Wilke M, Som SS, Bye AME, Bleasel A, Lawson JA. Voxel-based morphometry in the detection of dysplasia and neoplasia in childhood epilepsy: Limitations of grey matter analysis. *J Clin Neurosci.* 2009;**16**(6):780–5.

76. Henley S, Ridgway GR, Scahill RI, Kassubek J. Pitfalls in the use of voxel-based morphometry as a biomarker: examples from Huntington. *Am J Neuroradiol.* 2010;**31**:711–9.

77. Nersesyan H, Hyder F, Rothman DL, Blumenfeld H. Dynamic fMRI and EEG recordings during spike-wave seizures and generalized tonic-clonic seizures in WAG/Rij Rats. *J Cereb Blood Flow Metab.* 2004;**24**:589–99.

78. Mishra AM, Ellens DJ, Schridde U, et al. Where fMRI and electrophysiology agree to disagree: corticothalamic and striatal activity patterns in the WAG/Rij rat. *J Neurosci.* 2011;**31**(42):15053–64.

79. Meeren HKM, Pijn JPM, Van Luijtelaar ELJM, Coenen AML, Lopes da Silva FH. Cortical focus drives widespread corticothalamic networks during spontaneous absence seizures in rats. *J Neurosci.* 2002;**22**(4):1480–95.

80. Nersesyan H, Herman P, Erdogan E, Hyder F, Blumenfeld H. Relative changes in cerebral blood flow and neuronal activity in local microdomains during generalized seizures. *J Cereb Blood Flow Metab.* 2004;1057–68.

81. Tenney J, Duong T, King J. Corticothalamic modulation during absence seizures in rats: a functional MRI assessment. *Epilepsia.* 2003;**44**(9):1133–40.

82. Schridde U, Khubchandani M, Motelow JE, Sanganahalli BG, Hyder F, Blumenfeld H. Negative BOLD with large increases in neuronal activity. *Cereb Cortex.* 2008;**18**(8):1814–27.

83. DeSalvo MN, Schridde U, Mishra AM, et al. Focal BOLD fMRI changes in bicuculline-induced tonic-clonic seizures in the rat. *NeuroImage*. 2010;**50**(3):902–9.

84. Gotman J, Grova C, Bagshaw A, Kobayashi E, Aghakhani Y, Dubeau F. Generalized epileptic discharges show thalamocortical activation and suspension of the default state of the brain. *Proc Natl Acad Sci USA*. 2005;**102**(42):15236–40.

85. Hamandi K, Laufs H, Nöth U, Carmichael DW, Duncan JS, Lemieux L. BOLD and perfusion changes during epileptic generalised spike wave activity. *NeuroImage*. 2008;**39**(2):608–18.

86. Labate A, Briellmann RS, Abbott DF, Waites AB, Jackson GD. Typical childhood absence seizures are associated with thalamic activation. *Epileptic Disord*. 2005;**7**(4):373–7.

87. Biswal B, Yetkin FZ, Haughton VM, Hyde JS. Functional connectivity in the motor cortex of resting human brain using echo-planar MRI. *Magn Reson Med*. 1995;**34**(4):537–41.

88. Waites AB, Briellmann RS, Saling MM, Abbott DF, Jackson GD. Functional connectivity networks are disrupted in left temporal lobe epilepsy. *Ann Neurol*. 2006;**59**:335–43.

89. Muroi J, Okuno T, Kuno C, et al. An MRI study of the myelination pattern in West syndrome. *Brain Dev*. 1996;**18**:179–84.

90. Schropp C, Staudt M, Staudt F, et al. Delayed myelination in children with West syndrome: an MRI-study. *Neuropediatrics*. 1994;**25**:116–20.

91. Boska MD, Hasan KM, Kibuule D, et al. Quantitative diffusion tensor imaging detects dopaminergic neuronal degeneration in a murine model of Parkinson's disease. *Neurobiol Dis*. 2007;**26**(3):590–6.

92. Obenaus A, Jacobs RE. Magnetic resonance imaging of functional anatomy: use for small animal epilepsy models. *Epilepsia*. 2007;**48**(2002):11–7.

93. Song S-K, Kim JH, Lin S-J, Brendza RP, Holtzman DM. Diffusion tensor imaging detects age-dependent white matter changes in a transgenic mouse model with amyloid deposition. *Neurobiol Dis*. 2004;**15**(3):640–7.

94. Gulani V, Webb AG, Duncan ID, Lauterbur PC. Apparent diffusion tensor measurements in myelin-deficient rat spinal cords. *Magn Reson Med*. 2001;**45**(2):191–5.

95. Harsan LA, Poulet P, Guignard B, et al. Brain dysmyelination and recovery assessment by noninvasive in vivo diffusion tensor magnetic resonance imaging. *J Neurosci Res*. 2006;**83**(3):392–402.

96. Concha L, Livy D, Gross D, Wheatley B, Beaulieu C. Direct correlation between diffusion tensor imaging and electron microscopy of the fornix in humans with temporal lobe epilepsy. *Proc 16th Sci Meet Int Soc Magn Reson Med*. 2008;**566**.

97. Hui ES, Fu QL, So KF, Wu EX. Diffusion tensor MR study of optic nerve degeneration in glaucoma. *Proc Annu Int Conf IEEE Eng Med Biol*. 2007;4312–5.

Imaging Mechanisms of Drug Resistance in Experimental Models of Epilepsy

Jens P. Bankstahl and Marion Bankstahl

14.1 Introduction

Despite the development of numerous new anticonvulsant drugs, pharmacoresistance still affects about 20–30% of epilepsy patients.[1,2] The etiology of different types of epilepsy includes various causes like developmental disorders, brain insults, or genetic factors.[3] As drug resistance itself may also be caused by various mechanisms, a huge amount of variation between individual patients becomes obvious. This variation explains the need of diagnostic tools to evaluate potential mechanisms of drug resistance in order to provide adequate individual treatment. Preclinical molecular imaging provides the opportunity to evaluate proposed mechanisms of drug resistance in animal models of pharmacoresistant epilepsy with a high translational potential. Furthermore, new imaging concepts can be established and evaluated using imaging in combination with histology and molecular biology.

14.2 Animal Models of Drug-Resistant Epilepsy

Animal models of drug-resistant epilepsy are still crucial for evaluating mechanisms of drug resistance, as availability of diseased human brain tissue and brain tissue of corresponding healthy controls is extremely limited. A consensus definition of drug resistance[4] requires unsuccessful treatment with at least two antiepileptic drugs (AEDs) with different targets and at maximum tolerable dose before the diagnosis of drug resistance. Regarding the translational value of animal models, they should fulfil at least this requirement. One can distinguish two major types of rodent models of drug resistance: (1) models in which all animals are per se resistant to a variety of AEDs and (2) models in which certain percentages of animals are resistant (nonresponders) and others are responsive to AEDs (responders).[5,6] In principle, both types of models are valuable for evaluation of

mechanisms of drug resistance. However, an inevitable requirement for imaging studies is that animals underwent an epileptogenic process leading to brain alterations responsible for pharmacoresistance. Models that exhibit a priori drug-resistant induced seizures, like the 6-Hz psychomotor seizure model in mice,[7] have only limited value for imaging studies. In contrast, models with 100% pharmacoresistant chronically epileptic animals can serve for evaluation of potential new mechanisms of pharmacoresistance in comparison to control animals. Examples are lamotrigine-resistant kindled rats, showing also resistance to carbamazepine, phenytoin, and topiramate,[8] or methylazoxymethanolacetate-exposed rats, which are refractory to valproate, ethosuximide, or carbamazepine.[9] Nevertheless, this approach can give clear evidence that a certain mechanism is involved in pharmacoresistance only if a treatment strategy is successful and able to reverse drug resistance. The second type, i.e., models with responder and nonresponder subgroups, in contrast, gives the unique opportunity to directly compare both subgroups and to evaluate whether a certain resistance mechanism is differently expressed. Still, proof that a certain mechanism is responsible for drug resistance is given only if drug resistance can be counteracted in the nonresponder group. The criterion of resistance to at least two AEDs in the nonresponder group has been met by the amygdala-kindling rat model, in which pharmacoresistance to phenytoin extends to various other AEDs,[10,11] as well as post-status-epilepticus rat models in which resistance to phenobarbital extends to phenytoin (BLA model) or levetiracetam (pilocarpine model).[12–14] Notably, most published preclinical imaging studies did not use animal models that fulfil the requirements mentioned above. This means that results from these studies will give insight into brain changes during epileptogenesis or chronic epilepsy, but a direct relation to drug resistance remains vague and is strengthened only in combination with results of other, nonimaging studies.

14.3 Mechanisms of Drug Resistance in Epilepsy

Various mechanisms underlying drug-resistant epilepsy have been proposed. Still, the two major mechanistic concepts of drug-resistant epilepsies are represented by (1) the target hypothesis and (2) the transporter hypothesis.[15,16] The target hypothesis explains drug resistance by changes in drug target affinity and expression. For example, both in patients and in animal models, changes in drug sensitivity of voltage-gated sodium channels have been shown,[17,18] which might be caused by down-regulation of $\beta1$ and $\beta2$ subunits.[19] Another proposed mechanism of drug resistance is a changed $GABA_A$ receptor, the target of GABA-mediated inhibition. Changes in the subunit composition of $GABA_A$ receptors of epileptic rats correlated with profound alterations in pharmacosensitivity and receptor function.[20–22] For both targets, the $GABA_A$ receptor and voltage-gated sodium channels, molecular imaging approaches are available and first studies are published (see the "Imaging Drug Target Alterations" section).

However, a single change in one drug target alone cannot explain why a patient is resistant to different AEDs with different mechanisms of action, although it might be possible that multiple drug targets are changed within the same patient. In this context, the transporter hypothesis might provide further explanation for the existence of multiresistant patients. Over the last two decades, the transporter hypothesis has been the most extensively studied explanation for drug-resistant epilepsy.[15,23] Drug efflux transporters are expressed in various organs all over the body and particularly at tissue barriers like the blood-brain barrier. Here, they have a major influence on passage of their substrates. At the blood-brain barrier, several so-called multidrug transporters with partially overlapping substrate spectrum are physiologically expressed, of which P-glycoprotein has been by far most intensively studied until today. Furthermore, transport of various AEDs by P-glycoprotein has been shown both in vitro and in vivo.[24] Strong evidence is given for overexpression of drug transporters in pharmacoresistant epilepsy patients as well as in animal models of drug-resistant epilepsy.[25–28] As a proof of concept for the transporter hypothesis,

Brandt et al.[29] showed that pharmacoresistance toward the AED and P-glycoprotein substrate phenobarbital in chronically epileptic rats can be reversed by application of the P-glycoprotein inhibitor tariquidar. In parallel, van Vliet et al.[30] described that coadministration of phenytoin and tariquidar markedly increased seizure control.

14.4 Imaging Changes in Multidrug Transporter Expression

Over the last years, imaging of multidrug transporter expression has been a major research focus in preclinical molecular imaging of epilepsy models.[31,32] A large part of these studies has been performed within the European Union–funded research consortium EURIPIDES (European Research Initiative to Develop Imaging Probes for Early in Vivo Diagnosis and Evaluation of Response to Therapeutic Substances). For radio-isotope imaging of efflux transporters at the blood-brain barrier, two principal approaches were proposed: (1) radio-labeled transporter substrates for imaging transporter function and (2) radio-labeled nontransported inhibitors for imaging transporter expression.[33] Despite some evidence that apart from P-glycoprotein, other brain efflux transporters might be involved in pharmacoresistant epilepsy, imaging approaches for these transporters are still very limited and no studies in epilepsy models or patients are published.

The still most widely used transported radiotracer remains the P-glycoprotein substrate $[^{11}C]$verapamil, which was first evaluated with regard to P-glycoprotein function at the mouse blood-brain barrier by Hendrikse et al.[34] using the racemic formulation. They could demonstrate that in *mdr1a*$^{(-/-)}$ mice $[^{11}C]$-verapamil brain uptake was 9.5 times higher than in the wild type mice. Furthermore, after injection of a dose of 50 mg/kg of the P-glycoprotein inhibitor cyclosporine A, $[^{11}C]$verapamil brain uptake increased 10.6-fold in wild type mice whereas *mdr1a*$^{(-/-)}$ mice did not show any significant increase. However, many substrate tracers like *(R)*-$[^{11}C]$verapamil, described to be the superior enantiomer of $[^{11}C]$verapamil,[35] like the serotonin 5-HT_{1A} antagonist $[^{18}F]$MPPF[36] or like the more recently developed $[^{11}C]$-*N*-desmethyl-loperamide[37] are effectively transported (high-affinity substrates) resulting in very low brain uptake in rats (Figure 14.1). Brain uptake values for *(R)*-$[^{11}C]$verapamil in rats are well below 0.1% injected dose per gram

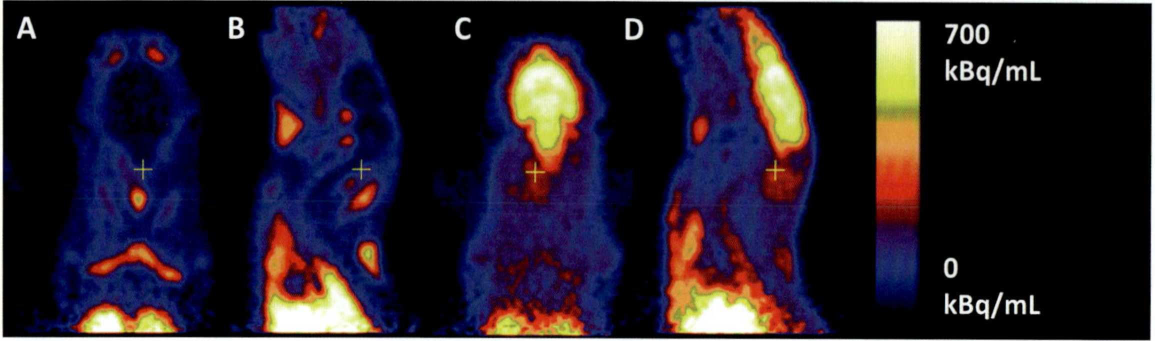

Figure 14.1. Horizontal (A and C) and sagittal (B and D) (R)-[^{11}C]verapamil PET summation images (0–60 minutes) of healthy rats recorded before (A and B) and 120 minutes after (C and D) administration of the P-glycoprotein inhibitor tariquidar. Complete inhibition of P-glycoprotein by 15 mg/kg tariquidar leads to a 12-fold increase in (R)-[^{11}C]verapamil brain uptake. Originally published in *J Nucl Med.* Bankstahl JP, Kuntner C, Abrahim A, et al. Tariquidar-induced P-glycoprotein inhibition at the rat blood–brain barrier studied with (R)-^{11}C-verapamil and PET. *J Nucl Med.* 2008;**49**:1328–35. © by the Society of Nuclear Medicine and Molecular Imaging, Inc.

tissue (%ID/g) at 60 min after tracer injection and <0.1 mean standardized uptake value (SUV) was reported for [^{18}F]MPPF and [^{11}C]-*N*-desmethyl-loperamide, whereas complete P-glycoprotein inhibition leads to up to 12-fold, 10-fold, and 5-fold increase, respectively (Figure 14.1).[36,38,39] Notably, the translational value of [^{18}F]MPPF for imaging transporter function in patients is rather limited as it appears not to be transported by human P-glycoprotein.[40]

As in epilepsy only increased expression of transporters will contribute to drug resistance, and thereby leads to even more reduced tracer brain uptake, any quantification is extremely challenging. Multiple approaches have been followed to overcome this limitation. First, less effectively transported substrates would lead to higher brain uptake, making increased transporter function more easily quantifiable. In this case decreased brain uptake would be indicative of increased transport activity. In this regard, the GABA$_A$ ligand [^{11}C]mephobarbital, a methylated analog of the AED phenobarbital, was evaluated for potential transport by P-glycoprotein,[41] as shown for phenobarbital both in vitro and in vivo.[24] Interestingly, insertion of a methyl function into phenobarbital completely abolished transporter substrate properties in [^{11}C]mephobarbital.[41] Furthermore, the sodium channel ligand [^{11}C]phenytoin was evaluated resulting in brain uptake levels of about 0.2 SUV in rats.[42] After maximal P-glycoprotein inhibition, an increase by only 45% is described,[42] confirming low-affinity substrate properties. The usefulness of this approach to identify disease-related increases in P-glycoprotein function has yet to be shown. Second, a prodrug-approach was suggested to make quantification of drug efflux easier. In this approach, a nontransported radiolabeled prodrug that can easily cross the blood-brain barrier is metabolized in the brain into a radiolabeled substrate, and then transported out of the brain. For P-glycoprotein, Sander et al.[43] could demonstrate high initial brain uptake and increased tracer levels after transporter inhibition by using such a radiolabeled prodrug, but no differences in brain clearance using pharmacokinetic modeling. More work is needed to prove the practicability and usefulness of this approach. Notably, some preclinical imaging studies using high-affinity substrates like (R)-[^{11}C]verapamil in epilepsy models were also performed after maximal P-glycoprotein inhibition. Without inclusion of baseline scans within the same subjects for data analysis, these studies will only allow to evaluate P-glycoprotein-transport-independent brain changes. Therefore, the third approach includes only partial P-glycoprotein inhibition in order to increase brain uptake of effectively transported substrates.[44] In this proof-of-concept study, the authors could demonstrate that P-glycoprotein expression levels correlate with efflux rate constant k_2 obtained by pharmacokinetic modeling after half-maximal P-glycoprotein inhibition (Figure 14.2). In a preceding dose finding study using the third generation P-glycoprotein inhibitor tariquidar in combination with (R)-[^{11}C]verapamil, the ED$_{50}$ value for tariquidar of 3.0 mg/kg was identified in rats.[45] Most strikingly, this dose leads to almost identical tariquidar

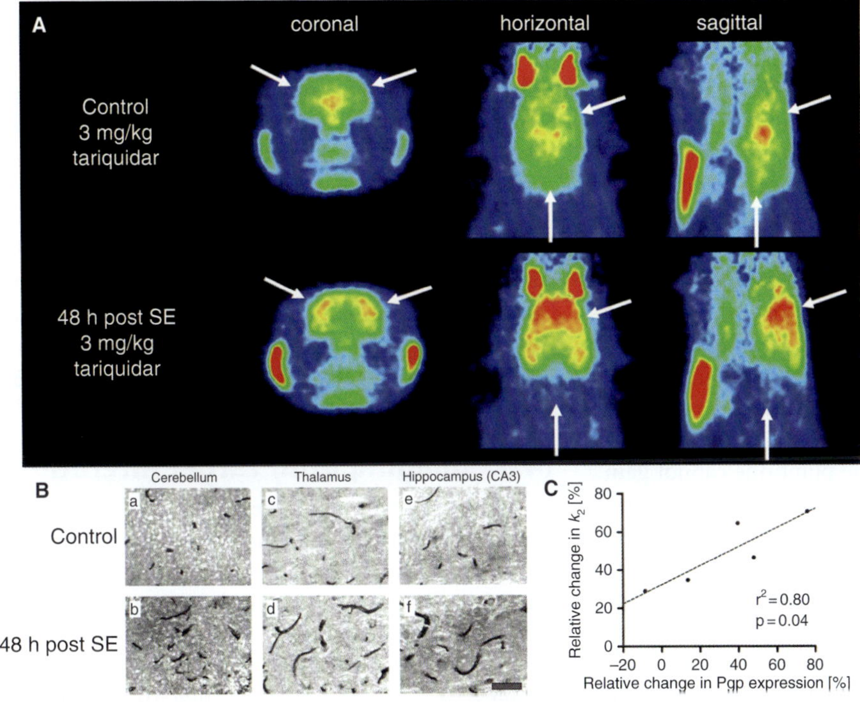

Figure 14.2. Changes in P-glycoprotein function and expression two days after status epilepticus (SE) in rats. (A) Coronal, horizontal, and sagittal (R)-$[^{11}C]$verapamil PET brain images (0–60 minutes) recorded 120 minutes after administration of 3 mg/kg tariquidar, i.e., half-maximal P-glycoprotein inhibition, in control and post-SE rats. White arrows indicate obvious differences in brain radioactivity uptake in cerebellum and cortical regions. (B) Representative examples of immuno-stained brain sections of a control rat (a, c, e) and a rat 48 hours after SE (b, d, f). P-glycoprotein expression is shown in brain capillaries of third cerebellar lobule (a, b), thalamus (c, d), and hippocampus (e, f). (C) Correlation analysis between SE-induced changes in P-glycoprotein expression and changes in compartmental-model-derived efflux rate constant k_2 relative to control group in five different brain regions after partial P-glycoprotein inhibition. From Bankstahl et al.[44]

plasma levels like the half-maximal inhibiting dose in human volunteers.[46] This partial-inhibition approach has already been successfully translated into clinical research.[47] Finally, Moerman et al.[48] described the very interesting idea of combining $[^{11}C]$-N-desmethyl-loperamide with the injection of different doses of AEDs in healthy rats for evaluation of potential transporter modulation. The extension of this approach to animal models of pharmacoresistance might be valuable.

Imaging of transporter expression rather than transporter function requires the availability of non-transported radio ligands. If such a tracer could be developed, increased brain signal would be indicative for increased transporter expression. P-glycoprotein inhibitors like tariquidar, elacridar, and laniquidar were described to bind to the transporter molecule at a nontransport binding site.[49] Surprisingly, during preclinical evaluation of both $[^{11}C]$tariquidar and $[^{11}C]$elacridar it became apparent that both inhibitors

are transported in tracer levels.[50] The same is true for $[^{11}C]$laniquidar, which also shows rather low brain uptake values that increase after P-glycoprotein inhibition by cyclosporine A, but, interestingly, not by the second-generation P-glycoprotein inhibitor valspodar.[51] This result is indicative of transport of $[^{11}C]$laniquidar, whereas it is unclear which other transporters might be involved. For all radio labeled transporter inhibitors, efforts to evaluate their potential for imaging of transporter expression are ongoing. Considering their unclear interaction with transport proteins, it is doubtful whether they can provide significant additional value over other transported tracers.

Most of the studies in epilepsy models have been performed using high-affinity substrate tracers. Two studies were performed as proof-of-concept studies using (R)-$[^{11}C]$verapamil PET shortly after a status epilepticus, which often acts as epileptogenic insult in rodent epilepsy models.[44,52] Syvänen et al.[52] performed

PET scans with and without transporter inhibition 7 days after kainic-acid-induced status epilepticus in rats. Unfortunately, at this time point P-glycoprotein levels were not changed, and therefore it was not very surprising that (R)-$[^{11}C]$verapamil uptake levels were also unchanged. Due to the known time profile of early P-glycoprotein overexpression in the pilocarpine post-status-epilepticus rat model,[53] Bankstahl et al.[44] performed their study at a time point more close to status epilepticus, i.e., after 48 hours, and after half-maximal inhibition of P-glycoprotein (Figure 14.2). During early epileptogenesis, an overexpression of P-glycoprotein in hippocampus, thalamus, and cerebellum measured by immunohistochemical analysis significantly correlated with increased efflux rate constant k_2. Of course, at this time point one cannot gain information about pharmacoresistance in chronic epilepsy. Therefore, more elaborate models are needed, in which AED responders and nonresponders undergo PET imaging. Bartmann et al.[54] performed PET scans in chronically epileptic phenobarbital responders and nonresponders (BLA model) using $[^{18}F]$MPPF. This study could not reveal any differences in influx rate constant K_1 or efflux rate constant k_2 without transporter inhibition. After administration of 5 mg/kg tariquidar, clear differences between responders and nonresponders became apparent. Interestingly, K_1 was significantly increased and k_2 significantly decreased in nonresponders as compared to responders while decreased influx and increased efflux would be expected. Although the authors explained this with a "greater tariquidar sensitivity of K_1," other factors like changes in 5HT1$_A$ receptor expression, the target of $[^{18}F]$MPPF, might have influenced the results. Syvänen et al.[55] performed another study in the same responder/nonresponder rat model using $[^{11}C]$quinidine and $[^{11}C]$laniquidar without and with complete transporter inhibition. Both $[^{11}C]$quinidine, another radiolabeled inhibitor for P-glycoprotein and $[^{11}C]$-laniquidar revealed, without inhibition, significantly higher hippocampal levels in nonresponders early after tracer injection as compared to controls, but not in responders. It was suggested that transporter inhibitors at tracer levels might bind at least to one transporter and another nontransport binding site at transport proteins.[50] This could explain why in addition to obvious transport of the radiolabeled inhibitors, increased signal without transporter inhibition might still be indicative for increased transporter expression, and is in very good agreement with data from Müllauer

et al.[56] showing a negative correlation between uptake of the radiolabeled P-glycoprotein inhibitor $[^{11}C]$tariquidar and volume of distribution V_t of (R)-$[^{11}C]$verapamil after half-maximal inhibition of P-glycoprotein. Notably, the difference shown by Syvänen et al.[55] disappeared at 40 minutes after tracer injection. Furthermore, after complete P-glycoprotein inhibition, $[^{11}C]$quinidine revealed significant differences between responders and nonresponders suggesting the involvement of other, not P-glycoprotein-related mechanisms. Further studies including alternative approaches for imaging transporter function as described above are still needed to elucidate the most promising method for translation to patients.

14.5 Imaging Drug Target Alterations

Changed expression of one or more drug targets might contribute to insufficient drug response in part of epilepsy patients. For most of the AEDs, the main seizure-preventing mechanism is relatively well understood.[2] While various radio tracers have been shown to visualize epilepsy-related brain changes in rodent models, like in dopamine receptors using $[^{18}F]$fallypride,[57] in glucose utilization using $[^{18}F]$fluor-deoxy-D-glucose,[58,59] or of neuroinflammation using TSPO ligands,[60,61] only few radioligands are available for AED targets. These include $[^{18}F]$flumazenil and $[^{11}C]$mephobarbital targeting the GABA$_A$ receptor,[41,62] and $[^{11}C]$phenytoin targeting voltage gated sodium channels.[42] Importantly, both $[^{18}F]$flumazenil and $[^{11}C]$phenytoin have been recently described as P-glycoprotein substrates in rodents.[42,63] Therefore, data analysis has to be performed very carefully, as brain uptake might be confounded by epilepsy-associated changes in transporter activity at the blood-brain barrier. To date, only $[^{18}F]$flumazenil has been evaluated in epilepsy models with regards to neuro-receptor expression. Liefaard et al.[64] could show a reduction of GABA$_A$ receptor density of 36% measured by $[^{18}F]$flumazenil PET in the amygdala kindling rat model, but did not distinguish between AED-responding and nonresponding animals. In parallel, the pharmacological total volume of distribution V_{Br} in the brain was increased by 78% suggesting a kindling-induced increase of transport at the blood-brain barrier. Syvänen et al.[65] used $[^{18}F]$flumazenil PET during epileptogenesis 7 days after a kainate-induced status epilepticus and before the development of spontaneous seizures.

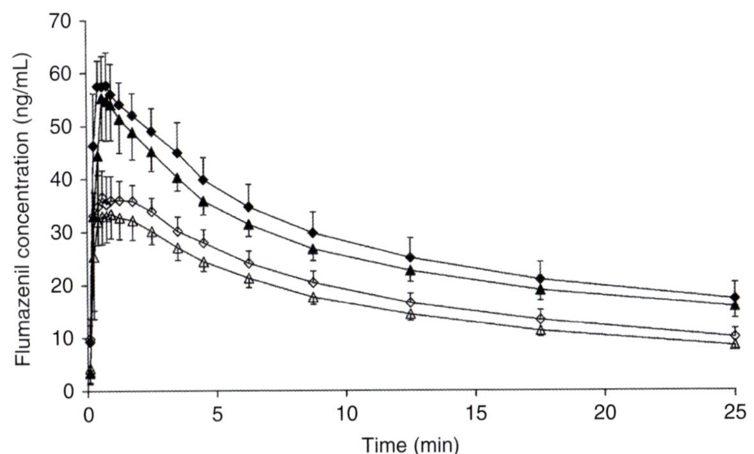

Figure 14.3. [^{11}C]Flumazenil PET quantification after a dose of approximately 4 mg of flumazenil in hippocampus in control rats (diamonds) and rats 7 days after kainate-induced status epilepticus (triangles). Open and closed symbols represent scans before and after P-glycoprotein inhibition by tariquidar treatment, respectively. Error bars indicate *SD*. Flumazenil concentrations were calculated by pharmacokinetic modeling and were lower in kainate-treated rats than in controls, both before and after tariquidar treatment. Originally published in *J Nucl Med*. Syvänen S, Labots M, Tagawa Y, et al. Altered GABA$_A$ receptor density and unaltered blood-brain barrier transport in a kainate model of epilepsy: an in vivo study using ^{11}C-flumazenil and PET. *J Nucl Med*. 2012;**53**: 1974–83. © by the Society of Nuclear Medicine and Molecular Imaging, Inc.

They could show a 12% decrease in GABA$_A$ receptor density in kainate-treated rats (Figure 14.3). They also evaluated potential epileptogenesis-associated differences in P-glycoprotein-mediated transport of [^{18}F]flumazenil at the blood-brain barrier, but did not find any changes,[52] which is consistent with unaltered P-glycoprotein expression at this time point described by the same group.[52] As yet, studies in chronically epileptic rats with or without selection of AED responders and nonresponders are not published.

14.6 Conclusion

Multiple approaches for radio-isotope imaging of mechanisms of drug resistance are available. Many of these protocols have been tested in rodent models of experimental epilepsy. Nevertheless, to date, only limited evidence is reported evaluating their suitability to distinguish between AED responding and nonresponding animals, but further studies are currently performed. Furthermore, there is an urgent need to further characterize available models of drug-resistant epilepsy, as complex mechanisms finally leading to drug resistance are only partially understood. For example, Bogdanovic et al.[61] showed higher brain uptake of the neuro-inflammation tracer [^{11}C]PK11195 in phenobarbital-resistant rats as compared to responsive animals, which might be related to a higher seizure frequency in nonresponders. To better judge the translational value of animal models of pharmacoresistance, ideally, data from pharmacoresistant and pharmacoresponsive patients would be needed for comparison.

References

1. Brodie MJ, Barry SJE, Bamagous GA, Norrie JD, Kwan P. Patterns of treatment response in newly diagnosed epilepsy. *Neurology*. 2012;78(20):1548–54.

2. Löscher W, Klitgaard H, Twyman RE, Schmidt D. New avenues for anti-epileptic drug discovery and development. *Nat Rev Drug Discov*. 2013;12(10):757–76.

3. Berg AT, Berkovic SF, Brodie MJ, et al. Revised terminology and concepts for organization of seizures and epilepsies: report of the ILAE Commission on Classification and Terminology, 2005–2009. *Epilepsia*. 2010;51(4):676–85.

4. Kwan P, Arzimanoglou A, Berg AT, et al. Definition of drug resistant epilepsy: consensus proposal by the ad hoc task force of the ILAE Commission on Therapeutic Strategies. *Epilepsia*. 2010;51(6):1069–77.

5. Löscher W. Critical review of current animal models of seizures and epilepsy used in the discovery and development of new antiepileptic drugs. *Seizure*. 2011;20(5):359–68.

6. Potschka H. Animal models of drug-resistant epilepsy. *Epileptic Disord*. 2012;14(3):226–34.

7. Barton ME, Klein BD, Wolf HH, White HS. Pharmacological characterization of the 6 Hz psychomotor seizure model of partial epilepsy. *Epilepsy Res*. 2001;47(3):217–27.

8. Srivastava AK, White HS. Carbamazepine, but not valproate, displays pharmacoresistance in lamotrigine-resistant amygdala kindled rats. *Epilepsy Res*. 2013;104(1–2):26–34.

9. Smyth MD, Barbaro NM, Baraban SC. Effects of antiepileptic drugs on induced epileptiform activity in a rat model of dysplasia. *Epilepsy Res*. 2002;50(3):251–64.

10. Löscher W. Animal models of drug-refractory epilepsy. In: Pitkänen A, Schwartzkroin P, Moshé S,

eds. *Models of Seizures and Epilepsy*. New York: Academic Press; 2006:551–67.

11. Löscher W, Rundfeldt C. Kindling as a model of drug-resistant partial epilepsy—selection of phenytoin-resistant and nonresistant rats. *J Pharmacol Exp Ther*. 1991;**258**(2):483–9.

12. Bankstahl M, Bankstahl JP, Löscher W. Inter-individual variation in the anticonvulsant effect of phenobarbital in the pilocarpine rat model of temporal lobe epilepsy. *Exp Neurol*. 2012;**234**(1):70–84.

13. Brandt C, Volk HA, Löscher W. Striking differences in individual anticonvulsant response to phenobarbital in rats with spontaneous seizures after status epilepticus. *Epilepsia*. 2004;**45**(12):1488–97.

14. Glien M, Brandt C, Potschka H, Löscher W. Effects of the novel antiepileptic drug levetiracetam on spontaneous recurrent seizures in the rat pilocarpine model of temporal lobe epilepsy. *Epilepsia*. 2002;**43**(4): 350–7.

15. Löscher W. Mechanisms of drug resistance. *Epileptic Disord*. 2005;**7**(suppl 1):3–9.

16. Potschka H, Brodie MJ. Pharmacoresistance. *Handbook Clin Neurol*. 2012;**108**:741–57.

17. Remy S, Gabriel S, Urban BW, et al. A novel mechanism underlying drug resistance in chronic epilepsy. *Ann Neurol*. 2003;**53**(4):469–79.

18. Remy S, Urban BW, Elger CE, Beck H. Anticonvulsant pharmacology of voltage-gated Na+ channels in hippocampal neurons of control and chronically epileptic rats. *Eur J Neurosci*. 2003;**17**(12): 2648–58.

19. Ellerkmann RK, Remy S, Chen J, et al. Molecular and functional changes in voltage-dependent Na+ channels following pilocarpine-induced status epilepticus in rat dentate granule cells. *Neuroscience*. 2003;**119**(2): 323–33.

20. Brooks-Kayal AR, Shumate MD, Jin H, Rikhter TY, Coulter DA. Selective changes in single cell GABA(A) receptor subunit expression and function in temporal lobe epilepsy. *Nat Med*. 1998;**4**(10):1166–72.

21. Coulter DA. Mossy fiber zinc and temporal lobe epilepsy: pathological association with altered "epileptic" gamma-aminobutyric acid A receptors in dentate granule cells. *Epilepsia*. 2000;**41**:S96–9.

22. Volk HA, Arabadzisz D, Fritschy JM, Brandt C, Bethmann K, Löscher W. Antiepileptic drug-resistant rats differ from drug-responsive rats in hippocampal neurodegeneration and GABA(A) receptor ligand binding in a model of temporal lobe epilepsy. *Neurobiol Dis*. 2006;**21**(3):633–46.

23. Löscher W, Potschka H. Drug resistance in brain diseases and the role of drug efflux transporters. *Nat Rev Neurosci*. 2005;**6**(8):591–602.

24. Zhang C, Kwan P, Zuo Z, Baum L. The transport of antiepileptic drugs by P-glycoprotein. *Adv Drug Deliv Rev*. 2012;**64**(10):930–42.

25. Kwan P, Brodie MJ. Potential role of drug transporters in the pathogenesis of medically intractable epilepsy. *Epilepsia*. 2005;**46**(2):224–35.

26. Potschka H, Volk HA, Löscher W. Pharmacoresistance and expression of multidrug transporter P-glycoprotein in kindled rats. *NeuroReport*. 2004;**15** (10):1657–61.

27. Tishler DM, Weinberg KI, Hinton DR, Barbaro N, Annett GM, Raffel C. Mdr1 gene-expression in brain of patients with medically intractable epilepsy. *Epilepsia*. 1995;**36**(1):1–6.

28. Volk HA, Löscher W. Multidrug resistance in epilepsy: rats with drug-resistant seizures exhibit enhanced brain expression of P-glycoprotein compared with rats with drug-responsive seizures. *Brain*. 2005;**128**: 1358–68.

29. Brandt C, Bethmann K, Gastens AM, Löscher W. The multidrug transporter hypothesis of drug resistance in epilepsy: proof-of-principle in a rat model of temporal lobe epilepsy. *Neurobiol Dis*. 2006;**24**(1): 202–11.

30. van Vliet EA, van Schaik R, Edelbroek PM, et al. Inhibition of the multidrug transporter P-glycoprotein improves seizure control in phenytoin-treated chronic epileptic rats. *Epilepsia*. 2006;**47**(4):672–80.

31. Löscher W, Langer O. Imaging of P-glycoprotein function and expression to elucidate mechanisms of pharmacoresistance in epilepsy. *Curr Top Med Chem*. 2010;**10**(17):1785–91.

32. Syvänen S, Eriksson J. Advances in PET imaging of P-glycoprotein function at the blood-brain barrier. *ACS Chem Neurosci*. 2013;**4**(2):225–37.

33. Bankstahl JP. What does a picture tell? In vivo imaging of ABC transporter function. *Drug Discov Today Technol*. 2014;**12**:e113–9.

34. Hendrikse NH, Schinkel AH, De Vries EGE, et al. Complete in vivo reversal of P-glycoprotein pump function in the blood-brain barrier visualized with positron emission tomography. *Br J Pharmacol*. 1998;**124**(7):1413–8.

35. Luurtsema G, Molthoff CFM, Windhorst AD, et al. (R)- and (S)-[C-11]verapamil as PET-tracers for measuring P-glycoprotein function: in vitro and in vivo evaluation. *Nucl Med Biol*. 2003;**30**(7):747–51.

36. Passchier J, van Waarde A, Doze P, Elsinga PH, Vaalburg W. Influence of P-glycoprotein on brain uptake of [18F]MPPF in rats. *Eur J Pharmacol*. 2000;**407**(3):273–80.

37. Lazarova N, Zoghbi SS, Hong J, et al. Synthesis and evaluation of [N-methyl-C-11]N-desmethyl-

loperamide as a new and improved PET radiotracer for imaging P-gp function. *J Med Chem.* 2008;**51**(19): 6034–43.

38. Bankstahl JP, Kuntner C, Abrahim A, et al. Tariquidar-induced P-glycoprotein inhibition at the rat blood-brain barrier studied with (R)-C-11-verapamil and PET. *J Nucl Med.* 2008;**49**(8):1328–35.

39. Farwell MD, Chong DJ, Iida Y, Bae SA, Easwaramoorthy B, Ichise M. Imaging P-glycoprotein function in rats using C-11-N-desmethyl-loperamide. *Ann Nucl Med.* 2013;**27**(7):618–24.

40. Tournier N, Cisternino S, Peyronneau MA, et al. Discrepancies in the P-glycoprotein-mediated transport of F-18-MPPF: a pharmacokinetic study in mice and non-human primates. *Pharm Res.* 2012;**29** (9):2468–76.

41. Mairinger S, Bankstahl JP, Kuntner C, et al. The antiepileptic drug mephobarbital is not transported by P-glycoprotein or multidrug resistance protein 1 at the blood-brain barrier: a positron emission tomography study. *Epilepsy Res.* 2012;**100** (1–2):93–103.

42. Verbeek J, Eriksson J, Syvanen S, et al. 11C phenytoin revisited: synthesis by 11C CO carbonylation and first evaluation as a P-gp tracer in rats. *EJNMMI Res.* 2012;**2** (1):36.

43. Sander K, Galante E, Gendron T, et al. Development of fluorine-18 labeled metabolically activated tracers for imaging of drug efflux transporters with positron emission tomography. *J Med Chem.* 2015;**58**(15): 6058–80.

44. Bankstahl JP, Bankstahl M, Kuntner C, et al. A novel positron emission tomography imaging protocol identifies seizure-induced regional overactivity of P-glycoprotein at the blood-brain barrier. *J Neurosci.* 2011;**31**(24):8803–11.

45. Kuntner C, Bankstahl JP, Bankstahl M, et al. Dose-response assessment of tariquidar and elacridar and regional quantification of P-glycoprotein inhibition at the rat blood-brain barrier using (R)-[11C]verapamil PET. *Eur J Nucl Med Mol Imaging.* 2010;**37**(5):942–53.

46. Bauer M, Zeitlinger M, Karch R, et al. Pgp-mediated interaction between (R)-C-11 verapamil and tariquidar at the human blood-brain barrier: a comparison with rat data. *Clin Pharmacol Ther.* 2012;**91**(2):227–33.

47. Feldmann M, Asselin MC, Liu J, et al. P-glycoprotein expression and function in patients with temporal lobe epilepsy: a case-control study. *Lancet Neurol.* 2013;**12** (8):777–85.

48. Moerman L, Wyffels L, Slaets D, Raedt R, Boon P, De Vos F. Antiepileptic drugs modulate P-glycoproteins in the brain: A mice study with 11C-desmethylloperamide. *Epilepsy Res.* 2011;**94**(1–2): 18–25.

49. Martin C, Berridge G, Mistry P, Higgins C, Charlton P, Callaghan R. The molecular interaction of the high affinity reversal agent XR9576 with P-glycoprotein. *Br J Pharmacol.* 1999;**128**(2):403–11.

50. Bankstahl JP, Bankstahl M, Römermann K, et al. Tariquidar and elacridar are dose-dependently transported by P-glycoprotein and Bcrp at the blood-brain barrier: a small-animal positron emission tomography and in vitro study. *Drug Metab Dispos.* 2013;**41**(4):754–62.

51. Luurtsema G, Schuit RC, Klok RP, et al. Evaluation of C-11 laniquidar as a tracer of P-glycoprotein: radiosynthesis and biodistribution in rats. *Nucl Med Biol.* 2009;**36**(6):643–9.

52. Syvänen S, Luurtsema G, Molthoff C, et al. (R)-[11C]Verapamil PET studies to assess changes in P-glycoprotein expression and functionality in rat blood-brain barrier after exposure to kainate-induced status epilepticus. *BMC Med Imaging.* 2011;**11**(1):1.

53. Bankstahl JP, Löscher W. Resistance to antiepileptic drugs and expression of P-glycoprotein in two rat models of status epilepticus. *Epilepsy Res.* 2008;**82** (1):70–85.

54. Bartmann H, Fuest C, la Fougère C, et al. Imaging of P-glycoprotein-mediated pharmacoresistance in the hippocampus: Proof-of-concept in a chronic rat model of temporal lobe epilepsy. *Epilepsia.* 2010;**51**(9): 1780–90.

55. Syvänen S, Russmann V, Verbeek J, et al. C-11 quinidine and C-11 laniquidar PET imaging in a chronic rodent epilepsy model: Impact of epilepsy and drug-responsiveness. *Nucl Med Biol.* 2013;**40**(6): 764–75.

56. Müllauer J, Karch R, Bankstahl JP, et al. Assessment of cerebral P-glycoprotein expression and function with PET by combined [C-11]inhibitor and [C-11]substrate scans in rats. *Nucl Med Biol.* 2013;**40**(6):755–63.

57. Yakushev IY, Dupont E, Buchholz H-G, et al. In vivo imaging of dopamine receptors in a model of temporal lobe epilepsy. *Epilepsia.* 2010;**51**(3):415–22.

58. Goffin K, Van Paesschen W, Dupont P, Van Laere K. Longitudinal microPET imaging of brain glucose metabolism in rat lithium-pilocarpine model of epilepsy. *Exp Neurol.* 2009;**217**(1):205–9.

59. Guo Y, Gao F, Wang S, et al. In vivo mapping of temporospatial changes in glucose utilization in rat brain during epileptogenesis: an [18F]-fluordeoxyglucose-small animal positron emission tomography study. *Neuroscience.* 2009;**162** (4):972–9.

60. Amhaoul H, Hamaide J, Bertoglio D, et al. Brain inflammation in a chronic epilepsy model: Evolving pattern of the translocator protein during epileptogenesis. *Neurobiol Dis.* 2015;**82**:526–39.

61. Bogdanovic RM, Syvänen S, Michler C, et al. (R)-C-11 PK11195 brain uptake as a biomarker of inflammation and antiepileptic drug resistance: evaluation in a rat epilepsy model. *Neuropharmacology.* 2014;**85**:104–12.

62. Maziere M, Hantraye P, Prenant C, Sastre J, Comar D. Synthesis of ethyl 8-fluoro-5,6-dihydro-5-C11 methyl-6-oxo-4h-imidazo 1,5-a 1,4 benzodiazepine-3-carboxylate (RO 15.1788-11C)—a specific radioligand for the in vivo study of central benzodiazepine receptors by positron emission tomography. *Int J Appl Radiat Isot.* 1984;**35**(10):973–6.

63. Froklage F, Syvanen S, Hendrikse NH, et al. [11C]Flumazenil brain uptake is influenced by the blood-brain barrier efflux transporter P-glycoprotein. *EJNMMI Res.* 2012;**2**(1):12.

64. Liefaard LC, Ploeger BA, Molthoff CFM, et al. Changes in GABA(A) receptor properties in amygdala kindled animals: In vivo studies using [C-11]flumazenil and positron emission tomography. *Epilepsia.* 2009;**50**(1):88–98.

65. Syvänen S, Labots M, Tagawa Y, et al. Altered $GABA_A$ receptor density and unaltered blood-brain barrier transport in a kainate model of epilepsy: an in vivo study using 11C-flumazenil and PET. *J Nucl Med.* 2012;**53**(12):1974–83.

15

Biomarkers of Drug Response and Pharmacoresistance to Epilepsy

Britta Wandschneider, Maria Feldmann, and Matthias Koepp

15.1 Introduction

Despite advances in antiepileptic drug (AED) therapy, about one third of patients with epilepsy are resistant to several, if not all, AEDs, even though these drugs act by diverse mechanisms.[1] Pharmacoresistant epilepsy is defined as failure of adequate response to two tolerated and appropriately prescribed AED schedules (whether as monotherapies or in combination) to achieve sustained seizure freedom.[2] Drug response in epilepsy applies not only to seizure control, albeit the defining symptom of the disease, but also to affective, behavioral, and cognitive symptoms, related to the disease and/or antiepileptic treatment. Many patients rank cognitive impairment as the most relevant complaint and, vice versa, consider the side-effect profile of specific AEDs as more important than achieved seizure control by these drugs.[3,4] Currently, we have no means to predict how a patient will respond to a drug, in terms of both efficacy and side effects. Rather than waiting for the next seizure or side effects to occur in order to assess efficacy and tolerability through trial and error, we ultimately need clinically meaningful surrogate markers for therapeutic guidance that measure how a patient feels and functions, and predict the medication effect. This chapter focuses on the currently available imaging biomarkers for assessment and prediction of drug response and refractoriness.

15.2 Imaging Markers of General Treatment Response

A limited number of studies investigated the relationship of metabolic and structural imaging markers and general treatment response in epilepsy, i.e., not related to specific AEDs. These studies mainly investigated patients with temporal lobe epilepsy (TLE) and concluded that more extensive damage, probably related to an epilepsy precipitating injury or to active epilepsy itself, is associated with poorer treatment response.

A proton magnetic resonance spectroscopy (MRS) study contrasting patients with TLE who responded to the first AED to those who continued to have seizures reported significantly reduced N acetyl aspartate/creatinine (NAA/Cr) ratios in the treatment failure group. NAA/Cr ratios are considered markers of axonal damage and/or dysfunction.[5] A voxel-based morphometry (VBM) study[6] in a large TLE cohort ($n = 165$) identified more widespread gray matter atrophy, extending beyond the medial temporal lobe, in treatment-resistant and "relapsing-remitting" cases when compared to treatment-responders. However, seizure frequency was higher and age of onset younger in treatment-resistant cases, thus structural changes may not be specific to treatment response but attributed to overall worse epilepsy. Seizure recurrence after AED withdrawal has also been associated with structural markers, e.g., the degree of hippocampal atrophy.[7]

15.3 Pharmacological fMRI Studies

Regional network effects of distinct AEDs have mainly been studied with functional MRI (fMRI). Over the last years, fMRI studies have led to major advances in understanding epilepsy and hence provide the appropriate framework to study specific AED effects in different epilepsy syndromes.

The following commonalities have been identified using fMRI in several epilepsy syndromes:

1. Dysfunctional networks appear to extend far beyond the presumed main disease focus.[8–10]
2. These regions include major brain networks relevant to cognitive function, such as the working memory and default mode network.[10–12]
3. The default mode network has been identified as a key network to be implicated in both cognitive dysfunction and seizure manifestation.[12,13]

In addition, critical disease hubs colocalize with major subsystems of the default mode network, i.e., the medial temporal lobe; hence function within these networks is directly modulated by disease factors, such as seizure frequency and disease duration.[14]

Despite these advances in understanding disease processes in epilepsy and its comorbidities, studies assessing the effect of and response to AEDs, so called pharmaco-fMRI (ph-MRI) studies, are rare and usually of small sample sizes.[15] This may be due to several methodological difficulties: The signal change in fMRI related to the drug is low; hence drug effects are generally studied as an interaction effect in task-related fMRI, i.e., task-related activation patterns for a drug are compared to those without the drug or placebo. Drugs can influence the blood-oxygen-level-dependent (BOLD) signal both at a neuronal and vascular level, complicating the interpretation of the effects observed. As the BOLD signal is contaminated by low-frequency noise, detection of slowly evolving medication effects can be challenging.[16] A more epilepsy-specific problem is that patients usually are already on AED treatment, therefore the study design has to control for effects of comedication in addition to other confounders, such as disease activity, syndrome, and comorbidities.[17] Nevertheless, ph-MRI has the major advantage that it can investigate effects of pharmacological agents at a network level and remotely from regions of highest target receptor densities, whereas PET and molecular studies can define target receptor occupancy and affinity without necessarily translating effects to large-scale networks.[16] Hence fMRI enables a system evaluation of networks underlying behavioral effects of a drug, independent of its biochemical mechanism of action. As AEDs often target several receptor subtypes with varying regional distribution and AED efficacy differs across these targets, fMRI can monitor the combined effect of these interactions across multiple brain regions.[18] A further advantage is that fMRI does not use ionizing radiation and has no known biological side effects.

Though not in epilepsy, ph-MRI has been widely applied in other CNS diseases, mainly affective disorders, addiction, and schizophrenia.[19] Overall, these studies show that pharmacological agents with known clinical efficacy have consistent effects on disease relevant neural networks and modulation of brain activation at treatment baseline is a potential surrogate marker for long-term efficacy.[19] Similar conclusions could potentially be drawn for anticonvulsive treatment with a larger body of studies present.

In the following, effects of specific antiepileptic agents on brain networks will be discussed.

15.3.1 Carbamazepine and Oxcarbazepine

Jokeit and colleagues[20] were the first to employ ph-MRI in an epilepsy cohort. They studied the relationship of mesiotemporal fMRI activation and carbamazepine (CBZ) concentrations in 21 patients with refractory TLE. Most patients were on monotherapy, and a visuospatial memory retrieval task was employed. The extent of task- and syndrome-specific fMRI activation within the medial temporal lobe was negatively correlated with the CBZ serum levels. The effect was most marked with close to toxic drug levels. However, there were no psychometric data available to relate effects to memory function in these patients.

Employing a graph theoretical approach and resting-state fMRI in a TLE cohort treated with CBZ or oxcarbazepine (OXC) and comparing to patients who were on other AEDs, Haneef and colleagues[21] report altered hubness in those on CBZ/OXC, i.e., less highly connected nodes connecting distant parts of the brain. Whereas betweenness centrality, or hubness, was reduced within the limbic circuit and thalamus with CBZ/OXC use, it was increased in default mode regions, i.e., cingulate and posterior cingulate/precuneus.

Previous data in TLE suggest a "redistribution" of hub regions with high betweenness centrality to mainly paralimbic and temporal association cortices.[22] It is therefore tempting to speculate that CBZ/OXC may have a region-specific effect on disease-related network changes.

15.3.2 Valproate

In a placebo-controlled, combined transcranial magnetic stimulation (TMS) and fMRI study, valproate (VPA) and lamotrigine (LTG) demonstrated network-specific effects. When TMS was applied over the motor region, both agents reduced TMS-specific effective connectivity between the primary motor and premotor cortex and the primary motor and supplementary motor area (SMA). Only LTG-treatment was associated with increased effective connectivity between the left dorsolateral prefrontal cortex and anterior cingulate when TMS was applied over the prefrontal cortex.[23]

For VPA, similar effects are seen in juvenile myoclonic epilepsy (JME). JME has been associated with increased functional and structural connectivity between central motor and prefrontal cognitive networks, likely accounting for cognitively triggered jerks, a reflex trait highly associated with the syndrome.[11,24,25] Abnormal motor cortex coactivation with cognitive networks during an fMRI working memory task in JME was shown to be modulated by disease factors, i.e., the trait was enhanced with more frequent seizures and in the morning, when seizures occur more frequently due to the typical chronodependency of the syndrome. Vice versa, abnormal coactivation was attenuated with increasing VPA dose, consistent with the clinical impression that VPA is effective in JME, particularly in treatment of myoclonic jerks, and not necessarily associated with cognitive side effects.[26]

An EEG-fMRI study attributes VPA treatment response among patients with idiopathic generalized epilepsy (IGE) to regional differences in generalized spike and wave discharge (GSWD) generators.[27] Spike-related activation in VPA-resistant patients was increased in the medial frontal cortex and anterior insula bilaterally when compared to VPA responders. The effect remained apparent in the medial frontal cortex even after controlling for spike frequency, but there was no difference between groups in the thalamus. Hence in the VPA-resistant group, seizure generators may be more extensively cortically distributed.

15.3.3 Levetiracetam

Several studies attribute a favorable cognitive profile to levetiracetam (LEV),[28] which has been described superior to CBZ by some authors.[29]

Consistent with neurobehavioral data, fMRI studies show a beneficial effect of LEV to cognitive networks. Wandschneider and colleagues[30] studied fMRI activation and deactivation patterns during a verbal and visuospatial working memory task in 53 left and 54 right TLE patients, comparing those treated with ($n = 59$) to those without ($n = 48$) LEV. The effect of LEV on functional network activations was task- and syndrome-specific, as well as dose-dependent: patients on LEV showed an augmentation of task-related deactivation in the diseased temporal lobe compared to patients without LEV; more specifically, this effect was seen in the left midtemporal gyrus in left TLE during the verbal, and the right hippocampus in right TLE during the visuospatial task and became more apparent with increasing LEV dose. As patients on LEV showed similar task-related deactivation patterns to healthy controls, LEV appears to be associated with restoration of normal fMRI activation.

Though out-of-scanner neurobehavioral data, including frontal lobe cognitive measures, did not differ between those treated with LEV and those without, LEV-modulated fMRI deactivation patterns can be interpreted as a beneficial drug effect. This is corroborated by data in healthy subjects and epilepsy patients where progressive deactivation of mesial temporal structures during cognitive tasks is observed with improved performance.[12,31]

Functional MRI data from individuals with amnestic mild cognitive impairment, which is associated with a risk of Alzheimer's disease, demonstrated that dysfunctional, increased hippocampal activation in the dentate gyrus/CA3 was normalized by low-dose LEV treatment with improvement of memory performance.[32,33] Hence across different disease entities, LEV appears to have a localization-specific effect, which is relevant to both seizure generation and propagation, and cognitive performance via a major hub of the default mode network.

15.3.4 Topiramate and Zonisamide

Both topiramate (TPM) and zonisamide (ZNS) are broad-spectrum AED and TPM is further approved for migraine prophylaxis.[34,35] Both contain a sulfamate moiety responsible for its weak carbonic anhydrase inhibitor mechanism.[34,35]

For TPM, cognitive dysfunction has been described in patients with epilepsy, those with migraine, and healthy controls, characterized by reduced attention, psychomotor speed, short-term memory, and, more specifically, impairment of expressive language and working memory. These deficits are noted even after single-dose administration and on steady dose in mono- or combination therapy, irrespective of seizure control. Psychometric measures consistently improve after significant dose reduction or discontinuation.[37–41] Zonisamide treatment leads to similar, probably less pronounced neurocognitive impairment.[41,42]

Topiramate is the AED most studied in ph-MRI trials, though in relatively small cohorts. Five fMRI studies employed expressive language tasks in 2 healthy subjects, 5 to 16 epilepsy patients, and 10 migraine patients after a single dose or on steady-

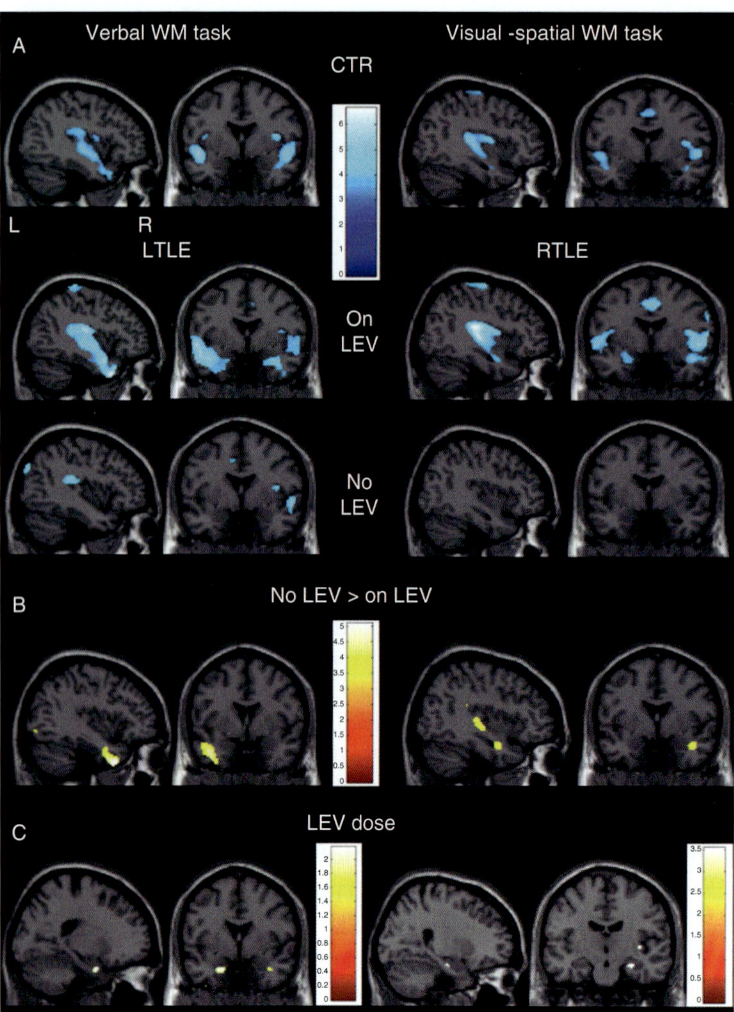

Figure 15.1. Group comparisons between patients with and without levetiracetam (LEV) during two working memory (WM) fMRI paradigms. Whereas healthy controls and patients on LEV show similar patterns of deactivation, patients without LEV show less deactivation in the medial temporal lobe areas than both controls and patients on LEV in either lateralizing task (Figure 15.1A). During the verbal WM task, left TLE patients without LEV significantly fail to deactivate the left midtemporal gyrus (Figure 15.1B; left TLE without LEV > left TLE with LEV, $p < .001$, 20 voxels threshold extent). During the right-lateralizing visuospatial task, patients with right TLE who are not treated with LEV fail to deactivate the right hippocampus (Figure 15.1B; right TLE without LEV > right TLE with LEV, $p < .001$, 20 voxels threshold extent). A post hoc analysis in patients treated with LEV demonstrated a dose-dependent effect of mesial temporal lobe deactivation through LEV. The lower the LEV dose, the lesser the right hippocampus is deactivated during the visuospatial WM task (Figure 15.1C; $p < .001$, 20 voxels threshold extent). A similar dose effect is observed in left TLE patients during the verbal WM task at a lower level of significance (Figure 15.1C; $p < .05$, uncorrected): the left > right hippocampus becomes less strongly deactivated with lower LEV dose. Figure 15.1C was inclusively masked for task-related deactivation networks ($p < .05$). CTR = healthy controls; LTLE = left temporal lobe epilepsy; RTLE = right temporal lobe epilepsy. Reproduced from Wandschneider et al.[29]

state TPM treatment. Combined results convey a pattern of reduced activation in language relevant regions, i.e., dominant inferior and middle frontal gyri, superior temporal gyrus,[43–45] and a failure to deactivate task-negative regions, including the default mode network.[44,46,47] Successful task execution in general has been associated with effective deactivation of task-negative areas.[48,49]

A recent study[50] investigated commonalities and differences of TPM and ZNS effects on fMRI activation patterns during a verbal fluency task in a larger cohort of focal epilepsy patients (TPM = 32, ZNS = 51). Consistent with the previous studies and behavioral data reporting similar cognitive impairment patterns of TPM and ZNS, both AEDs led to decreased activations in language-associated frontal and working memory and sustained attention parietal lobe networks compared to a patient control group on LEV. In addition, TPM appeared to be specifically associated with attenuation of task-related deactivation, including subsystems of the default mode network implicated in language processing (i.e., right inferior parietal lobule, supramarginal gyrus).

Both AEDs contain a sulfamate moiety, and specific detrimental effects on verbal intellectual abilities have also been described in a related drug, sulthiame.[51] Sulfa-compound containing drugs share the carbonic anhydrase inhibition mechanism and the mechanism underlying the similar fMRI changes in TPM and ZNS may be comparable to acetazolamide, another carbonic anhydrase inhibitor, which has been shown to increase blood flow with constant oxygen consumption, leading to an enhancement of the resting BOLD signal and decrease of the activation BOLD.[52,53]

15.4 Molecular Imaging Biomarker of Drug Resistance

There are several hypotheses that explain the mechanisms associated with pharmacoresistance in epilepsy. Current theories on the causes of pharmacoresistance in epilepsy include the transporter hypothesis, the target hypothesis, the network hypothesis, the gene variant/methylation hypothesis, and the intrinsic severity hypothesis. However, none of these hypotheses is currently a stand-alone theory that is able to convincingly explain how drug resistance arises in human epilepsy.[54]

15.4.1 The Transporter Hypothesis

The drug transporter hypothesis proposes that pharmacoresistance is related to increased expression of multidrug efflux transporter proteins such as P-glycoprotein (Pgp). These proteins are thought to prevent AED entry by actively extruding AEDs from their target site.[55] Multidrug efflux transporters are highly expressed in capillary endothelial cells and astrocytic foot processes that form the blood-brain barrier (BBB). They limit intracellular concentration of substrates by pumping them out of the cell through an active energy-dependent mechanism. Pgp (encoded by the adenosine triphosphate [ATP]-binding cassette subfamily B member 1 gene [ABCB1]) was discovered more than 30 years ago.[56]

15.4.2 Imaging Biomarker of Multidrug Transporter Function

Noninvasive brain imaging of multidrug efflux transporter function in pharmacoresistant and seizure-free epilepsy patients is a strategy to evaluate whether the overexpression of multidrug efflux transporters at the BBB as postulated in the transporter hypothesis has any functional consequences that underlie pharmacoresistance in epilepsy. Additionally, PET tracers for multidrug efflux transporters could be useful to identify epilepsy patients with increased multidrug efflux transporter activity who will benefit from treatment with multidrug efflux transporter modulation drugs and therefore hold great promise for individualized medicine.[57]

Almost all efforts to develop multidrug efflux transporters ligands have focused on Pgp expressed at the BBB or in tumors. To date only radiolabeled transporter substrates have been described to measure multidrug efflux transporters in vivo in humans.

15.4.3 [^{11}C]verapamil

[^{11}C]verapamil (VPM) is the best validated PET tracer to image Pgp function to date. (R)-[^{11}C]verapamil is a high-affinity substrate of Pgp and is therefore very effectively transported by Pgp at the BBB. This results in low brain uptake of this radiotracer thus making it difficult to detect regional differences in cerebral Pgp function. A possible strategy to overcome the limitation of the low brain uptake of Pgp substrate radiotracer is to perform PET scans after partial blockade by Pgp modulating drugs such as cyclosporine A (CsA) or tariquidar (TQD).[58] The third-generation Pgp inhibitor TQD is safer than the nonselective CsA for use in human subjects and was shown to lack interaction with metabolism and plasma protein binding of (R)-[^{11}C]verapamil.[59,60] Blocking Pgp with an inhibitor allows the radiotracer to enter the BBB and hence increase its uptake and signal in the brain.

The function of Pgp can be quantified in vivo from dynamic PET data using standard compartmental models. Pgp mediates movement of substrates, such as verapamil, from brain to plasma. Since we administer the Pgp substrate PET tracer [^{11}C]-verapamil intravenously, its entry into the brain is limited by Pgp activity, and the net effect is measured as K_1, the transport rate constant from plasma to brain; a low net influx from plasma to brain, indicated by a low K_1 value, therefore indicates high efflux from the brain and, thus, high Pgp activity. As shown in rodent models of Pgp overactivity,[61] partial Pgp inhibition increases (R)-[^{11}C]verapamil K_1, but this increase is attenuated in areas of high P-glycoprotein activity, since a fixed dose of Pgp inhibitor (2 or 3 mg/kg) inhibits a lower proportion of binding sites in areas of high Pgp activity than in areas of low activity. Therefore partial Pgp inhibition should be used because if Pgp glycoprotein were fully inhibited, brain uptake of [^{11}C]-verapamil K_1 would be driven by passive diffusion only and differences in Pgp activity could not be detected.

Bartels et al.[62] studied the Pgp function using [^{11}C]-verapamil PET in 17 healthy volunteers aged 18–86. Analysis with statistical parametric mapping showed significantly decreased Pgp function in older subjects

concentrations are lower within the epileptic focus was not addressed in this study.

15.4.6 Other Radiolabeled Pgp Substrates

Several more radiolabeled drugs and radioligands have been investigated as PET tracers for Pgp. The radioligand [18F]MPPF has been developed as an alternative to short-lived (11C)-labeled tracers for PET studies of serotonin 5-HT$_{1A}$ receptors. [18F]-MPPF has been used in animal studies together with the third-generation Pgp inhibitor TQD to study Pgp activity. Preliminary results obtained in a clinical study with 10 mTLE patients showed that the Pgp inhibitor CsA significantly increased the [18F]MPPF binding potential (the ratio of receptor density and radioligand affinity) by 14% in most brain regions, regardless of their involvement in seizure generation or propagation.[75]

15.5 Conclusions

Neuroimaging, particularly fMRI, has demonstrated syndrome and regionally specific effects of distinct AEDs in epilepsy. So far, as a potentially common theme, AEDs appear to lead to attenuation of either task-related activation and/or deactivation in colocalized hubs critical to both the specific epilepsy syndrome and the network relevant to the cognitive function studied.

It has been argued that general conclusions on specific AED effects from the studies available are difficult to draw since effects are seen in the context of a specific task.[53] However, others argue that evaluation of CNS drug effects are necessarily investigated with tasks as activation of specific brain networks defines the consequences of a drug effect on cognition and behavior and/or on disease processes. Activation patterns in the context of a task are stable and highly reproducible, and their alteration by disease and medication may therefore serve as distinct biomarkers.[18] Translating from other areas of fMRI research, e.g., pain drug discovery, fMRI can be envisioned as a tool to integrate several aspects of CNS drug research, such as functional pathology due to the disease, the contribution of genes to both functional networks and drug response, and early surrogate markers of drug efficacy and side effects.[18]

Molecular imaging techniques have helped to investigate mechanisms of pharmacoresistance in epilepsy. In particular, PET has been used to evaluate the transporter hypothesis and demonstrated that Pgp is increased in pharmacoresistant epilepsy. Several open questions remain and molecular imaging techniques can help to enhance our understanding of underlying mechanisms for pharmacoresistance in epilepsy.

Using noninvasive imaging with PET enables us to identify individual patients where pharmacoresistance is caused by Pgp overactivity and potentially individualize treatment. Moreover, comparative studies between pharmacoresistant and seizure-free epilepsy patients can enable testing for a correlation between Pgp function and the pharmacoresponse. This imaging technique can also guide patient selection for future clinical studies. In particular, PET imaging of Pgp function may allow individualized application of approaches to overcome Pgp-associated pharmacoresistance. In the future combined imaging and clinical trials of novel treatment strategies, such as Pgp inhibitors or modulators of overexpression in patients who have Pgp overactivity on VPM-PET could be employed. Targeting Pgp by modulators can enhance the efficacy of AEDs. The compound verapamil is a substrate for Pgp at low concentrations (which is used in PET), but, like many substrates, verapamil is also an inhibitor for Pgp at high concentrations.[76] Verapamil was among the first identified inhibitors of Pgp and may function to block Pgp–modulated efflux of AEDs in the brain, thereby raising the intracellular concentration of AEDs.[77] The third-generation modulator TQD increased the efficacy of phenytoin in a chronic rat epilepsy model and helped to overcome pharmacoresistance to phenobarbital in chronic rat epilepsy models.[78] Third-generation inhibitors are considered fairly specific. But there is recent evidence for the third-generation inhibitor TQD that it can also affect the efflux transporter BCRP/ABCG2. Moreover, long-term inhibition of this transporter needs to take into account that Pgp serves as a protective mechanism and gatekeeper in several blood tissue barriers as well as hematopoietic cells.[79] In addition to limiting access of harmful xenobiotics to sensitive tissues or cells, Pgp also accelerates extrusion of xenobiotics based on its efflux function in the liver and kidneys. Therefore, alternate approaches that leave basal transporter expression and function unaffected might offer advantages for tolerability and safety issues. Preventing seizure-associated transporter upregulation might offer an intriguing alternate approach to overcoming transporter associated pharmacoresistance.[80,81]

Although recent studies provide the proof of concept for the transporter hypothesis and show that there is functionally relevant Pgp overactivity in pharmacoresistant mTLE,[67] the critical question remains whether it is sufficient to overcome Pgp overexpression as one putative mechanism of a multifactorial problem or whether other mechanisms such as intrinsic disease severity, alterations in targets, different gene variants, or network alterations need to be taken into account and further investigated in the future. Given the complexity of epilepsy, it is unlikely that pharmacoresistant epilepsy is caused by a single mechanism but instead is due to several mechanisms that may even occur in the same patient. Overcoming pharmacoresistance in epilepsy represents a challenge and will necessitate a multifactorial approach and the combined efforts of basic and clinical epilepsy researchers.[54]

References

1. Kwan P, Brodie MJ. Early identification of refractory epilepsy. *N Engl J Med.* 2000;**342**:314–9.

2. Kwan P, Arzimanoglou A, Berg A, et al. Definition of drug resistant epilepsy: consensus proposal by the ad hoc task force of the ILAE Commission on Therapeutic Strategies. *Epilepsia.* 2010;**51**:1069–77.

3. Fisher RS, Vickrey BG, Gibson P, et al. The impact of epilepsy from the patient's perspective I. Descriptions and subjective perceptions. *Epilepsy Res.* 2000;**41**:39–51.

4. Bootsma HP, Ricker L, Hekster YA, et al. The impact of side effects on long-term retention in three new antiepileptic drugs. *Seizure.* 2009;**18**:327–31.

5. Campos BAG, Yasuda CL, Castellano G, Bilevicius E, Li LM, Cendes F. Proton MRS may predict AED response in patients with TLE. *Epilepsia.* 2010;**51**:783–8.

6. Bilevicius E, Yasuda CL, Silva MS, Guerreiro CAM, Lopes-Cendes I, Cendes F. Antiepileptic drug response in temporal lobe epilepsy: a clinical and MRI morphometry study. *Neurology.* 2010;**75**: 1695–701.

7. Cardoso TAM, Coan AC, Kobayashi E, Guerreiro CAM, Li LM, Cendes F. Hippocampal abnormalities and seizure recurrence after antiepileptic drug withdrawal. *Neurology.* 2006;**67**:134–6.

8. Stretton J, Winston G, Sidhu M, et al. Neural correlates of working memory in temporal lobe epilepsy—an fMRI study. *NeuroImage.* 2012;**60**: 1696–703.

9. Centeno M, Vollmar C, O'Muircheartaigh J, et al. Memory in frontal lobe epilepsy: an fMRI study. *Epilepsia.* 2012;**53**:1756–64.

10. de Campos BM, Coan AC, Lin Yasuda C, Casseb RF, Cendes F. Large-scale brain networks are distinctly affected in right and left mesial temporal lobe epilepsy. *Hum Brain Mapp.* 2016;**37**(9):3137–52.

11. Vollmar C, O'Muircheartaigh J, Barker GJ, et al. Motor system hyperconnectivity in juvenile myoclonic epilepsy: a cognitive functional magnetic resonance imaging study. *Brain J Neurol.* 2011;**134**:1710–9.

12. Stretton J, Winston GP, Sidhu M, et al. Disrupted segregation of working memory networks in temporal lobe epilepsy. *NeuroImage Clin.* 2013;**2**: 273–81.

13. Danielson NB, Guo JN, Blumenfeld H. The default mode network and altered consciousness in epilepsy. *Behav Neurol.* 2011;**24**:55–65.

14. Andrews-Hanna JR, Reidler JS, Sepulcre J, Poulin R, Buckner RL. Functional-anatomic fractionation of the brain's default network. *Neuron.* 2010;**65**:550–62.

15. Koepp MJ. Gender and drug effects on neuroimaging in epilepsy. *Epilepsia.* 2011;**52**(suppl 4):35–7.

16. Mehta MA, O'Daly OG. Pharmacological application of fMRI. *Methods Mol Biol.* 2011;**711**: 551–65.

17. Beltramini GC, Cendes F, Yasuda CL. The effects of antiepileptic drugs on cognitive functional magnetic resonance imaging. *Quant Imaging Med Surg.* 2015;**5**: 238–46.

18. Borsook D, Becerra L, Hargreaves R. A role for fMRI in optimizing CNS drug development. *Nat Rev Drug Discov.* 2006;**5**:411–24.

19. Nathan PJ, Phan KL, Harmer CJ, Mehta MA, Bullmore ET. Increasing pharmacological knowledge about human neurological and psychiatric disorders through functional neuroimaging and its application in drug discovery. *Curr Opin Pharmacol.* 2014;**14**:54–61.

20. Jokeit H, Okujava M, Woermann FG. Carbamazepine reduces memory induced activation of mesial temporal lobe structures: a pharmacological fMRI-study. *BMC Neurol.* 2001;**1**:6.

21. Haneef Z, Levin HS, Chiang S. Brain graph topology changes associated with anti-epileptic drug use. *Brain Connect.* 2015;**5**:284–91.

22. Bernhardt BC, Chen Z, He Y, Evans AC, Bernasconi N. Graph-theoretical analysis reveals disrupted small-world organization of cortical thickness correlation networks in temporal lobe epilepsy. *Cereb Cortex.* 2011;**21**:2147–57.

23. Li X, Large CH, Ricci R, et al. Using interleaved transcranial magnetic stimulation/functional magnetic resonance imaging (fMRI) and dynamic causal modeling to understand the discrete circuit specific changes of medications: lamotrigine and valproic acid

changes in motor or prefrontal effective connectivity. *Psychiatry Res.* 2011;**194**:141–8.

24. Vollmar C, O'Muircheartaigh J, Symms MR, et al. Altered microstructural connectivity in juvenile myoclonic epilepsy: the missing link. *Neurology.* 2012;**78**:1555–9.

25. Yacubian EM, Wolf P. Praxis induction. Definition, relation to epilepsy syndromes, nosological and prognostic significance. A focused review. *Seizure.* 2014;**23**:247–51.

26. Wandschneider B, Thompson PJ, Vollmar C, Koepp MJ. Frontal lobe function and structure in juvenile myoclonic epilepsy: a comprehensive review of neuropsychological and imaging data. *Epilepsia.* 2012;**53**:2091–8.

27. Szaflarski JP, Kay B, Gotman J, Privitera MD, Holland SK. The relationship between the localization of the generalized spike and wave discharge generators and the response to valproate. *Epilepsia.* 2013;**54**: 471–80.

28. Helmstaedter C, Witt J-A. The effects of levetiracetam on cognition: a non-interventional surveillance study. *Epilepsy Behav.* 2008;**13**:642–9.

29. Helmstaedter C, Witt J-A. Cognitive outcome of antiepileptic treatment with levetiracetam versus carbamazepine monotherapy: a non-interventional surveillance trial. *Epilepsy Behav.* 2010;**18**:74–80.

30. Wandschneider B, Stretton J, Sidhu M, et al. Levetiracetam reduces abnormal network activations in temporal lobe epilepsy. *Neurology.* 2014;**83**:1508–12.

31. Cousijn H, Rijpkema M, Qin S, van Wingen GA, Fernández G. Phasic deactivation of the medial temporal lobe enables working memory processing under stress. *NeuroImage.* 2012;**59**:1161–7.

32. Bakker A, Krauss GL, Albert MS, et al. Reduction of hippocampal hyperactivity improves cognition in amnestic mild cognitive impairment. *Neuron.* 2012;**74**: 467–74.

33. Bakker A, Albert MS, Krauss G, Speck CL, Gallagher M. Response of the medial temporal lobe network in amnestic mild cognitive impairment to therapeutic intervention assessed by fMRI and memory task performance. *NeuroImage Clin.* 2015;**7**: 688–98.

34. Abou-Khalil BW. Antiepileptic drugs. *Continuum.* 2016;**22**:132–56.

35. Mula M. Topiramate and cognitive impairment: evidence and clinical implications. *Ther Adv Drug Saf.* 2012;**3**:279–89.

36. Schulze-Bonhage A. The safety and long-term efficacy of zonisamide as adjunctive therapy for focal epilepsy. *Expert Rev Neurother.* 2015;**15**:857–65.

37. Martin R, Kuzniecky R, Ho S, et al. Cognitive effects of topiramate, gabapentin, and lamotrigine in healthy young adults. *Neurology.* 1999;**52**:321–7.

38. Thompson PJ, Baxendale SA, Duncan JS, Sander JW. Effects of topiramate on cognitive function. *J Neurol Neurosurg Psychiatry.* 2000;**69**:636–41.

39. Meador KJ, Loring DW, Vahle VJ, et al. Cognitive and behavioral effects of lamotrigine and topiramate in healthy volunteers. *Neurology.* 2005;**64**:2108–14.

40. Bootsma HPR, Ricker L, Diepman L, et al. Long-term effects of levetiracetam and topiramate in clinical practice: a head-to-head comparison. *Seizure.* 2008;**17**:19–26.

41. Mula M, Trimble MR. Antiepileptic drug-induced cognitive adverse effects: potential mechanisms and contributing factors. *CNS Drugs.* 2009;**23**:121–37.

42. Ojemann LM, Ojemann GA, Dodrill CB, Crawford CA, Holmes MD, Dudley DL. Language disturbances as side effects of topiramate and zonisamide therapy. *Epilepsy Behav.* 2001; **2**:579–84.

43. Jansen JFA, Aldenkamp AP, Marian Majoie HJ, et al. Functional MRI reveals declined prefrontal cortex activation in patients with epilepsy on topiramate therapy. *Epilepsy Behav.* 2006;**9**:181–5.

44. Szaflarski JP, Allendorfer JB. Topiramate and its effect on fMRI of language in patients with right or left temporal lobe epilepsy. *Epilepsy Behav.* 2012;**24**:74–80.

45. De Ciantis A, Muti M, Piccolini C, et al. A functional MRI study of language disturbances in subjects with migraine headache during treatment with topiramate. *Neurol Sci.* 2008;**29**(suppl 1):S141–3.

46. Yasuda CL, Centeno M, Vollmar C, et al. The effect of topiramate on cognitive fMRI. *Epilepsy Res.* 2013;**105**: 250–5.

47. Tang Y, Xia W, Yu X, et al. Altered cerebral activity associated with topiramate and its withdrawal in patients with epilepsy with language impairment: an fMRI study using the verb generation task. *Epilepsy Behav.* 2016;**59**:98–104.

48. Raichle ME, MacLeod AM, Snyder AZ, Powers WJ, Gusnard DA, Shulman GL. A default mode of brain function. *Proc Natl Acad Sci USA.* 2001;**98**:676–82.

49. Seghier ML, Price CJ. Functional heterogeneity within the default network during semantic processing and speech production. *Front Psychol.* 2012;**3**:281.

50. Wandschneider B, Burdett J, Townsend L, et al. Effect of topiramate and zonisamide on fMRI cognitive networks. *Neurology.* 2017;**88**(12):1165–71.

51. Dodrill CB. Effects of sulthiame upon intellectual, neuropsychological, and social functioning abilities

among adult epileptics: comparison with diphenylhydantoin. *Epilepsia*. 1975;**16**:617–25.

52. Bruhn H, Kleinschmidt A, Boecker H, Merboldt KD, Hänicke W, Frahm J. The effect of acetazolamide on regional cerebral blood oxygenation at rest and under stimulation as assessed by MRI. *J Cereb Blood Flow Metab*. 1994;**14**:742–8.

53. van Veenendaal TM, IJff DM, Aldenkamp AP, et al. Metabolic and functional MR biomarkers of antiepileptic drug effectiveness: a review. *Neurosci Biobehav Rev*. 2015;**59**:92–9.

54. Löscher W, Klitgaard H, Twyman R, Schmidt D. New avenues for anti-epileptic drug discovery and development. *Nat Rev Drug Discov*. 2013;**12**:757–76.

55. Sisodiya SM, Lin WR, Harding BN, Squier MV, Thom M. Drug resistance in epilepsy: expression of drug resistance proteins in common causes of refractory epilepsy. *Brain*. 2002;**125**:22–31.

56. Juliano R, Ling V. A surface glycoprotein modulating drug permeability in Chinese hamster ovary cell mutants. *Biochim Biophys Acta*. 1976;**455**:152–62.

57. Sisodiya SM, Bates SE. Treatment of drug resistance in epilepsy: one step at a time. *Lancet Neurol*. 2006;**5**: 380–1.

58. Löscher W, Langer O. Imaging of P-glycoprotein function and expression to elucidate mechanisms of pharmacoresistance in epilepsy. *Curr Top Med Chem*. 2010;**10**:1785–91.

59. Bankstahl J, Kuntner C, Abrahim A, et al. Tariquidar-induced P-glycoprotein inhibition at the rat blood-brain barrier studied with (R)-11C-verapamil and PET. *J Nucl Med*. 2008;**49**:1328–35.

60. Wagner C, Feurstein T, Karch R, et al. A pilot study to assess the efficacy of tariquidar to inhibit P-glycoprotein at the human blood-brain barrier with (R)[11C]verapamil and PET. *J Nucl Med*. 2009;**50**:1192.

61. Bankstahl J, Bankstahl M, Kuntner C, et al. A novel positron emission tomography imaging protocol identifies seizure-induced regional overactivity of P-glycoprotein at the blood-brain barrier. *J Neurosci*. 2011;**31**:8803–11.

62. Bartels AL, Kortekaas R, Bart J, et al. Blood-brain barrier P-glycoprotein function decreases in specific brain regions with aging: a possible role in progressive neurodegeneration. *Neurobiol Aging*. 2009;**30**:1818–24.

63. Brunner M, Langer O, Sunder-Plassmann R, et al. Influence of functional haplotypes in the drug transporter gene ABCB1 on central nervous system drug distribution in humans. *Clin Pharmacol Ther*. 2005;**78**:182–90.

64. Bauer M, Karch R, Neumann F, et al. Assessment of regional differences in tariquidar-induced

65. Eyal S, Ke B, Muzi M, et al. Regional P-glycoprotein activity and inhibition at the human blood-brain barrier as imaged by positron emission tomography. *Clin Pharmacol Ther*. 2010;**87**:579–85.

66. Langer O, Bauer M, Hammers A, et al. Pharmacoresistance in epilepsy: a pilot PET study with the P-glycoprotein substrate R-[(11)C]verapamil. *Epilepsia*. 2007;**48**: 1774–84.

67. Feldmann M, Asselin M-C, Liu J, et al. P-glycoprotein expression and function in patients with temporal lobe epilepsy: a case-control study. *Lancet Neurol*. 2013;**12**: 777–85.

68. Bauer M, Karch R, Zeitlinger M, et al. In vivo P-glycoprotein function before and after epilepsy surgery. *Neurology*. 2014;**83**:1326–31.

69. Jensen S, DiPaolo A, Lastella M, et al. Pharmacogenetics of ABCB1 and brain kinetics of 99 m-Tc-MIBI in epilepsy patients. *Epilepsia*. 2006;**47** (suppl 3):88–9.

70. Basic S, Hajnsek S, Bozina N, et al. The influence of C3435 T polymorphism of ABCB1 gene on penetration of phenobarbital across blood-brain barrier in patients with generalized epilepsy. *Seizure Eur J Epilepsy*. 2008;**17**:524–30.

71. Seneca N, Zoghbi SS, Liow J-S, et al. Human brain imaging and radiation dosimetry of 11C-N-desmethyl-loperamide, a PET radiotracer to measure the function of P-glycoprotein. *J Nucl Med*. 2009;**50**: 807–13.

72. Kreisl W, Liow J, Kimura N, et al. P-glycoprotein function at the blood-brain barrier in humans can be quantified with the substrate radiotracer 11C-N-desmethyl-loperamide. *J Nucl Med*. 2010;**51**: 559–66.

73. Luna-Tortós C, Fedrowitz M, Löscher W. Several major antiepileptic drugs are substrates for human P-glycoprotein. *Neuropharmacology*. 2008;**55**: 1364–75.

74. Baron J, Roeda D, Munari C, Crouzel C, Chodkiewicz J, Comar D. Brain regional pharmacokinetics of 11C-labeled diphenylhydantoin: positron emission tomography in humans. *Neurology*. 1983;**33**:580–5.

75. Hammers A, Bouvard S, Costes N, et al. Impact of P-glycoprotein on the distribution of [18 F]-MPPF in pharmacoresistant temporal lobe epilepsia. *Epilepsia*. 2010;**51**:48.

76. Kannan P, John C, Zoghbi SS, et al. Imaging the function of P-glycoprotein with radiotracers:

pharmacokinetics and in vivo applications. *Clin Pharmacol Ther.* 2009;86:368–77.

77. Summers MA, Moore JL, McAuley JW. Use of verapamil as a potential P-glycoprotein inhibitor in a patient with refractory epilepsy. *Ann Pharmacother.* 2004;38:1631–4.

78. Van Vliet EA, Van Schaik R, Edelbroek PM, et al. Inhibition of the multidrug transporter p-glycoprotein improves seizure control in phenytoin-treated chronic epileptic rats. *Epilepsia.* 2006;47:672–80.

79. Huls M, Russel FGM, Masereeuw R. The role of ATP binding cassette transporters in tissue defense and organ regeneration. *J Pharmacol Exp Ther.* 2009;328:3–9.

80. Potschka H. Modulating P-glycoprotein regulation: future perspectives for pharmacoresistant epilepsies? *Epilepsia.* 2010;51:1333–47.

81. Potschka H. Animal and human data: where are our concepts for drug-resistant epilepsy going? *Epilepsia.* 2013;54:29–32.

Chapter

16

Predicting the Outcome of Surgical Interventions for Epilepsy Using Imaging Biomarkers

Clarissa Lin Yasuda, Ana Carolina Coan, Marina K. Alvim, and Fernando Cendes

16.1 Introduction

It is still unknown why outcome is suboptimal in approximately 30–50% of operated patients.[1] Unfortunately, even a comprehensive presurgical investigation cannot yet distinguish candidates who will benefit from surgical intervention from those who will persist with seizures.[2] Yet, the unpredictability of ultimate seizure control may thwart more broad indication of surgical treatment. Although several high-quality studies have investigated possible outcome predictors in temporal lobe epilepsy (TLE), including combinations of clinical, electrophysiological, and multimodal imaging findings, a conclusive and definite presurgical biomarker capable of stratifying the most probable surgical outcome is still unavailable.

Considering extratemporal epilepsies, and more specifically focal cortical dysplasias, the uncertainties regarding surgical prognosis are even greater, since for most patients qualitative MRI is unremarkable.[3] The preoperative investigation is usually more complex and relies on the combination of multimodal neuroimaging and electrophysiological methods.[4]

16.2 Magnetic Resonance Imaging

The available studies suggest quantitative MRI alterations of multiple limbic and extralimbic networks rather than a circumscribed lesion in the hippocampus are related to surgical outcome.[2] Studies are conveying evidence that a favorable outcome is more likely to occur when cerebral abnormalities are particularly confined to the targeted area to be resected or disconnected.[4]

16.2.1 MR Volumetry

As hippocampal sclerosis (HS) has been considered the main source of seizure activity in TLE, different studies have attempted to extract morphological characteristics that could be associated with surgical outcome. While some studies associated good surgical outcome with presence of predominantly unilateral hippocampal atrophy,[5] in a study of 87 patients, volumes of both ipsilateral and contralateral mesial structures (hippocampus and entorhinal cortex) were not associated with surgical outcome; neither was the volume asymmetry.[6]

It is now clear that the presence of a unilateral atrophic hippocampus cannot reliably predict surgical outcome for TLE-HS patients, regardless of concordant clinical and electroencephalography (EEG) data.

16.2.2 Hippocampal Shape Analysis

The investigation of more complex morphological characteristics of hippocampus, shape analysis, consistently revealed that patients with poorer surgical outcomes also presented with atrophy of anterior/posterior lateral aspect of the contralateral hippocampus.[7] In a study of surface-shape analysis of mesiotemporal structures, the accelerated progression of atrophy of contralateral entorhinal cortex predicted poorer surgical outcome in TLE patients.[8]

The application of an unsupervised pattern-learning algorithm on subregional surface data of mesiotemporal structures (hippocampus, amygdala, and entorhinal cortex) in a large cohort of TLE patients yielded four homogeneous classes, each with a unique profile of abnormalities (TLE I = marked bilateral atrophy [68% Engel I], TLE II = ipsilateral atrophy [89% Engel I], TLE III = mild bilateral atrophy [65% Engel I], TLE IV = hypertrophy [44% Engel I]).[9] By submitting these data to a linear discriminant analysis algorithm (LDA), it was possible to accurately predict surgical outcome in 81% of all patients (class-blind model, 89% of seizure-free patients and 59% of patients with persistent seizures), and in 92±1% of patients of each class separately (class-based models).[9] Surface-based LDA performed better than conventional volumetry of hippocampus

(73% of correct outcome prediction), entorhinal cortex (72% of correct outcome prediction), amygdala (71% of correct outcome prediction), and their combination (71% of correct outcome prediction).

16.2.3 Extra-hippocampal Morphometry

Previous morphometric studies have associated unfavorable outcomes with extrahippocampal abnormalities.[2] From nonspecific findings of extrahippocampal abnormalities to quantification of gray matter atrophy with voxel-based morphometry (VBM),[10–12] persistence of seizures has been repeatedly associated with a widespread, extrahippocampal pattern of alterations. In the analysis of 49 left TLE patients with VBM of preoperative T1-weighted images, support vector classification was applied to the extracted maps and accurately predicted surgical outcome (94% for men, 96% for women). While in the male cohort larger WM volumes were associated to a good surgical outcome, in the female cohort relatively larger WM volumes were associated to a suboptimal surgical outcome.[11] A logistic regression model showed that presurgical GM volume reduction outside the temporal lobes associated with presence or absence of mesial pathology classified correctly 85.3% of TLE patients with good outcome and 70% of those with poor outcome.[10]

Both morphometric[13] and visual[14] analyses have been helpful to predict surgical outcome in patients with neocortical epilepsies associated with MRI abnormalities suggestive of FCD with up to 90% sensitivity and 67% specificity,[13] with better prognosis being associated with complete excision of the MRI lesion.[14]

16.2.4 Large-scale Network Abnormalities

Cortical analyses with thickness measurements in TLE-HS patients revealed an association between poorer outcomes and atrophy in temporopolar/insular cortices, while surgical failure in patients without HS was related to atrophy in the posterior quadrant.[15] The investigation of large-scale whole-brain networks with graph theory applied to cortical thickness correlated suboptimal postoperative seizure outcome with increased disruption of network topological organization (characterized by increased path length).[16] Interestingly, the restricted analysis of topological properties of mesiotemporal structures preoperatively (based on surface-shape information collected from hippocampus, entorhinal cortex, and amygdala) revealed that a more segregated ipsilateral hippocampus (characterized by increased path

length) predicted long-term seizure control. While restricted topological alterations of mesiotemporal networks may be characterized by an "isolated hippocampus" (easier to be surgically "neutralized"), the presence of large-scale network abnormalities may reflect a more widespread pattern of structural alterations that facilitates seizure relapse after surgery.[17]

16.3 Diffusion Tensor Imaging

As diffusion images provide information regarding directionality and magnitude of water diffusion in each voxel, they have been used to assess white matter properties and organization (more specifically about the fiber tracts and connections);[18] it is therefore not surprising that a number of studies applying different methods (either voxel-based or tractography-based) uncovers temporal and extratemporal abnormalities in epilepsy.[19] Although the association between white matter alterations and seizure frequency has not been fully explored, there has been increased interest in investigating its potential as a biomarker of seizure control.

Regression models revealed that both larger hippocampal asymmetry indices and ipsilateral apparent diffusion coefficient (ADC) measurement predicted optimal surgical results.[20] Another study showed suboptimal surgical outcome in patients with significant bilateral fractional anisotropy (FA) reduction in thalamotemporal paths, even after correction for extent of resection.[6]

Despite the limitations of diffusion imaging to fully depict microstructural properties in white matter in both healthy and diseased states, diffusion MRI is currently the only method capable to assess fiber trajectories in vivo. Given the ability to appraise white matter architecture, it is not surprising that diffusion MRI tractography has been used to study structural connectivity, and ultimately applied in the construction of structural brain connectomes.[21]

In an analysis of preoperative diffusion MRI from 20 TLE-HS patients, the poor outcome group presented with altered graph theory properties in the subnetwork within ipsilateral temporal lobe (higher integration with lower segregation).[22] In addition, they also presented with increased connectivity between ipsilateral medial temporal and both lateral temporal/parietal lobes, as well as between contralateral temporal pole and parietal lobe. In a more recent study of TLE patients (with and without HS),[23] subnetworks from structural brain connectome (based on preoperative diffusion MRI) linking ipsilateral temporal with extratemporal regions were used to estimate surgical outcome. Good seizure control was more likely in patients with fewer alterations within this subnetwork (ipsilateral hippocampus, amygdala, thalamus, superior frontal region, lateral temporal gyri, insula, orbitofrontal cortex, cingulate and lateral occipital gyrus). Predictive models revealed that neural architecture predicted seizure-free outcome with 90% specificity (83% accuracy), while the combination with clinical data provided 94% specificity (88% accuracy). In another study of DTI brain connectome in TLE, a two-step connectome-based prediction model was applied to evaluate the potential of machine learning algorithm to predict surgical outcome.[24] In addition to correctly separate patients from controls with 80% accuracy, the model predicted surgical outcome with 70% of accuracy (similar to "expert-based" clinical decision).

Presurgical DTI brain connectome in left TLE has recently been used in computational modeling of epilepsy and surgical simulation (with removal of nodes from patient networks and comparison of likelihood of postoperative seizure control). Persistence of seizures after simulated surgery was associated with contralateral and frontal abnormalities; in addition, a more patient-specific surgical approach (rather than a predefined standard temporal lobectomy) would potentially yield better outcomes.[25]

These results provide evidence for a potential role of presurgical connectome (combined with clinical data) as a biomarker of surgical outcome.

16.4 Magnetic Resonance Spectroscopy

Proton magnetic resonance spectroscopy (MRS) assesses neuronal integrity by quantifying the peak of N-acetyl-aspartate (NAA), a marker of neuronal integrity, usually by comparing its concentrations with choline or creatine peaks. Unlike MRI, single photon emission tomography (SPECT), and positron emission tomography (PET) techniques, only a few relatively large voxels are sampled with proton MRS. The relatively poor signal-to-noise ratio of proton MRS and the relatively long time required to obtain spectra makes this technique of limited clinical use.[26]

Comparisons with the EEG localization and surgical results have demonstrated that the reduced signal intensity of NAA can lateralize and localize the epileptogenic focus in patients with focal epilepsies, in particular TLE. However, these changes are often bilateral and the relative concentration of NAA can normalize after successful surgery in patients with mesial TLE.[26]

16.5 Positron Emission Tomography

Preoperative investigation in about one-quarter of refractory patients remains difficult,[27] in particular when MRI is normal and/or discordant with EEG and clinical information. In this scenario, functional imaging like PET has an important role.[28] Given PET's "unique capability of imaging brain metabolism,"[29] it has been used to aid localization of the epileptogenic zone,[30] based on the differences of regional brain metabolism between ictal and interictal phases.[30]

The first image method employed in TLE evaluation was 18F-fluoro-2-deoxyglucose (FDG) PET,[31] which measures regional glucose metabolism. It is still the most commonly used PET tracer in epilepsy and typically shows a reduced uptake inter-ictally.[32]

FDG-PET hypometabolism agrees in 56–90% with seizure onset (determined by intracranial EEG), and has a moderate to high sensitivity in localizing the seizure focus (71–89%). In patients with either normal MRI or conflicting data (MRI discordant with clinical/EEG data) FDG-PET indicated the correct localization of the seizure focus in 63% of TLE and 38% of frontal lobe epilepsy patients.[29]

While the presence of unilateral temporal hypometabolism[33] or asymmetric glucose consumption in both temporal and uncal regions[34] predicted better surgical outcome in TLE patients, the presence of extratemporal and bilateral temporal cortical hypometabolism were correlated with poor outcome.[33] Wong et al. investigated the occurrence of extratemporal glucose hypometabolism in TLE patients and noted that it was more common in frontal lobe and insula, and more widespread in patients with secondarily generalized seizures, although not necessarily contiguous with the temporal lobe hypometabolism.[35] These areas were named as "remote hypometabolism" by the author, and its extension was related to poor outcome.[35] One study with nonlesional TLE patients revealed a good surgical outcome associated with greater proportion of resection of FDG-PET hypometabolism volume.[36]

Although MRI seems to be the most important exam to influence the surgical decision in TLE, in the follow-up, only FDG-PET yielded significant predictive value for surgical outcome in the study of Struck el al.[28] In this study with 34 operated patients, multivariate analysis with logistic regression showed that FDG-PET was an independent predictor of surgical outcome, but not MRI or EEG. MRI-negative PET-positive TLE patients presented with very good surgical outcome, comparable to those with unilateral HS on MRI.[37]

Given the difficulties in detecting epileptogenic focus in extratemporal epilepsy, a multimodal approach is generally necessary for investigation; therefore, prognostic factors usually derive from combinations of different methods. For the investigation of epilepsy in tuberous sclerosis complex (TSC), the combination of FDG-PET/MRI (coregistered) with higher values of ADC presented higher accuracy in detecting epileptogenic tubers and adjacent cortical dysplasia, compared to the information extracted exclusively from structural MRI (tuber size) and DTI (lower FA values). This multimodal approach (FDG-PET, MRI, ADC) also yielded a good surgical outcome (9 of 11 patients became seizure-free after tuber and adjoining cortex removal).[38] In a large study of neocortical epilepsy, the correct lesion localization by FDG-PET was an independent predictor of good outcome with an odds ratio (OR) of 2.49 (minimum of 2 years free of seizures), along with focal lesion on MRI (OR = 2.54) and localized ictal onset on EEG (OR = 2.37).[39]

In patients with focal cortical dysplasias, FDG-PET had a sensitivity of 95% to detect lesions and improved the surgical prognosis of patients with negative MRI.[40] Localized FDG-PET hypometabolism was a significant and independent predictor of good surgical result.[41] FDG-PET was also associated with better surgical outcome in patients with cryptogenic neocortical epilepsy when concordant with interictal EEG findings.[42]

Another useful radiotracer in epilepsy investigation is [11C]flumazenil (FMZ). It binds to the benzodiazepine site of g-aminobutyric acid (GABA) receptors[32] and reflects the functional integrity of GABAergic system.[43] Previous studies showed superiority of [11C]FMZ compared to FDG-PET to detect epileptic foci, marked by a more localized reduction of [11C]FMZ uptake in the epileptic foci. A suboptimal surgical outcome in TLE-HS patients was associated with increased ligand binding of FMZ in periventricular white matter (suggesting the presence of heterotopic neurons undetected by visual inspection),[43] which was later confirmed in an independent cohort.[32]

In extratemporal refractory epilepsy poorer surgical outcome correlated with higher number of cortical areas with decreased FMZ binding, which were not removed with surgery. Although resection may not be required for all these remote lesions, a careful investigation of such cortical areas (even with intracranial EEG) may eventually add information for an optimized surgical outcome.[44]

α-[11C]methyl-L-tryptophan (AMT) PET was originally developed to evaluate the brain synthesis of serotonin. AMT-PE has been useful to detect the epileptogenic cortex even in the absence findings on MRI or FDG-PET. Early studies in patients with TSC showed an increased uptake of AMT in epileptogenic tubers,[45] probably resulting from an increased metabolism via the inflammatory kynurenine pathway, rather than abnormalities in serotonin synthesis, suggesting an immune activation during the epileptogenic process.[45] The resection of tubers with increased AMT uptake yielded better prognosis. Favorable outcome was also observed in patients with focal cortical dysplasia and AMT uptake, which was more associated with Type IIB FCD.[46]

16.6 Single-Photon Emission Computed Tomography

Single-photon emission computed tomography (SPECT) is useful to detect the epileptogenic zone, by determining the cerebral perfusion during ictal and interictal phases.[30] The tracers used are ^{99m}Tc-hexamethyl-propyleneamine oxime (^{99m}Tc-HMPAO) or ^{99m}Tc-ethylcysteinate dimer (^{99m}Tc-ECD), which typically shows areas of hyperperfusion during ictal acquisition, reflecting a hypometabolism in areas associated with seizure onset or propagation.[47]

The comparison between interictal and ictal SPECT images is traditionally performed by visual analysis, although the accuracy of this method in extratemporal epilepsy is less satisfactory.[48] An improved analysis method to analyze is based on the subtraction of interictal SPECT from the ictal image (coregistered with MRI [SISCOM]), which has consistently demonstrated better accuracy to localize seizure onset,[48] in both temporal (70–90%)[49] and extratemporal epilepsy,[47,49] including focal cortical malformations.[48,50] SISCOM analysis has been a predictor of good surgical outcome for both temporal and extratemporal epilepsies with unremarkable MRI findings.[41]

Despite ictal SPECT localizing cortical malformations (53–72% of efficacy) (including focal cortical dysplasia),[48,50] it is still contraversial how much of the ictal hyperperfusion has to be resected for optimal outcome. While one SISCOM analysis (15 patients) considered it unnecessary to completely resect the hyperperfusion cluster,[47] another study reported better surgical outcome after complete removal of the hyperperfusion zone.[50] The contradictory results may derive from difference in sample size and mostly from methodological differences, given a better delineation of epileptogenic zone by SISCOM.[50]

Despite the validation of SISCOM, it does not account for normal variation of voxel intensities between sequential scans from a subject. To compensate this random variation between images, some groups have modified the analysis performing postprocessing techniques with SPM (http://www.fil.ion.ucl.ac.uk/spm/), either coregistered with MRI (STATISCOM—statistical ictal SPECT coregistered to MRI)[51] or without (ISAS—ictal-interictal SPECT analyzed by SPM).[52] In a study with TLE patients (with and without IIS), STATISCOM was superior to SISCOM for seizure localization.[51] In addition, a higher probability of good surgical outcome was associated with a correct determination of TLE subtype by STATISCOM, 81% of patients with correct localizing STATISCOM were seizure-free, compared to 53% of patients with incorrect or indeterminate STATISCOM.[51] A study of refractory epilepsy with normal MRI (mixed temporal and extratemporal epilepsies) confirmed superiority of both SPM-based analysis (ISAS and STATISCOM) to correctly localize surgical site, compared to SISCOM.[53]

16.7 Functional MRI

Functional MRI (fMRI) is based on the ability of paramagnetic deoxyhemoglobin molecules to create the BOLD (blood-oxygenation-level-dependent) contrast related to increased oxygen consumption during neuronal activity.[54] fMRI can be used for both task-related and task-free (resting-state) acquisitions. In epilepsy surgery, task-related fMRI (i.e., motor or visual paradigms) has been widely utilized in presurgical evaluation with the aim to localize eloquent cortex.[27] Functional connectivity extracted from resting-state fMRI is a method that evaluates interactions among brain regions at a time the individual is not performing any specific task.[55] This allows the observation of patterns of connectivity of large-scale brain networks, which may be compromised in various neurological disorders, including epilepsy.[56]

To date, the major problem of resting-state fMRI use in the clinical practice of patients with epilepsy is the difficulty to obtain patient-specific results. Promising results have shown that single-subject resting-state fMRI connectivity can be concordant with the presumed epileptogenic zone as determined by clinical, EEG or other imaging information.[56] However, no studies have adequately compared surgical outcome with fMRI connectivity at the individual level. Recently, in a group-level analysis, one study[57] used clusters of hemodynamic changes correlated with interictal EEG discharges to perform fMRI connectivity analysis in patients with different types of epilepsies before surgery. They associated seizure recurrence after surgery with a less lateralized functional connectivity, suggesting that a high laterality of connectivity of the epileptogenic zone could be a predictor of surgical outcome. Although encouraging, these results have no clinical impact up to this point.

On the other hand, the combined use of multimodal neuroimaging techniques, such as electroencephalography (EEG) and fMRI (EEG-fMRI), allows individual-level analysis.[58] EEG-fMRI provides noninvasive simultaneous evaluation of neuronal activity and cerebral hemodynamics through the variation of the BOLD signal, detecting hemodynamic changes related to pathological ictal and interictal neuronal activity; therefore, it aids the determination of brain region responsible for the initiation and propagation of seizures.[59,60] Distinctly from resting-state fMRI, some studies published in the last years have tried to evaluate the value of EEG-fMRI in predicting surgical outcome in patients with focal refractory epilepsies. However, it is noteworthy that so far the number of studies and patients evaluated are still limited and currently EEG-fMRI is not widely used in the clinical practice. Additionally, systematic evaluation of the potential of EEG-fMRI to define the surgical outcome has been performed only through retrospective studies.

16.7.1 Functional MRI in TLE

In TLE patients, one recent study tried to compare fMRI connectivity of patients with or without seizure recurrence after surgery at both group and individual levels.[61] TLE patients with recurring seizures after surgery had differences in the connectivity patterns of the right hippocampus with the ipsilateral thalamus compared with seizure-free patients. However, this difference could be either an increase or a decrease in the connectivity values and the small sample size did not allow further conclusions for individual patients.

In a recent study of TLE, Coan et al.[62] retrospectively evaluated the value of EEG-fMRI to predict the long-term outcome (mean follow-up of 46 months), demonstrating that the presence of interictal-related BOLD changes in the area of the surgical resection was associated with good surgical outcome (seizure freedom, except for auras); it provided sensitivity and specificity of 81% and 79%, respectively, and positive and negative predictive values of 81% and 79%, respectively (Figure 16.1). In a multivariate analysis, EEG-fMRI results were associated with good surgical outcome independently of etiology, epilepsy duration or the duration of the follow-up.[62]

16.7.2 Functional MRI in Extra-temporal Lobe Epilepsy

No studies comparing single-subject fMRI connectivity and surgical outcome in patients with extra-temporal lobe epilepsy have been published to date, but a few EEG-fMRI reports have combined both TLE and ETE showing that individuals with preoperative interictal EEG-fMRI results discordant with the surgical resection presented with poor seizure control after surgery.[60] However, it appears that interictal EEG-fMRI has a low sensitivity to detect the epileptogenic zone.[63]

An et al.[64] evaluated the role of EEG-fMRI to predict surgical outcome in 35 consecutive temporal and extratemporal epilepsy patients and found that the concurrence of both maximum BOLD change and surgical resected area, compared to the absence of any BOLD signal inside or in the vicinity of the surgical lacuna, could predict seizure-free outcome with high sensitivity (87.5%), specificity (76.9%), positive predictive value (70%), and negative predictive value of 90.9%.

16.8 Electric Source Imaging

EEG measures the neuronal activity on a sub-millisecond time scale. However, it lacks the ability to accurately localize the responsible brain source due to the so-called inverse problem: different source arrangements can generate the same distribution of electrical potentials. The development of different inverse solutions algorithms in the last decade has allowed us to implement source localization of the electrical potentials obtained by EEG (electric source imaging [ESI]).[65] ESI can be used to localize the generators associated with both interictal or ictal epileptiform discharges, although the majority of studies are based on interictal activity. Although ESI has an interesting appeal to be used as a clinical tool, especially because of its low cost comparable with other modalities, its interpretation requires caution due to the variability of methodologies used in the different studies. The models used to generate the source maps are based on complex mathematical assumptions that vary between different studies.[66] In addition, the accuracy of the source localization is significantly increased with 128 or more electrodes, in comparison with the standard 31 electrodes montage.[65]

Various studies have demonstrated that ESI can estimate the source of epileptiform potentials, with good accuracy to localize the epileptogenic zone defined either by noninvasive[65] or invasive data,[67] in both adults and children with focal epilepsies. The concordance of EEG source localization and intracranial EEG results has been demonstrated in both temporal and extra-temporal lobe epilepsy.[68,69] More recently, different studies have addressed the role of ESI in predicting the surgical outcome in patients with refractory epilepsies.

Few ESI series focused specifically on TLE patients. Recently, TLE patients underwent investigation with high resolution (256-channels) ESI.[70] The results showed high sensitivity (91.4% for sublobar localization and 97.1% for lobar localization) and specificity (75%) for localization of the epileptogenic zone. Patients with sources localized within the brain area resected had better surgical outcome (Engel I or II) than those who did not.[70] In ESI studies combining both TLE and ETE patients, while some failed to show significant difference in its accuracy to predict the surgical outcome between TLE and ETE,[71] others showed higher localization power for TLE patients.[72]

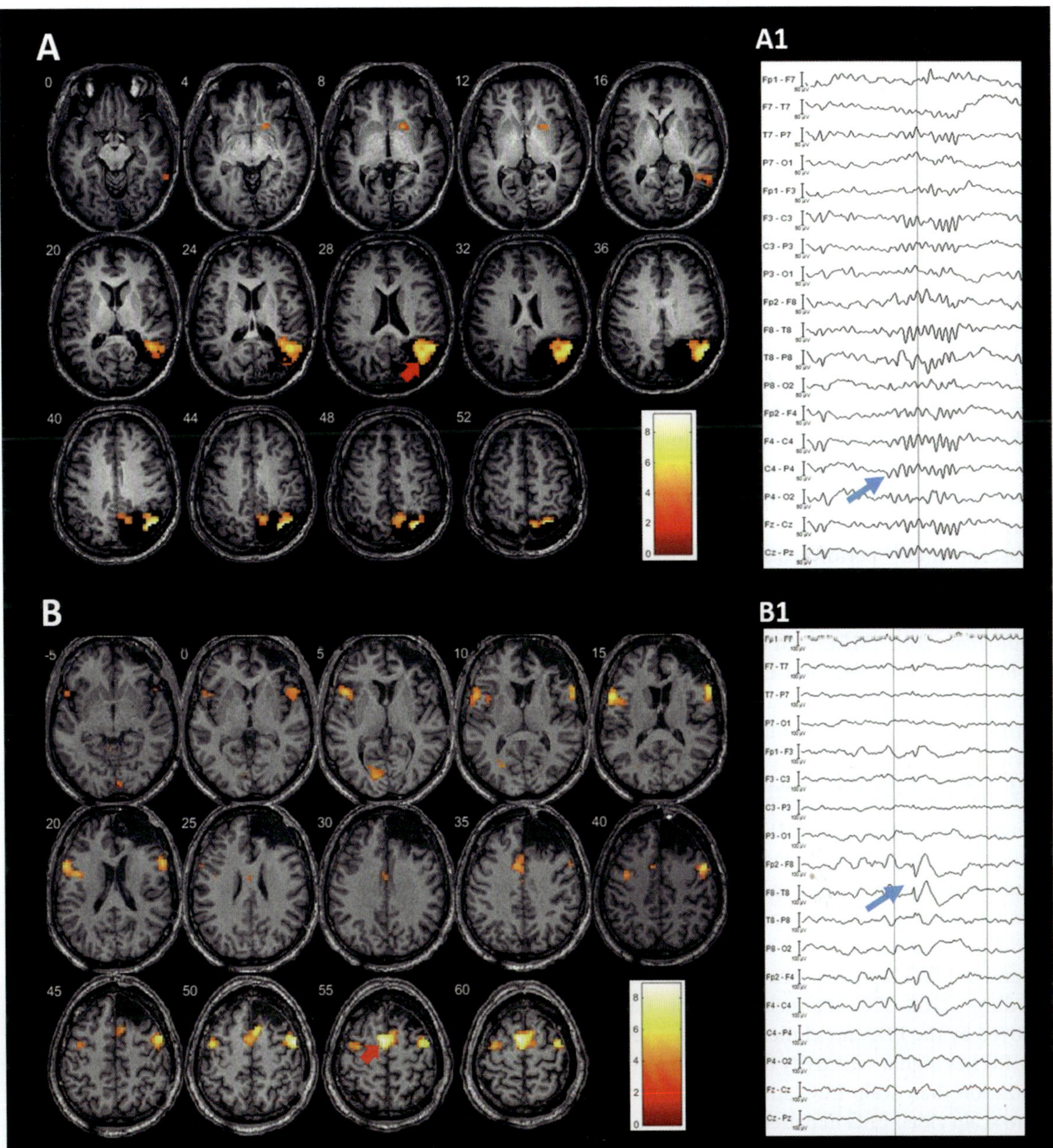

Figure 16.1. Example of interictal-related EEG-fMRI activations acquired during the presurgical evaluation of two distinct patients with pharmacoresistant epilepsies. In both cases, the presurgical BOLD maps are shown coregistered with the postoperative MRI and one example of the interictal discharge observed during the MRI acquisition is shown (A1 and B1; blue arrow). In case A, the area of maximum BOLD activation (red arrow) coincides with the surgical resection and the patient is seizure-free after a follow-up of three years. In case B, there is no BOLD activation inside the surgical resection and the area of maximum BOLD activation (red arrow) was not removed; this patient remained with frequent seizures after surgery.

A series including both temporal and extra-temporal lobe epilepsy patients evaluated for surgery showed that high resolution (128-channels) ESI detected a focal epileptogenic region in 32 patients, which was highly correlated with the defined epileptogenic zone (94% of cases); in addition, the identified source colocalized with the resected region in 79% of patients (19 of 30 operated patients).[65]

A large prospective study included 152 operated patients with focal epilepsies (minimum 1 year of follow-up) who underwent ESI during the presurgical evaluation.[73] Comparing the concordance of the source localization with the resected area and the surgical outcome, the authors observed that high resolution (128–256 electrodes) ESI had sensitivity and specificity of 84% and 88%, respectively.[73]

More recently, Russo et al. also compared the use of three-dimensional low resolution ESI in the presurgical evaluation of 60 children with refractory epilepsies (mixed temporal and extratemporal) with other neuroimaging techniques.[72] In addition to information for surgical planning as an adjunctive tool, ESI showed a sensitivity of 65.9% and a specificity of 36.8% to determine favorable outcome (Engels I and II) at 1-year follow-up and 60.6% and a specificity of 50% at 2-year follow-up.[72]

Lascano et al.[74] performed a prospective study of 58 consecutive operated patients who had been submitted to four different modalities to localize the epileptogenic zone (MRI, high density ESI, PET, and ictal SPECT). Only MRI and ESI were predictors of favorable outcome, with increasing predictive values if both modalities presented with positive results.[74]

16.9 Magnetoencephalography

Magnetoencephalography (MEG) is a noninvasive technique that measures the magnetic field produced by the electric currents flow generated by the brain neuronal activity.[75,76] The superimposed MEG data on an MRI is called magnetic source imaging (MSI). Currently, MEG systems provides high-resolution recordings (100–300 channels) and it has been approved by the FDA for clinical use since 1997.

MEG has higher sensitivity for spike detection and localization than scalp EEG and it can add information to scalp video EEG in patients with nonlocalizing findings.[77] Studies comparing MSI with other modalities have demonstrated its ability to localize the epileptogenic zone, as well as its relationship with the structural lesions, in both adults and children with epilepsy.[75] MEG can also map the motor and sensory cortex adding information to presurgical evaluation. To date, most MEG studies in epilepsy are based on interictal findings, although some reports show that ictal MEG might provide superior localization information than interictal MEG in some cases.[78]

The comparison of MEG and ictal intracranial EEG demonstrates that MEG exams with well-localized sources are highly concordant with the seizure onset zone determined by the intracranial recordings.[79] Indeed, seizure-free outcome is more commonly observed if a single cluster is identified by MSI and is concordant with the ictal onset zone determined by intracranial EEG. Differently, multiple clusters most often correlate with extensive ictal onset zone and poor seizure control after surgery.[80] Focusing more often in well-localized MEG clusters, different studies have compared MEG/MSI results with surgical outcome in patients with refractory epilepsies.

According to the current data, MSI greatest value is to localize the epileptogenic zone in patients with lateral TLE, whereas for mesial TLE its accuracy is reduced.[41] However, different studies have focused specifically on TLE patients, showing MEG can add information to the surgical plan of these patients.[81] Anterior temporal localization of MEG spikes is associated with good surgical outcome after anterior temporal lobectomy, while spikes localized in regions other than the anterior temporal lobe or with lobar spike localization appear to have worse surgical results.[82]

Another study evaluated MEG results of 131 patients submitted to epilepsy surgery. Considering only the group of patients with good surgical outcome (Engel I or II), MEG correctly identified the epileptic lobe in 87.3% of TLE (patients) and in 100% of ETE.[83]

In a group of five patients with lesional frontal lobe epilepsy submitted to epilepsy surgery, Genow et al. showed that those with the majority of MEG spikes localized within the resected area presented good surgical outcome.[84]

References

1. de Tisi J, Bell GS, Peacock JL, et al. The long-term outcome of adult epilepsy surgery, patterns of seizure remission, and relapse: a cohort study. *Lancet.* 2011;**378**:1388–95.

2. Bonilha L, Keller SS. Quantitative MRI in refractory temporal lobe epilepsy: relationship with surgical outcomes. *Quant Imaging Med Surg.* 2015;5: 204–24.

3. Bernasconi A, Bernasconi N, Bernhardt BC, et al. Advances in MRI for "cryptogenic" epilepsies. *Nat Rev Neurol.* 2011;7:99–108.

4. Yasuda CL, Cendes F. Neuroimaging for the prediction of response to medical and surgical treatment in epilepsy. *Expert Opin Med Diagn*. 2012;6:295–308.

5. West S, Nolan SJ, Cotton J, et al. Surgery for epilepsy. *Cochrane Database Syst Rev*. 2015;7:CD010541.

6. Keller SS, Richardson MP, Schoene-Bake JC, et al. Thalamotemporal alteration and postoperative seizures in temporal lobe epilepsy. *Ann Neurol*. 2015;77:760–74.

7. Lin JJ, Salamon N, Dutton RA, et al. Three-dimensional preoperative maps of hippocampal atrophy predict surgical outcomes in temporal lobe epilepsy. *Neurology*. 2005;65:1094–7.

8. Bernhardt BC, Kim H, Bernasconi N. Patterns of subregional mesiotemporal disease progression in temporal lobe epilepsy. *Neurology*. 2013;81:1840–7.

9. Bernhardt BC, Hong SJ, Bernasconi A, et al. Magnetic resonance imaging pattern learning in temporal lobe epilepsy: classification and prognostics. *Ann Neurol*. 2015;77:436–46.

10. Doucet GE, He X, Sperling M, et al. Frontal gray matter abnormalities predict seizure outcome in refractory temporal lobe epilepsy patients. *NeuroImage Clin*. 2015;9:458–66.

11. Feis DL, Schoene-Bake JC, Elger C, et al. Prediction of post-surgical seizure outcome in left mesial temporal lobe epilepsy. *NeuroImage Clin*. 2013;2:903–11.

12. Yasuda CL, Valise C, Saude AV, et al. Dynamic changes in white and gray matter volume are associated with outcome of surgical treatment in temporal lobe epilepsy. *NeuroImage*. 2010;49:71–9.

13. Wang ZI, Jones SE, Jaisani Z, et al. Voxel-based morphometric magnetic resonance imaging (MRI) postprocessing in MRI-negative epilepsies. *Ann Neurol*. 2015;77:1060–75.

14. Leach JL, Miles L, Henkel DM, et al. Magnetic resonance imaging abnormalities in the resection region correlate with histopathological type, gliosis extent, and postoperative outcome in pediatric cortical dysplasia. *J Neurosurg Pediatr*. 2014;14:68–80.

15. Bernhardt BC, Bernasconi N, Concha L, et al. Cortical thickness analysis in temporal lobe epilepsy: reproducibility and relation to outcome. *Neurology*. 2010;74:1776–84.

16. Bernhardt BC, Chen Z, He Y, et al. Graph-theoretical analysis reveals disrupted small-world organization of cortical thickness correlation networks in temporal lobe epilepsy. *Cereb Cortex*. 2011;21:2147–57.

17. Bernhardt BC, Bernasconi N, Hong SJ, et al. Subregional mesiotemporal network topology is altered in temporal lobe epilepsy. *Cereb Cortex*. 2016;26:3237–48.

18. Beaulieu C. The basis of anisotropic water diffusion in the nervous system—a technical review. *NMR Biomed*. 2002;15:435–55.

19. Gross DW. Diffusion tensor imaging in temporal lobe epilepsy. *Epilepsia*. 2011;52(suppl 4):32–4.

20. Goncalves Pereira PM, Oliveira E, Rosado P. Apparent diffusion coefficient mapping of the hippocampus and the amygdala in pharmaco-resistant temporal lobe epilepsy. *AJNR Am J Neuroradiol*. 2006;27:671–83.

21. Bernhardt BC, Bonilha L, Gross DW. Network analysis for a network disorder: the emerging role of graph theory in the study of epilepsy. *Epilepsy Behav*. 2015;50:162–70.

22. Bonilha L, Helpern JA, Sainju R, et al. Presurgical connectome and postsurgical seizure control in temporal lobe epilepsy. *Neurology*. 2013;81:1704–10.

23. Bonilha L, Jensen JH, Baker N, et al. The brain connectome as a personalized biomarker of seizure outcomes after temporal lobectomy. *Neurology*. 2015;84:1846–53.

24. Munsell BC, Wee CY, Keller SS, et al. Evaluation of machine learning algorithms for treatment outcome prediction in patients with epilepsy based on structural connectome data. *NeuroImage*. 2015;118:219–30.

25. Hutchings F, Han CE, Keller SS, et al. Predicting surgery targets in temporal lobe epilepsy through structural connectome based simulations. *PLOS Comput Biol*. 2015;11:e1004642.

26. Cendes F, Knowlton Robert C, Novotny E, et al. Magnetic resonance spectroscopy in epilepsy: clinical issues. *Epilepsia*. 2002;43:32–9.

27. Duncan JS. Imaging in the surgical treatment of epilepsy. *Nat Rev Neurol*. 2010;6:537–50.

28. Struck AF, Hall LT, Floberg JM, et al. Surgical decision making in temporal lobe epilepsy: a comparison of [(18)F]FDG-PET, MRI, and EEG. *Epilepsy Behav*. 2011;22:293–7.

29. Burneo JG, Poon R, Kellett S, et al. The utility of positron emission tomography in epilepsy. *Can J Neurol Sci*. 2015;42:360–71.

30. Wehner T, Lüders H. Role of neuroimaging in the presurgical evaluation of epilepsy. *J Clin Neurol*. 2008;4:1–16.

31. LoPinto-Khoury C, Sperling MR, Skidmore C, et al. Surgical outcome in PET-positive, MRI-negative patients with temporal lobe epilepsy. *Epilepsia*. 2012;53:342–8.

32. Yankam Njiwa J, Gray KR, Costes N, et al. Advanced [(18)F]FDG and [(11)C]flumazenil PET analysis for individual outcome prediction after temporal lobe epilepsy surgery for hippocampal sclerosis. *NeuroImage Clin*. 2015;7:122–31.

33. Choi JY, Kim SJ, Hong SB, et al. Extratemporal hypometabolism on FDG PET in temporal lobe epilepsy as a predictor of seizure outcome after temporal lobectomy. *Eur J Nucl Med Mol Imaging*. 2003;30:581–7.

34. Dupont S, Semah F, Clemenceau S, et al. Accurate prediction of postoperative outcome in mesial temporal lobe epilepsy: a study using positron emission tomography with 18fluorodeoxyglucose. *Arch Neurol*. 2000;57:1331–6.

35. Wong CH, Bleasel A, Wen L, et al. The topography and significance of extratemporal hypometabolism in refractory mesial temporal lobe epilepsy examined by FDG-PET. *Epilepsia*. 2010;51:1365–73.

36. Vinton AB, Carne R, Hicks RJ, et al. The extent of resection of FDG-PET hypometabolism relates to outcome of temporal lobectomy. *Brain*. 2007;130: 548–60.

37. Carne RP, O'Brien TJ, Kilpatrick CJ, et al. MRI-negative PET-positive temporal lobe epilepsy: a distinct surgically remediable syndrome. *Brain*. 2004;127:2276–85.

38. Chandra PS, Salamon N, Huang J, et al. FDG-PET/MRI coregistration and diffusion-tensor imaging distinguish epileptogenic tubers and cortex in patients with tuberous sclerosis complex: a preliminary report. *Epilepsia*. 2006;47:1543–9.

39. Yun CH, Lee SK, Lee SY, et al. Prognostic factors in neocortical epilepsy surgery: multivariate analysis. *Epilepsia*. 2006;47:574–9.

40. Chassoux F, Rodrigo S, Semah F, et al. FDG-PET improves surgical outcome in negative MRI Taylor-type focal cortical dysplasias. *Neurology*. 2010;75:2168–75.

41. Knowlton RC, Elgavish RA, Bartolucci A, et al. Functional imaging: II. Prediction of epilepsy surgery outcome. *Ann Neurol*. 2008;64:35–41.

42. Lee SK, Lee SY, Kim KK, et al. Surgical outcome and prognostic factors of cryptogenic neocortical epilepsy. *Ann Neurol*. 2005;58:525–32.

43. Koepp MJ, Woermann FG. Imaging structure and function in refractory focal epilepsy. *Lancet Neurol*. 2005;4:42–53.

44. Juhasz C, Asano E, Shah A, et al. Focal decreases of cortical GABAA receptor binding remote from the primary seizure focus: what do they indicate? *Epilepsia*. 2009;50:240–50.

45. Chugani HT, Luat AF, Kumar A, et al. alpha-[^{11}C]-methyl-L-tryptophan–PET in 191 patients with tuberous sclerosis complex. *Neurology*. 2013;81:674–80.

46. Chugani HT, Kumar A, Kupsky W, et al. Clinical and histopathologic correlates of 11C-alpha-methyl-L-tryptophan (AMT) PET abnormalities in children with intractable epilepsy. *Epilepsia*. 2011;52:1692–8.

47. Dupont P, Van Paesschen W, Palmini A, et al. Ictal perfusion patterns associated with single MRI-visible focal dysplastic lesions: implications for the noninvasive delineation of the epileptogenic zone. *Epilepsia*. 2006;47:1550–7.

48. O'Brien TJ, So EL, Cascino GD, et al. Subtraction SPECT coregistered to MRI in focal malformations of cortical development: localization of the epileptogenic zone in epilepsy surgery candidates. *Epilepsia*. 2004;45: 367–76.

49. Tepmongkol S, Tangtrairattanakul K, Lerdlum S, et al. Comparison of brain perfusion SPECT parameters accuracy for seizure localization in extratemporal lobe epilepsy with discordant pre-surgical data. *Ann Nucl Med*. 2015;29:21–8.

50. Krsek P, Kudr M, Jahodova A, et al. Localizing value of ictal SPECT is comparable to MRI and EEG in children with focal cortical dysplasia. *Epilepsia*. 2013;54:351–8.

51. Kazemi NJ, Worrell GA, Stead SM, et al. Ictal SPECT statistical parametric mapping in temporal lobe epilepsy surgery. *Neurology*. 2010;74:70–6.

52. McNally KA, Paige AL, Varghese G, et al. Localizing value of ictal-interictal SPECT analyzed by SPM (ISAS). *Epilepsia*. 2005;46:1450–64.

53. Sulc V, Stykel S, Hanson DP, et al. Statistical SPECT processing in MRI-negative epilepsy surgery. *Neurology*. 2014;82:932–9.

54. Ogawa S, Tank DW, Menon R, et al. Intrinsic signal changes accompanying sensory stimulation: functional brain mapping with magnetic resonance imaging. *Proc Natl Acad Sci USA*. 1992;89:5951–5.

55. Biswal BB. Resting state fMRI: a personal history. *NeuroImage*. 2012;62:938–44.

56. Zhang CH, Lu Y, Brinkmann B, et al. Lateralization and localization of epilepsy related hemodynamic foci using presurgical fMRI. *Clin Neurophysiol*. 2015;126:27–38.

57. Negishi M, Martuzzi R, Novotny EJ, et al. Functional MRI connectivity as a predictor of the surgical outcome of epilepsy. *Epilepsia*. 2011;52:1733–40.

58. Bagshaw AP, Kobayashi E, Dubeau F, et al. Correspondence between EEG-fMRI and EEG dipole localisation of interictal discharges in focal epilepsy. *NeuroImage*. 2006;30:417–25.

59. Kobayashi E, Hawco CS, Grova C, et al. Widespread and intense BOLD changes during brief focal electrographic seizures. *Neurology*. 2006;66: 1049–55.

60. Thornton R, Laufs H, Rodionov R, et al. EEG correlated functional MRI and postoperative outcome

in focal epilepsy. *J Neurol Neurosurg Psychiatry.* 2010;81:922–7.

61. Morgan VL, Sonmezturk HH, Gore JC, et al. Lateralization of temporal lobe epilepsy using resting functional magnetic resonance imaging connectivity of hippocampal networks. *Epilepsia.* 2012;53:1628–35.

62. Coan AC, Chaudhary UJ, Grouiller F, et al. EEG-fMRI in the presurgical evaluation of temporal lobe epilepsy. *J Neurol Neurosurg Psychiatry.* 2016;87: 642–9.

63. Elshoff L, Groening K, Grouiller F, et al. The value of EEG-fMRI and EEG source analysis in the presurgical setup of children with refractory focal epilepsy. *Epilepsia.* 2012;53:1597–606.

64. An D, Fahoum F, Hall J, et al. Electroencephalography/ functional magnetic resonance imaging responses help predict surgical outcome in focal epilepsy. *Epilepsia.* 2013;54:2184–94.

65. Michel CM, Lantz G, Spinelli L, et al. 128-channel EEG source imaging in epilepsy: clinical yield and localization precision. *J Clin Neurophysiol.* 2004;21:71–83.

66. Plummer C, Harvey AS, Cook M. EEG source localization in focal epilepsy: where are we now? *Epilepsia.* 2008;49:201–18.

67. Zumsteg D, Friedman A, Wieser HG, et al. Propagation of interictal discharges in temporal lobe epilepsy: correlation of spatiotemporal mapping with intracranial foramen ovale electrode recordings. *Clin Neurophysiol.* 2006;117:2615–26.

68. Gavaret M, Badier JM, Marquis P, et al. Electric source imaging in temporal lobe epilepsy. *J Clin Neurophysiol.* 2004;21:267–82.

69. Gavaret M, Badier JM, Marquis P, et al. Electric source imaging in frontal lobe epilepsy. *J Clin Neurophysiol.* 2006;23:358–70.

70. Feng R, Hu J, Pan L, et al. Application of 256-channel dense array electroencephalographic source imaging in presurgical workup of temporal lobe epilepsy. *Clin Neurophysiol.* 2016;127:108–16.

71. Megevand P, Spinelli L, Genetti M, et al. Electric source imaging of interictal activity accurately localises the seizure onset zone. *J Neurol Neurosurg Psychiatry.* 2014;85:38–43.

72. Russo A, Jayakar P, Lallas M, et al. The diagnostic utility of 3D electroencephalography source imaging in pediatric epilepsy surgery. *Epilepsia.* 2016;57:24–31.

73. Brodbeck V, Spinelli L, Lascano AM, et al. Electroencephalographic source imaging: a prospective study of 152 operated epileptic patients. *Brain.* 2011;134:2887–97.

74. Lascano AM, Perneger T, Vulliemoz S, et al. Yield of MRI, high-density electric source imaging (HD-ESI), SPECT and PET in epilepsy surgery candidates. *Clin Neurophysiol.* 2016;127:150–5.

75. Heers M, Hedrich T, An D, et al. Spatial correlation of hemodynamic changes related to interictal epileptic discharges with electric and magnetic source imaging. *Hum Brain Mapp.* 2014;35:4396–414.

76. Tanaka N, Peters JM, Prohl AK, et al. Clinical value of magnetoencephalographic spike propagation represented by spatiotemporal source analysis: correlation with surgical outcome. *Epilepsy Res.* 2014;108:280–8.

77. Knowlton RC. Can magnetoencephalography aid epilepsy surgery? *Epilepsy Curr.* 2008;8:1–5.

78. Eliashiv DS, Elsas SM, Squires K, et al. Ictal magnetic source imaging as a localizing tool in partial epilepsy. *Neurology.* 2002;59:1600–10.

79. Knowlton RC. The role of FDG-PET, ictal SPECT, and MEG in the epilepsy surgery evaluation. *Epilepsy Behav.* 2006;8:91–101.

80. Oishi M, Kameyama S, Masuda H, et al. Single and multiple clusters of magnetoencephalographic dipoles in neocortical epilepsy: significance in characterizing the epileptogenic zone. *Epilepsia.* 2006;47:355–64.

81. Wheless JW, Willmore LJ, Breier JI, et al. A comparison of magnetoencephalography, MRI, and V-EEG in patients evaluated for epilepsy surgery. *Epilepsia.* 1999;40:931–41.

82. Iwasaki M, Nakasato N, Shamoto H, et al. Surgical implications of neuromagnetic spike localization in temporal lobe epilepsy. *Epilepsia.* 2002;43:415–24.

83. Stefan H, Hummel C, Scheler G, et al. Magnetic brain source imaging of focal epileptic activity: a synopsis of 455 cases. *Brain.* 2003;126:2396–405.

84. Genow A, Hummel C, Scheler G, et al. Epilepsy surgery, resection volume and MSI localization in lesional frontal lobe epilepsy. *NeuroImage.* 2004;21:444–9.

Chapter 17

Imaging Neural Excitability and Networks in Genetic Absence Epilepsy Models

Grygoriy Tsenov, Giuseppe Bertini, Michele Pellitteri, Elena Nicolato, Pasquina Marzola, Paolo Francesco Fabene, and Gilles van Luijtelaar

17.1 Introduction

Epilepsies and epileptic syndromes are complex disease states that are accompanied by abnormal hyper excitability and/or hyper synchronicity within the central nervous system (CNS). The classifications of the International League against Epilepsy distinguish generalized from focal epilepsies (http://www.ilae.org/visitors/centre/Definition-2014.cfm), although the classification for some of them has been debated. Absence epilepsies are among the most well-characterized seizure types in patients and in animal models, and represent a most fascinating disease group, classically considered as prototypical generalized epilepsies.[1–6]

Genetic rat models of absence epilepsy have been studied since the 1980s and are believed to mimic more accurately the spontaneous seizures of human epilepsy than drug-induced animal models do.[1,7–9] The GAERS (Genetic Absence Epilepsy Rats from Strasbourg) and WAG/Rij (Wistar Albino Glaxo, originating from the city of Rijswijk) strains are two commonly used and accepted genetic models for childhood absence epilepsy. Rats of both strains are endowed with spontaneously occurring spike-wave discharges (SWDs), with frequencies from 7 to 11 Hz, amplitudes of 200 to 1,000 µV, lasting 1 to 45 seconds, concomitant to mild facial myoclonus.[10,11]

Historically, the GAERS colony was derived from Wistar rats and selectively bred for the seizure phenotype so that 100% of progeny spontaneously develop epilepsy. A nonepileptic control (NEC) strain was also derived from the original colony by selectively breeding for the lack of seizure expression, providing a powerful control strain, since any differences between the two strains would have a high a priori chance of being etiologically associated with consequences the epilepsy trait.

The WAG/Rij and GAERS rats share the same electroclinical phenotype in spite of some minor differences, similar in magnitude to those observed

within the various sublines of GAERS.[12,13] SWDs in WAG/Rij become manifest in the cortical EEG at two to three months of age. At an age of 75 days, one out of six subjects shows SWDs. At six months, all animals are affected, and both male and female rats show about 16–20 discharges per hour, with an average duration of about 5 s, which amounts to several hundred discharges per day.[14,15]

The WAG/Rij and GAERS strains also exhibit a range of behaviors indicative of affective disturbance, such as depressive-like behavior;[16,17] in addition, GAERS show higher anxiety levels, compared to NEC rats.[18] These complex phenotypes bear some resemblance with mood disturbances observed in clinical populations.[19–22] Interestingly, early and chronic anti-epileptogenesis treatment could prevent the depressive-like behavior.[23] Other behavioral characteristics of WAG/Rij rats include a short latency to emerge from the home cage into familiar and novel environments, low open-field defecation, and high open-field ambulation. In the same animals, under stressful circumstances, ambulation is low. They also have a low apomorphine-induced gnawing score, a high running-wheel activity, a low amount of REM sleep and interrupted non-REM-REM sleep cycle, a circadian distribution of the SWDs, good two-way, active shock-avoidance acquisition, and poor spatial reference memory in a hole-board, normal working memory scores in two spatial memory tasks, normal prepulse inhibition, but clearly abnormal auditory sensory gating.[9,10,17,24–26]

17.2 On the Origin of Spike-wave Discharges

Outcomes of early mapping, neurophysiologic, lesion, and pharmacologic studies in the genetic models confirmed the involvement of the thalamus[27–32] in the occurrence of SWDs. Moreover, a pacemaker role of the reticular thalamic nucleus (RTN) and a different role of the ventral basal complex and the RTN in the occurrence of SWDs was suggested. Ibotenic lesions of the lateral thalamus, including the rostral RTN, confirmed that an intact thalamus is a prerequisite for SWD occurrence. Thalamic lesions including the caudal pole of the RTN, however, enhanced SWDs, suggesting that the rostral and caudal poles of the RTN seemed to antagonize each other sustaining SWDs.[33]

The role of the cortex became more obvious after a comprehensive study of network mechanisms responsible for the immediate onset, widespread generalization, and high synchrony of SWDs in the genetic models.[34–37] Based on field potentials simultaneously recorded from multiple cortical and thalamic sites, as well as the analyses of cortico-cortical, intrathalamic, and cortico-thalamic interrelationships between these field potentials with various network analysis techniques, a focal area in the facial somatosensory cortex was identified, followed in time by an extremely fast and large-scale synchronization across cortex and thalamus. These results are incompatible with the classical assumption that the thalamus acts as the primary driving source for the discharges.[5] Instead, network analyses indicate that a cortical focus plays a leading role in the origin of generalized SWDs characteristic of absence seizures in the genetic absence models. The existence of a focal zone as the initiation site has been confirmed and extended in the GAERS model through the identification of hyper-excitable cells in the deep layers of the facial area of the somatosensory cortex, and by inactivation studies with tetrodotoxine.[38,39] Since SWDs can be evoked by cortical and thalamic low frequency electrical stimulation,[34,40,41] it seems rather likely that absence epilepsy should be understood as originating in and engaging a cortico-thalamo-cortical (CTC) network.[36,37,42] A trigger pulse applied either within or outside the network may initiate oscillations in the tightly and reciprocally interconnected CTC network, and SWDs may occur when the brain is in an appropriate state. Reviewing the literature over the last 50 years, it can be concluded that much of the historical controversies concerning the exact mechanisms and sites of origin of the SWDs can be ascribed to the usage of different experimental models[5,43] and that it is risky to interpret oscillations in slices as genuine SWDs.

17.3 Cortical Excitability

Evidence for an excitable focal cortical region is suggested by an increased expression of Nav1.1 and Nav1.6 selectively at the focal region in the somatosensory cortex and only in 5- to 6-month-old WAG/Rij rats.[44] Other evidence comes from neurophysiologic studies, both in vitro and in vivo. In vitro studies showed an increased excitability in the somatosensory cortex in WAG/Rij rats: NMDA-sensitive late EPSP led to action potential discharges in 44% of regular bursting cells in deep neocortical layers, vs. 8% in control rats.[45] Whole cell patch-clamp recording from layer II–III cortical pyramidal neurons in WAG/Rij rats was complemented by immunohistochemical, Western blot, and PCR studies of HCN1–HCN4 subunits of the I_h channel. The fast component of I_h activation in neurons of WAG/Rij rats showed a 50% decrease in current density, and was four times slower than in neurons of non-epileptic control Wistar rats. The results of Western blot and PCR analysis corresponded to a decreased I_h current. A 34% decrease was found in HCN1 subunit protein levels in the cerebral cortex of WAG/Rij rats as compared to Wistar rats, but HCN1 mRNA expression was not different. The protein and mRNA levels of the other three I_h channel subunits (HCN2–HCN4) were not altered.[46] The rapid age-dependent decline in expression of HCN1 channels preceded the onset of SWDs, suggesting that the loss of HCN1 channel expression is inherited rather than acquired.[47] An in vivo electrophysiologic study in free-moving animals with electrical evoked potentials in the somatosensory and motor cortex in WAG/Rij and in nonepileptic, age-matched control rats confirmed the increased excitability in the somatosensory cortex since only the epileptic rats showed an increased response toward cortical electrical stimulation. Moreover, stimulation induced SWD-like afterdischarges, suggesting not only that the focal area was more excitable, but also that the whole CTC network became more prone to SWDs.[40] It has been proposed that this loss of neocortical HCN1 function but also many neurophysiologic and biochemical changes contribute to an increased cortical excitability.[48]

Besides increased excitability, impairment of inhibitory processes was found in the frontal cortex of WAG/Rij rats in vivo when they were compared to three other rat strains in a paired pulse inhibition paradigm (sensory gating). The amplitude of the auditory evoked potentials toward the second stimulus was enhanced,[17,25] suggesting functional inhibitory disturbances in the neocortex of WAG/Rij rats. Other evidence for diminished cortical inhibition was obtained from a pharmacological study in which tiagabine, a GABA reuptake blocker, was locally applied in the somatosensory cortex. A dose-dependent decrease in SWD occurrence was found, suggesting that decreased cortical inhibition might also play a role in the pathogenesis of absence epilepsy.[45] It is not always clear, however, whether the differences between strains are found exclusively in the perioral region of the somatosensory cortex or if other cortical regions differ between the epileptic and control rats. The latter cannot be established without using a complete high-resolution cortical grid. The electrophysiologic studies lack a sufficient spatial resolution and therefore imaging techniques might be helpful to better delineate putative focal cortical zones, and to elucidate the involvement of and changes in wider brain regions than the classically assumed cortex and thalamus. In the next paragraph, first the structural imaging data will be reviewed, followed by some new MRI data collected in WAG/Rij rats. Next, fMRI studies will be reviewed, both regarding the interictal state in the absence models, as well as regarding the hemodynamic response in parallel with the occurrence of SWDs.

17.4 Imaging Absence Epilepsy

17.4.1 Structural MRI Studies

In a structural volumetric analysis (T2-w MRI, 4.7 T) of several brain structures in 14-week-old epileptic GAERS ($n = 11$) and NEC rats, all GAERS rats were reported to display SWDs.[49] It was found that the somatosensory cortex was thicker in GAERS than in NEC rats. Furthermore, the analysis of ROIs showed larger brain structures bilaterally, including the amygdalae, cortices, and cerebral ventricles in GAERS relative to NEC animals. The differences were not generalized, as no differences were observed in other regions, such as the striatum. On the other hand, GAERS rats showed also a hippocampal volume loss, undetectable with classical analyses, but observed with high-dimensional mapping, a postacquisition analysis technique that is sensitive to subtle changes in shape of the structure. The enlarged cortical width and volume could be explained by aberrant neuronal branching and arborization in this area, such as that described in the WAG/Rij model.[50]

Diffusion tensor imaging (DTI), a method based on the detection of water diffusion in biological tissues, allows characterization of long-range white matter networks and their modifications with neuropathology and treatment.[51] WAG/Rij rats at two different developmental stages (1.7 and 8 months) were studied before and after the onset of epilepsy to identify DTI changes related to epileptogenesis, compared to age-matched nonepileptic (control) Wistar rats.[52] The results were compared with 4.2-month GAERS and NEC rats to determine the specificity of these changes. Adult WAG/Rij exhibited a localized decrease in FA (fractional anisotropy, a measure of structural integrity of white matter) in the anterior part of the corpus callosum, compared to controls. The decreased FA in the anterior corpus callosum was not seen in young WAG/Rij rats before the onset of SWDs. Also GAERS exhibited a marked decrease in FA in the anterior corpus callosum vs. age-matched NEC. This decrease was more extensive than in WAG/Rij rats. Symptomatic WAG/Rij and GAERS also have an increased perpendicular diffusivity ($\lambda\perp$) in the anterior corpus callosum, which could be the cause of the reduced FA observed in the epileptic animals. Tractography was used to identify the white matter pathways of the anterior corpus callosum; it showed that the fibers interconnect the facial region of the somatosensory cortex between the two hemispheres. All this suggests that SWDs lead to microstructural changes in white matter pathways interconnecting the regions where the SWDs have their site of origin. These ex vivo DTI results in both absence epilepsy models[52] are an important step for understanding neurological difficulties in children suffering from absence epilepsy, suggesting that white matter abnormalities could contribute to chronic dysfunction in what has classically been considered exclusively a gray matter disorder.

Since treatment of WAG/Rij rats with ethosuximide (ESX) initiated before the developmental onset of SWDs and continued through adulthood was found to suppress seizures even 3 months after the medication was suspended,[53] the possibility that early and sustained blockade of SWDs by early and chronic administration of ESX could prevent white matter alterations was investigated by means of ex vivo, posttreatment DTI.[54] Cortical excitability in the form of electrical evoked potentials, but also CTC network activity by measuring SWDs and by electrical stimulation-induced 8 Hz afterdischarges, and depressive-like behavior were investigated. Four months of treatment with ESX suppressed SWDs during treatment, and 6 days and 2 months posttreatment. Also the duration of afterdischarges was reduced 6 days post-treatment. Increased FA in corpus callosum and internal capsula on DTI was found, an increased amplitude of the evoked potential (P8, a positive component with a latency of 8 ms), and a decreased immobility in the forced swim test, the latter suggesting an anti-depressant-like response. Shorter treatments with ESX had no large effects on any of the above parameters. Chronic ESX has widespread effects not only within but also outside the circuitry in which SWDs are initiated and generated, including preventing epileptogenesis, reducing depressive-like symptoms in WAG/Rij rats[54] and anxiogenic responses in GAERS.[55] It is thought that the treatment of patients before symptom onset might prevent many of the adverse consequences of chronic epilepsy.

17.4.2 Structural MRI: Original Data

We investigated structural changes in the major cerebral key players in SWD initiation and maintenance (lateral thalamus and somatosensory cortex) and in noninvolved control areas such as the hippocampus. We report here an original structural MRI study carried out to establish whether 4-month-old WAG/Rij rats (with a lower incidence of SWDs) and 9-month-old WAG/Rij rats (large amounts of SWDs) differ in T2-w values, indicating alterations in either water content or protein extravasation in the brain tissue. Given that the thalamus is not homogenous with respect to its role in SWD initiation and maintenance,[33,37] we also investigated whether the lateral thalamus including the VPM might differ from the medial thalamus. Since hyperexcitable cells in GAERS have been identified in the subgranular layers of the somatosensory cortex, differences between the sub- and supragranular layers were investigated as well, thanks to the spatial resolution of our scanner.

The T2 values in the motor cortex, hippocampus, and medial section of the thalamus did not show differences between 4-month-old (4 m, $n = 6$, mildly symptomatic) and 9-month-old (9 m, $n = 5$, fully symptomatic) WAG/Rij rats, suggesting that becoming 5 months older does not induce large structural differences in presumably noninvolved brain areas. In contrast, age-dependent decrease (all $ps < .05$) were found exclusively for the three regions involved in the occurrence of SWDs: the supra-somatosensory cortex (supra SS: 4 m 87.2 ± 0.8, 9 m 83.4 ± 0.9), the sub-somatosensory cortex (sub SS: 4 m 83.6 ± 0.7, 9 m 81.3 ± 0.7), and the lateral thalamus (lat thal: 4 m 79.2 ± 0.3, 9 m 76.4 ± 0.4). The data are presented in Figure 17.1 and Table 17.1.

Differences in T2 between supra SS and sub SS ($p < .05$) were found in the young WAG/Rij group only, suggesting that the difference disappeared when the number of seizures increase.

The reviewed data show an extended set of structural changes, mostly in the CTC network, although the limbic system also seemed affected in rats with hundreds of SWDs per day. The structural changes in the cortex were rather selective for the somatosensory cortex as they did not occur in the motor cortex and also selective for the lateral thalamus containing the VPM and VPL, as they did not occur in the limbic thalamus. An increase in FA was also observed in the anterior commissure and in the anterior part of the callosum. The latter connects the focal regions in the two hemispheres. All these fit well with the cortical focus theory. The volumetric changes in the amygdala could be related to the increased anxiety in GAERS. Treatment aimed at the prevention of expression of SWDs prevented the DTI changes in WAG/Rij rats, suggesting that at least some of the structural changes were caused by the frequently occurring SWDs.[54]

17.5 Functional Imaging of the Dynamic Response Accompanying SWDs

Tenney et al.[56] used EEG-triggered fMRI (T2*-weighted echo planar imaging at 4.7 T) of SWDs in the WAG/Rij model to determine the activity of several thalamic nuclei and cortical areas that are supposed to be involved in the genesis and maintenance of SWD activity. Rats were first acclimated to the restrainer and next surgically implanted with MR-compatible epidural EEG electrodes, and Antisedan was administered so the combined EEG and MR study was done in fully conscious rats. BOLD activation associated with SWDs (34 episodes with a mean duration of 7 s) was clustered throughout the primary somatosensory, parietal association, and temporal cortices, along with several thalamic nuclei including the RTN, MD, VPM/VPL, and Po. No significant negative BOLD activation was seen for any seizures.

Also others were investigating whether the increases in oxygen delivery during SWDs are truly generalized throughout the entire brain, or whether the increases occurred preferentially in focal regions where neuronal firing is most intense. A second question was whether neuronal activity was matched by changes in local cerebral blood flow.[57] WAG/Rij were anesthetized with a mix of neurolept-fentanyl anesthesia, which permits the occurrence of SWDs, closely mimicking those that occur spontaneously. A total of 256 SWDs obtained from 9 rats were investigated. The time course of the BOLD fMRI signal changed in the presence of SWDs. Clear increases in the BOLD fMRI signal were found in the facial part of the somatosensory cortex and in the thalamus. On the other hand, no significant changes related to SWD activity were observed in the primary visual cortex.

The most comprehensive fMRI study was done by Mishra et al.[58] A total of 1,856 SWD events were analyzed in 22 neurolept-fentanyl anesthetized WAG/Rij rats. BOLD activation, cerebral blood volume (CBV), as well as laser Doppler cerebral blood flow (CBF) were determined with a 9.4 T scanner. Local field potential (LFP) and multiunit activity (MUA) recordings were obtained in order to establish whether neuronal activity was matched by changes in local cerebral blood flow. The study confirmed and extended the previous results that BOLD signal changes during SWDs do not involve the whole brain uniformly. Rather, it showed both increases and decreases in specific cortical and subcortical brain regions, including intense bilateral increases in BOLD response in the facial zone of the somatosensory cortex, thalamus, anterior cingulate, posterior cingulate/retrosplenial cortex. Furthermore, BOLD increases in the superior colliculi were found in about 40% of the animals. The somatosensory cortex and thalamus showed increased fMRI, CBV, CBF, LFP, and MUA signals accompanying SWDs. However, the caudate-putamen showed fMRI, CBV, and CBF decreases despite increases in LFP and MUA signals. A dissociation between electrophysiological signals and fMRI activation also exists in the cortex, as SWDs can be found in frontal, central, and parietal cortex while the cortical BOLD increase is selective for the facial region of the parietal cortex.[57]

Table 17.1 Summary of T2 Values of Four Cortical and Three Subcortical Regions for the 4- and 9-Month-Old WAG/Rij Rats

Region	4 months	9 months
Motor cortex supragranular	88.35 ± 3.39	86.83 ± 3.06
Motor cortex subgranular	89.01 ± 4.51	84.01 ± 2.84
Somatosensory cortex supragranular	87.82 ± 2.90	84.14 ± 2.68
Somatosensory cortex subgranular	84.33 ± 3.32	81.22 ± 2.39
Hippocampus	90.09 ± 3.10	88.35 ± 4.42
Lateral thalamus	79.08 ± 2.18	76.31 ± 1.05
Medial thalamus	87.47 ± 3.99	84.85 ± 1.52

Data are expressed as mean ± standard deviation.

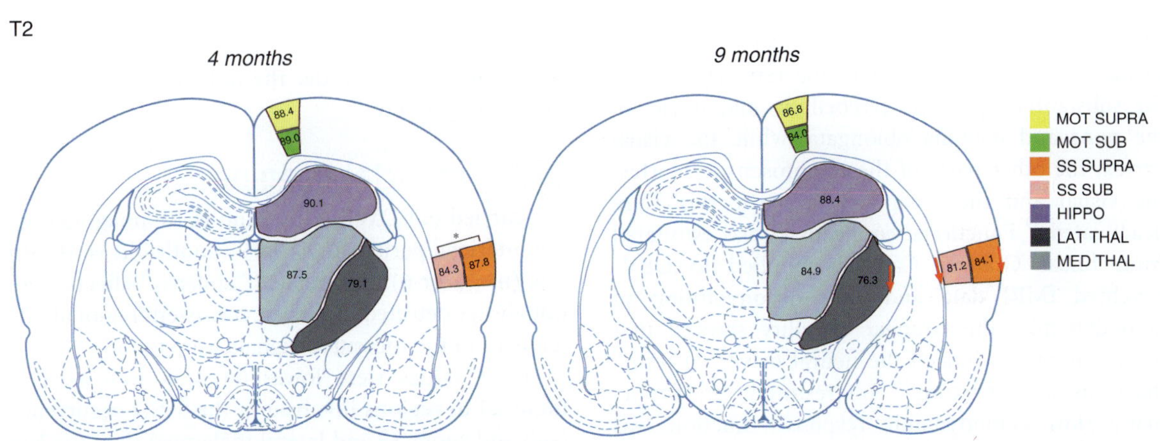

Figure 17.1. Schematic representation of T2 values comparing mild symptomatic (4 months old) versus fully symptomatic (9 months) animals. Asterisks denote a difference for multiple areas in the same group of animals ($p < .05$). Arrows indicate differences between 4- and 9-month-old animals. Modified from Ott et al.[21]

Near-infrared spectroscopy (NIRS) is a recently developed technique for hemodynamic studies that is particularly suitable for dynamic recordings in humans and animals. It uses the particular absorption properties of living tissues in the near-infrared range to measure changes in the concentrations of oxy-, deoxy-, and total hemoglobin (HbO_2, HHb, and HbT, respectively) in tissues. In order to assess whether SWDs are accompanied by a cortical hemodynamic response, 444 SWD episodes from 6 GAERS rats in a quiet waking state were used as triggers for the analyses of the NIRS measurements.[26] It was found that the concentration of HbO_2 starts to decrease 13 s before SWD onset (20% of baseline), then increases (25% of baseline) from 1 s before the onset of SWDs, reaches a maximum 7 s after the onset, decreases (22% of baseline) from 1 s before the end of the EEG SWD, reaches a trough 9 s after the end, and finally returns to baseline. The concentration of HHb shows an inverse pattern of similar amplitude, while the concentration of HbT closely follows HbO_2.

Next, whether the cortical epileptic focus in the genetic models can be interpreted as a driving source was investigated by means of fMRI data collected in a 2.35 T magnet.[59] Standard techniques estimating functional connectivity measures are jeopardized by the variation of blood flow dynamics between regions. The authors were able to remove hemodynamic effects by using appropriate modeling techniques. Highly significant increases and decreases in CBV were found during SWDs at the group level in neurolept anesthetized GAERS in the facial area of the somatosensory cortex, the ventrolateral thalamus, but also in the central medial and mediodorsal thalamus, next to the reticular part of the substantia nigra, the cerebellum, and nuclei of the pons and medulla oblongata, while the visual cortex, the other parts of the somatosensory cortex, the visual and the secondary motor cortex were deactivated.[59] Functional connectivity was estimated from linear Granger Causality applied to CBV-weighted fMRI data and after deconvolution of hemodynamics the Granger causality showed that the facial area is indeed the neural driver, affecting thalamus and striatum.[59] Interestingly, an abnormally slow hemodynamic response function was also found in the focal region. EEG data collected from various subcortical regions in free moving GAERS indicated that the averaged cortical SWDs preceded the thalamic one and this validated the concept that the focal region is indeed driving the thalamus and striatum.[59]

It seems, therefore, that the increases in oxygen delivery during SWDs are not truly generalized throughout the entire brain. Rather, the data point to the facial area of the somatosensory cortex and the thalamus as the major actors in SWD occurrence. The increase in hemodynamic variables parallels the appearance of SWD, and in most regions an increase in CBV and CBF. The driving function of the facial region in the somatosensory cortex to the thalamus throughout the whole SWD was additionally proposed.

Brain networks were investigated in WAG/Rij rats by means of fMRI-based resting functional connectivity measures. Simultaneous EEG-fMRI data were acquired at 9.4 T in epileptic WAG/Rij rats compared to nonepileptic Wistar and WAG/Rij controls. Two focal regions were identified: one in the left and one in the right part of the facial zone of the somatosensory cortex. Both showed an fMRI increase during SWDs. These two regions were used for connectivity analysis to investigate whether chronic seizure activity is associated with changes in network resting functional connectivity. High degrees of interictal cortical-cortical correlations were found in all WAG/Rij rats, but not in NEC WAG/Rij rats. Strongest connectivity was seen between the bilateral somatosensory and adjacent cortices. This result, together with the DTI studies described above, suggest that the two focal regions in the left and right hemispheres are strongly functionally connected despite the decrease in FA. Both seemingly opposite changes occur, most likely, under the influence of hundreds of SWDs per day.[58]

17.6 Interictal Imaging

We carried out DWI, rCBV, and rCBF imaging during interictal periods in 4- vs. 9-month-old WAG/Rij rats ($n = 5$ or 6). Age-related "epilepsy" effects were not always detected, however, consistent regional differences in the measured signals were found.

DWI values were lower (at 4 as well as at 9 months) in regions involved in the SWD occurrence (sub and supra SS and lateral thalamus) than the less involved regions (motor cortex, medial thalamus, hippocampus). Next, the subgranular cortical regions were always lower that the supragranular layers, and the difference was significant for both cortical regions sub SS (0.00063 ± 0.00003) < supra SS (0.00065 ± 0.00004) and sub MOT (0.00068 ± 0.00005) < supra MOT (0.00075 ± 0.00005). Thus, WAG/Rij rats have higher supra than sub DWI values in the cortex as a whole and lower DWI values in the SS compared to the motor cortex. The data are presented in Figure 17.2.

rCBV data obtained from images acquired with a gradient-echo sequence before and after USPIO administration showed no significant effects of age on rCBV in any of the brain regions investigated. However, we found regional differences and an interaction between region and age. In particular, at 4 months of age, the motor cortex supragranular layer had higher ($p < .05$) rCBV values than the subgranular layer, at 9 months the subgranular layer had increased ($p < .05$) its rCBV value and it was now higher ($p < .05$) than the supragranular layer (Table 17.2). This demonstrated that the difference between supra- and subgranular layers and how they change over time are typical for the motor cortex, our control region, since they were not present in the somatosensory cortex.

Age and region effects were found for rCBF in all regions: 9-month-old rats had higher values ($8.14 ± 0.53$) than 4-month-old rats ($5.34 ± 1.90$; see Figure 17.3 and Table 17.3). Post hoc tests regarding the regional differences showed lower ($p < .01$) rCBF values in the somatosensory cortex than in the motor cortex and higher ($p < .05$ and $p < .01$) rCBF values in the sub- than in the supragranular cortical layers in the somatosensory and motor cortex, respectively. This all implies that the age-related increases in rCBF were not found in the SS cortex.

We also found subcortical regional differences in rCBF. The highest values were in the hippocampus, lower ($ps < .05$) for the medial and lateral thalamus. Thus, rCBF data demonstrate cortical and subcortical regional differences and higher values for the older animals in both cortical and subcortical regions. However, this subcortical pattern does not clearly points toward selective changes rCBF in the networks involved in absence epilepsy.

17.7 Cortico-thalamic-cortical and Cortico-cortical Networks

A wealth of evidence strongly suggests that in the genetic rodent absence models the facial region of the somatosensory cortex contains a hyperexcitable zone. The fMRI data reviewed here, describing the hemodynamic changes accompanying SWDs, confirm a selective involvement of cortical and subcortical regions (lateral thalamus, but also other medial thalamic nuclei, the striatum, the cerebellum, and some brainstem nuclei) and are all in line with the cortical focus theory of absence epilepsy, which states that absence epilepsy in these genetic models is due to an increased cortical excitability in a focal region, accompanied by a reduced cortical inhibition. The foci in the left and right hemispheres form a strongly interconnected network. The regional selective increased BOLD, CBF, and CBV values accompanying the SWDs agree with the increased neuronal activity in most parts of brain. An interesting observation remains that SWDs are generalized over the cortex, while the hemodynamic responses are restricted, which suggests that the

Table 17.2 Summary of rCBV Results for both Experimental Groups

Region	4 months	9 months
Motor cortex supragranular	7440 ± 1235	6122 ± 454
Motor cortex subgranular	5205 ± 678	7566 ± 315
Somatosensory cortex supragranular	5550 ± 2619	5050 ± 781
Somatosensory cortex subgranular	4611 ± 812	5238 ± 768
Hippocampus	7203 ± 1505	6269 ± 640
Lateral thalamus	5025 ± 1720	5121 ± 924
Medial thalamus	6310 ± 1472	4889 ± 318

Data are expressed as mean ± standard deviation.

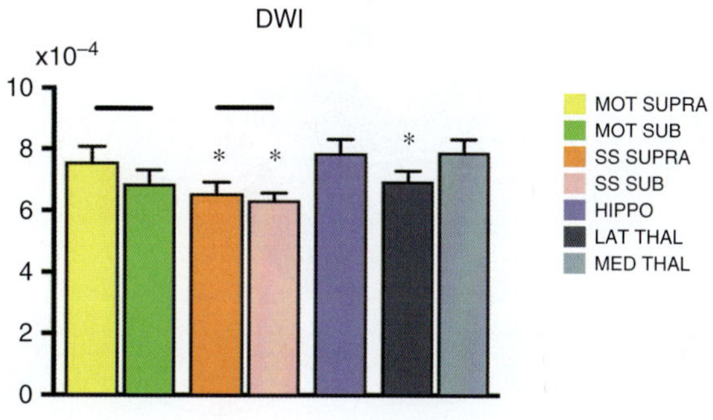

Figure 17.2. Summary of DWI results (pooled groups). Data are expressed as mean ± *SD*. Asterisks denote a difference with the pooled noninvolved brain regions ($p < .05$). Black horizontal lines indicate differences between supra- and subgranular layer ($p < .05$).

presence of SWDs is not sufficient to increase oxygen demands. A driving function, as previously established,[35,37,59] is a better candidate for an increased demand of oxygen in this area. The early changes (13 s pre SWD onset) as observed with NIRS deserve to be investigated in more detail. The increased hemodynamic response in the thalamus can be ascribed to the SWD accompanying increased tonic inhibition.[60]

The functional interictal connectivity changes point toward an increase in white matter tracts, connecting the bilateral foci. Although DWI reductions

reflect pathological conditions in brain tissue that are only partially understood, and involve changes in the diffusion characteristics of intra- and extracellular water compartments and water exchange across permeable boundaries,[61] they have been associated with acute cytotoxic edema.[61] The observation that changes were exclusively present in the absence epilepsy brain circuitry may have some consequences for local changes in excitability or vice versa. And this merits further investigation.

The age-related changes in rCBV in control regions in 4- vs. 9-month-old WAG/Rij rats did not occur in the focal region, and the latter area was characterized by a lower rCBF than the cortical control region, suggesting that absence epilepsy is accompanied not only by an interictal but also by an ictal change in the hemodynamic response function.[59]

The age-related increase in rCBF in all the cortical and subcortical regions and the higher rCBF values might be epilepsy related, but it can also be a general aging effect. It might point toward an increased glucose metabolism, resulting from increased inflammatory cell activation,[62] commonly seen in aging.

The structural imaging data (T2 and DTI) are also in line with a special and perhaps unique role of the bilateral foci in the facial region of the somatosensory cortex, including the morphological changes expressed in the anterior corpus callosum, interconnecting the bilateral foci. It might be a reason, together with the functional changes that were reported, why the pathological activity seems

Table 17.3 Summary of rCBF Results for All Experimental Groups

Region	4 months	9 months
Motor cortex supragranular	5.62 ± 1.67	8.26 ± 0.31
Motor cortex subgranular	6.01 ± 1.15	8.67 ± 0.49
Somatosensory cortex supragranular	4.50 ± 2.54	7.73 ± 0.29
Somatosensory cortex subgranular	5.24 ± 2.16	7.92 ± 0.56
Hippocampus	6.11 ± 1.22	8.10 ± 0.30
Lateral thalamus	4.53 ± 2.70	7.26 ± 0.41
Medial thalamus	4.85 ± 2.20	7.38 ± 0.45

Data are expressed as mean ± standard deviation.

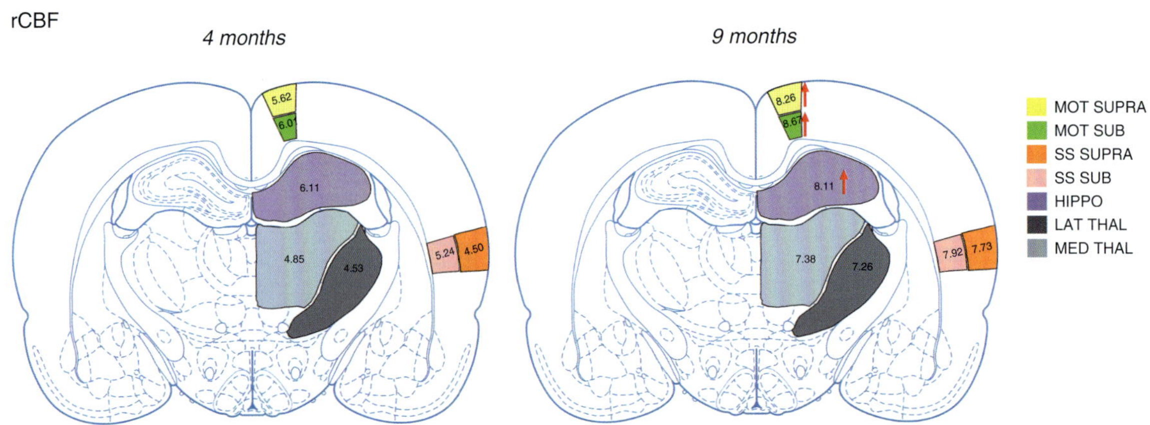

Figure 17.3. Schematic representation of rCBF results comparing 9-month-old with 4-month-old animals. Arrows indicate differences between age groups ($p < .05$).

generalized after being initiated in a single hemisphere. The structural data also point toward interictal changes in the lateral thalamus.

The decrease in T2 in the brain regions specifically involved in SWD occurrence might indicate a decline in extracellular water, and once more this might affect the excitability of neuronal tissue. As an alternative explanation, some authors[63] suggest that an increased concentration of deoxyhemoglobin causes a decrease in T2. This so-called negative BOLD effect has been shown to occur in MRI studies of hypoxia in the rat. Deoxyhemoglobin increases in situations in which the metabolic requirement for oxygen is not met by the delivery of oxygen.

The way in which structural and functional changes in local excitability and in the CTC and cortico-cortical networks relevant for absence epilepsy are related is by no means clear. What is clear, though, is that the loss of cortical HCN1 found in WAG/Rij rats preceded the onset of SWD, suggesting that the increased excitability is inherited, rather than required for SWD to occur.[47] Moreover, the same authors suggested that this loss of HCN1 provides a somatodendritic mechanism for increasing the synchronization of cortical output, and is therefore likely to play an important role in the generation of SWDs via the CTC and cortico-cortical networks.

An undisputed role is acknowledged for the corpus callosum in synchronizing the bilateral cortical and thalamic SWDs. Each hemisphere is able to initiate SWDs independently and SWDs become bilateral and symmetrical through the rapid interhemispheric communication via the monosynaptic callosal projections connecting homotopic regions of the somatosensory cortex. Interestingly, it has been demonstrated that the structural changes in the connectivity between the facial focal regions of the somatosensory cortex, as found by Hal Blumenfeld's group,[43] could be reduced by the prevention of SWDs, suggesting a causal relationship between the morphological callosal and cortical excitability changes. The reduction in excitability induced by antiepileptogenic treatment, as expressed by the enhancement of HCN1 channels, was accompained by a reduction in the amplitude of the elicited local electrical evoked potential, and the reduction of SWDs and by stimulation-induced afterdischarges.[53,54]

Changes in absence epileptic brains are not restricted to the mentioned networks, and functional changes in the limbic system are well documented both in WAG/Rij and in GAERS rats: both strains showed a resistance to electrical amygdala kindling.[64] Imaging studies showed structural changes in the amygdala and hippocampus in GAERS. Whether the behavioral phenotype showing depressive-like and anxiety behavior is related to that needs to be investigated. It has been suggested that also these changes might be due to the daily occurrence of hundreds of SWDs.

Considering the wealth of other cortical and subcortical changes in, among others, GABAergic and glutamatergic neurotransmission functions, it is likely that other cortical functions are disturbed, such as sensory gating or the organization of sleep by the hypothalamus-brainstem network.

17.8 Conclusions

Work in genetic absence epilepsy models (WAG/Rij and GAERS) has revolutionized the theories on SWD generation in absence epilepsy and, more in general, on genetic generalized epilepsies. The analysis of EEG signals from in vivo studies has demonstrated the existence of cortical foci in the facial region of the somatosensory cortex, from which SWD activity drives other cortical and thalamic regions. Different types of high-resolution structural and functional imaging studies in these models have yielded a wealth of new data, demonstrating a selective involvement of brain structures in SWD occurrence, and highlighting functional and structural changes in neuronal networks within the CTC and cortico-cortical networks, but also in other subcortical brain regions. These data leave no room for the classical axiom that absence epilepsy is a generalized type of epilepsy. The terms "focal epilepsy" and "network epilepsy" are a better fit for the experimental neurophysiologic and imaging data. Arguably, they might be helpful in the understanding of cognitive, emotional, and sleep-related problems often encountered in persons with absence epilepsy.

References

1. Danober L, Deransart C, Depaulis A, et al. Pathophysiological mechanisms of genetic absence epilepsy in the rat. *Prog Neurobiol.* 1998;55:27–57.

2. Khosravani H, Zamponi GW. Voltage-gated calcium channels and idiopathic generalized epilepsies. *Physiol Rev.* 2006;86:941–66.

3. Llinas RR, Steriade M. Bursting of thalamic neurons and states of vigilance. *J Neurophysiol.* 2006;**95**:3297–308.

4. Manning JP, Richards DA, Bowery NG. Pharmacology of absence epilepsy. *Trends Pharmacol Sci.* 2003;**24**: 542–9.

5. Meeren H, van Luijtelaar G, Lopes da Silva F, et al. Evolving concepts on the pathophysiology of absence seizures: the cortical focus theory. *Arch Neurol.* 2005;**62**:371–6.

6. Niedermeyer E. Primary (idiopathic) generalized epilepsy and underlying mechanisms. *Clin Electroencephalogr.* 1996;**27**:1–21.

7. Coenen AM, van Luijtelaar EL. Genetic animal models for absence epilepsy: a review of the WAG/Rij strain of rats. *Behav Genet.* 2003;**33**:635–55.

8. Depaulis A, van Luijtelaar G. Genetic models of absence epilepsy in the rat. In: Pitkänen A, Schwartzkroin PA, Moshé SL, eds. *Models of Seizures and Epilepsy.* New York: Academic Press; 2006:233–48.

9. van Luijtelaar EL, Drinkenburg WH, van Rijn CM, et al. Rat models of genetic absence epilepsy: what do EEG spike-wave discharges tell us about drug effects? *Methods Find Exp Clin Pharmacol.* 2002;**24** (suppl D):65–70.

10. van Luijtelaar EL, Coenen AM. Two types of electrocortical paroxysms in an inbred strain of rats. *Neurosci Lett.* 1986;**70**:393–7.

11. Vergnes M, Marescaux C, Micheletti G, et al. Spontaneous paroxysmal electroclinical patterns in rat: a model of generalized non-convulsive epilepsy. *Neurosci Lett.* 1982;**33**:97–101.

12. Akman O, Demiralp T, Ates N, et al. Electroencephalographic differences between WAG/ Rij and GAERS rat models of absence epilepsy. *Epilepsy Res.* 2010;**89**:185–93.

13. Powell KL, Tang H, Ng C, et al. Seizure expression, behavior, and brain morphology differences in colonies of genetic absence epilepsy rats from Strasbourg. *Epilepsia.* 2014;**55**:1959–68.

14. Coenen AM, van Luijtelaar EL. The WAG/Rij rat model for absence epilepsy: age and sex factors. *Epilepsy Res.* 1987;**1**:297–301.

15. Schridde U, van Luijtelaar G. The influence of strain and housing on two types of spike-wave discharges in rats. *Genes Brain Behav.* 2004;**3**:1–7.

16. Sarkisova KY, Midzianovskaia IS, Kulikov MA. Depressive-like behavioral alterations and c-fos expression in the dopaminergic brain regions in WAG/ Rij rats with genetic absence epilepsy. *Behav Brain Res.* 2003;**144**:211–26.

17. van Luijtelaar G. The prevention of behavioral consequences of idiopathic generalized epilepsy: evidence from rodent models. *Neurosci Lett.* 2011;**497**: 177–84.

18. Jones NC, Salzberg MR, Kumar G, et al. Elevated anxiety and depressive-like behavior in a rat model of genetic generalized epilepsy suggesting common causation. *Exp Neurol.* 2008;**209**:254–60.

19. Caplan R, Siddarth P, Gurbani S, et al. Depression and anxiety disorders in pediatric epilepsy. *Epilepsia.* 2005;**46**:720–30.

20. Davies S, Heyman I, Goodman R. A population survey of mental health problems in children with epilepsy. *Dev Med Child Neurol.* 2003;**45**:292–5.

21. Ott D, Siddarth P, Gurbani S, et al. Behavioral disorders in pediatric epilepsy: unmet psychiatric need. *Epilepsia.* 2003;**44**:591–7.

22. Tellez-Zenteno JF, Patten SB, Jette N, et al. Psychiatric comorbidity in epilepsy: a population-based analysis. *Epilepsia.* 2007;**48**:2336–44.

23. Sarkisova KY, Kuznetsova GD, Kulikov MA, et al. Spike-wave discharges are necessary for the expression of behavioral depression-like symptoms. *Epilepsia.* 2010;**51**:146–60.

24. de Bruin NM, van Luijtelaar EL, Cools AR, et al. Dopamine characteristics in rat genotypes with distinct susceptibility to epileptic activity: apomorphine-induced stereotyped gnawing and novelty/amphetamine-induced locomotor stimulation. *Behav Pharmacol.* 2001;**12**:517–25.

25. de Bruin NM, van Luijtelaar EL, Cools AR, et al. Auditory information processing in rat genotypes with different dopaminergic properties. *Psychopharmacology.* 2001;**156**:352–9.

26. Roche-Labarbe N, Zaaimi B, Mahmoudzadeh M, et al. NIRS-measured oxy- and deoxyhemoglobin changes associated with EEG spike-and-wave discharges in a genetic model of absence epilepsy: the GAERS. *Epilepsia.* 2010;**51**:1374–84.

27. Aker RG, Ozkara C, Dervent A, et al. Enhancement of spike and wave discharges by microinjection of bicuculline into the reticular nucleus of rats with absence epilepsy. *Neurosci Lett.* 2002; **322**:71–4.

28. Avanzini G, Vergnes M, Spreafico R, et al. Calcium-dependent regulation of genetically determined spike and waves by the reticular thalamic nucleus of rats. *Epilepsia.* 1993;**34**:1–7.

29. Inoue M, Duysens J, Vossen JM, et al. Thalamic multiple-unit activity underlying spike-wave discharges in anesthetized rats. *Brain Res.* 1993;**612**:35–40.

30. Liu Z, Vergnes M, Depaulis A, et al. Involvement of intrathalamic GABAB neurotransmission in the control of absence seizures in the rat. *Neuroscience.* 1992;**48**:87–93.

31. Marescaux C, Vergnes M, Depaulis A. Genetic absence epilepsy in rats from Strasbourg–a review. *J Neural Transm Suppl.* 1992;**35**:37–69.

32. Vergnes M, Marescaux C, Depaulis A. Mapping of spontaneous spike and wave discharges in Wistar rats with genetic generalized non-convulsive epilepsy. *Brain Res.* 1990;**523**:87–91.

33. Meeren HK, Veening JG, Moderscheim TA, et al. Thalamic lesions in a genetic rat model of absence epilepsy: dissociation between spike-wave discharges and sleep spindles. *Exp Neurol.* 2009;**217**:25–37.

34. Lüttjohann A, van Luijtelaar G. Thalamic stimulation in absence epilepsy. *Epilepsy Res.* 2013;**106**:136–45.

35. Meeren HK, Pijn JP, Van Luijtelaar EL, et al. Cortical focus drives widespread corticothalamic networks during spontaneous absence seizures in rats. *J Neurosci.* 2002;**22**:1480–95.

36. Lüttjohann A, van Luijtelaar G. The dynamics of cortico-thalamo-cortical interactions at the transition from pre-ictal to ictal LFPs in absence epilepsy. *Neurobiol Dis.* 2012;**47**:49–60.

37. Lüttjohann A, van Luijtelaar G. Dynamics of networks during absence seizure's on- and offset in rodents and man. *Front Physiol.* 2015;**6**:16.

38. Polack PO, Guillemain I, Hu E, et al. Deep layer somatosensory cortical neurons initiate spike-and-wave discharges in a genetic model of absence seizures. *J Neurosci.* 2007;**27**:6590–9.

39. Polack PO, Mahon S, Chavez M, et al. Inactivation of the somatosensory cortex prevents paroxysmal oscillations in cortical and related thalamic neurons in a genetic model of absence epilepsy. *Cereb Cortex.* 2009;**19**:2078–91.

40. Lüttjohann A, Zhang S, de Peijper R, et al. Electrical stimulation of the epileptic focus in absence epileptic WAG/Rij rats: assessment of local and network excitability. *Neuroscience.* 2011;**188**:125–34.

41. Zheng TW, O'Brien TJ, Morris MJ, et al. Rhythmic neuronal activity in S2 somatosensory and insular cortices contribute to the initiation of absence-related spike-and-wave discharges. *Epilepsia.* 2012;**53**:1948–58.

42. Stefan H, Lopes da Silva FH. Epileptic neuronal networks: methods of identification and clinical relevance. *Front Neurol.* 2013;**4**:8.

43. Blumenfeld H. Cellular and network mechanisms of spike-wave seizures. *Epilepsia.* 2005;**46**(suppl 9):21–33.

44. Klein JP, Khera DS, Nersesyan H, et al. Dysregulation of sodium channel expression in cortical neurons in a rodent model of absence epilepsy. *Brain Res.* 2004;**1000**:102–9.

45. D'Antuono M, Inaba Y, Biagini G, et al. Synaptic hyperexcitability of deep layer neocortical cells in a genetic model of absence seizures. *Genes Brain Behav.* 2006;**5**:73–84.

46. Strauss U, Kole MH, Brauer AU, et al. An impaired neocortical Ih is associated with enhanced excitability and absence epilepsy. *Eur J Neurosci.* 2004;**19**:3048–58.

47. Kole MH, Brauer AU, Stuart GJ. Inherited cortical HCN1 channel loss amplifies dendritic calcium electrogenesis and burst firing in a rat absence epilepsy model. *J Physiol.* 2007;**578**:507–25.

48. van Luijtelaar G, Zobeiri M. Progress and outlooks in a genetic absence epilepsy model (WAG/Rij). *Curr Med Chem.* 2014;**21**:704–21.

49. Bouilleret V, Hogan RE, Velakoulis D, et al. Morphometric abnormalities and hyperanxiety in genetically epileptic rats: a model of psychiatric comorbidity? *NeuroImage.* 2009;**45**:267–74.

50. Karpova AV, Bikbaev AF, Coenen AM, et al. Morphometric Golgi study of cortical locations in WAG/Rij rats: the cortical focus theory. *Neurosci Res.* 2005;**51**:119–28.

51. Basser PJ, Jones DK. Diffusion-tensor MRI: theory, experimental design and data analysis—a technical review. *NMR Biomed.* 2002;**15**:456–67.

52. Chahboune H, Mishra AM, DeSalvo MN, et al. DTI abnormalities in anterior corpus callosum of rats with spike-wave epilepsy. *NeuroImage.* 2009;**47**:459–66.

53. Blumenfeld H, Klein JP, Schridde U, et al. Early treatment suppresses the development of spike-wave epilepsy in a rat model. *Epilepsia.* 2008;**49**:400–9.

54. van Luijtelaar G, Mishra AM, Edelbroek P, et al. Anti-epileptogenesis: electrophysiology, diffusion tensor imaging and behavior in a genetic absence model. *Neurobiol Dis.* 2013;**60**:126–38.

55. Dezsi G, Ozturk E, Stanic D, et al. Ethosuximide reduces epileptogenesis and behavioral comorbidity in the GAERS model of genetic generalized epilepsy. *Epilepsia.* 2013;**54**:635–43.

56. Tenney JR, Duong TQ, King JA, et al. FMRI of brain activation in a genetic rat model of absence seizures. *Epilepsia.* 2004;**45**:576–82.

57. Nersesyan H, Herman P, Erdogan E, et al. Relative changes in cerebral blood flow and neuronal activity in local microdomains during generalized seizures. *J Cereb Blood Flow Metab.* 2004;**24**:1057–68.

58. Mishra AM, Bai X, Motelow JE, et al. Increased resting functional connectivity in spike-wave epilepsy in WAG/Rij rats. *Epilepsia.* 2013;**54**:1214–22.

59. David O, Guillemain I, Saillet S, et al. Identifying neural drivers with functional MRI: an electrophysiological validation. *PLOS Biol.* 2008;**6**: 2683–97.

60. Crunelli V, Cope DW, Terry JR. Transition to absence seizures and the role of GABA(A) receptors. *Epilepsy Res.* 2011;**97**:283–9.

61. Gass A, Szabo K, Behrens S, et al. A diffusion-weighted MRI study of acute ischemic distal arm paresis. *Neurology.* 2001;**57**:1589–94.

62. Juhler M, Paulson OB. Regional cerebral blood flow in acute experimental allergic encephalomyelitis. *Brain Res.* 1986;**363**:272–8.

63. Lei H, Zhang Y, Zhu XH, et al. Changes in the proton T2 relaxation times of cerebral water and metabolites during forebrain ischemia in rat at 9.4 T. *Magn Reson Med.* 2003;**49**:979–84.

64. Onat FY, Aker RG, Gurbanova AA, et al. The effect of generalized absence seizures on the progression of kindling in the rat. *Epilepsia.* 2007;**48**(suppl 5):150–6.

William Davis Gaillard and Madison M. Berl

18 Network Excitability and Cognition in the Developing Brain

18.1 Introduction

This chapter reviews imaging data on normal brain development, and examines the effect of epilepsy on these systems. FDG-PET, SPECT, and an array of MR imaging provide insights into brain development. These data allow in vivo assessments that recapitulate the elegant anatomic studies of human cortical development[1] and myelination.[2] This chapter does not review the neurobiology of fetal brain development including cellular division and migration. The process in general is of the establishment of synaptic connections, synaptic pruning, and myelination, a process not "complete" until 25 years although learning, and its underlying biochemical, biophysical, and synaptic processes continue throughout the lifetime.

18.2 Control Data for Normal Brain Development

One challenge has been to obtain data from normal, typically developing "controls." Infant MRI can be obtained as children sleep, and children above four years can be easily imaged unless significantly developmentally delayed, anxious, hyperactive, or oppositional. One can more readily obtain reliable data with children with developmental and intellectual disabilities after nine years of age.[3] As sedation is not warranted, it has become more challenging to obtain data in children three months to four years of age, often obtained when asleep during nighttime studies. There is a large cohort of cross-sectional and longitudinal structural MRI data acquired on 1,500 children for the national NINDS, NIMH, NICHD initiative on 1.5 T scanners, and several other large datasets with children scanned every year for three to four years creating a matrix of small-scale, overlapping, longitudinal data for children ages 4/5 years to 18 years.[4–8]

For cognitive studies, children can perform tasks such as motor tapping nend listening to stories at age three (admittedly the more precocious children and

IQs are in the 120s). However, the paradigms must be developmentally and cognitively age appropriate. For example, children younger than seven find it difficult to perform verbal fluency tasks and semantic decision tasks. Younger children tend to have higher IQs; most fMRI development research has mean IQs in the high normal range (e.g., 115).[9–20] Thus, there is a slight selection bias for many functional imaging studies.

For infants and young children, passive listening tasks and motor tasks have been conducted when asleep or lightly sedated as deep anesthesia obliterates the blood-oxygen-level-dependent (BOLD) response.[21–24] Resting state can be obtained, but it is more difficult to know what occurs during resting state when asleep, lightly sedated, or in the very young, let alone in an older child or an adolescent.

FDG-PET and Xenon SPECT confer radiation exposure, which is, understandably, tightly regulated in patient populations and for clear reasons not permissible to perform in typically developing infants, children, and adolescents. Some "normative" data have been obtained to evaluate children thought to have or be at risk for developmental disorders (e.g., children with port wine birthmark in V1 distribution and suspected Sturge-Weber), or in children with focal abnormalities isolated to one hemisphere with the normal hemisphere used as the control or normal hemisphere (see below).

18.3 Metabolism and Blood Flow

Two FDG-PET studies examined local cerebral metabolic rates for glucose, a measure of cellular energy demand that is primarily synaptic in origin.[25,26] These studies show increasing cortical glucose uptake rapidly increasing to five years, then plateauing, then decreasing to adult levels in late adolescence.[25,26] Cerebral blood flow (CBF) is coupled (generally) to metabolic consumption but is an indirect measure of metabolism. Xenon SPECT studies show a similar pattern.[27] These studies are interpreted as showing the growth in cortical thickness, synaptogenesis, and subsequent pruning

(Figures 18.1, 18.2). These trends are global, though there are differences in some regions. These regional differences are most pronounced in the first year of life when most glucose uptake is found in motor cortex, less thalamus, and visual cortex, followed by cerebellum. Later in infancy glucose uptake is seen in most other cortical regions to varying degrees. This pattern helps explain why these regions are most vulnerable to perinatal hypoxic ischemia. This primary motor and sensory cortex achieve adult values earlier during development; association cortex achieves adult maturity latest.

18.4 Structural Imaging

High-resolution T1 weighted images show general brain development but can also be used to examine sulcation as well as cortical thickness. These investigations have been primarily performed with children aged five years and older. As the cortex is only two millimeters thick on average, high-resolution images are necessary to properly measure changes in the thickness of the cortical ribbon and clearly demarcate it in relation to underlying white matter. T2 images are necessary to assess best maturation of white matter. The main developmental process is thinning of the cortex with maturation (Figure 18.3). Once the cortical mantle is established, the developmental decreases in cortical thickness vary in degree and by

regions. There are regional differences in the time line of achieving cortical thickness seen in adults. The primary motor and sensory cortex are the first to mature. The association cortices, frontal and parietal, are the last to achieve adult thickness. The phenomenon of cortical thinning is thought to reflect synaptic pruning.[5–8,28,29] The consequence of these findings is that efforts to identify regional atrophy or other changes in structural measures caused by, or associated with, epilepsy need to be assessed in the context of normal measures and trajectories of structural development. Between age five years and young adulthood there is little change in brain volume, however there is a decrease in overall gray matter volume due to the cortical thinning described above, and an increase in white matter volume due to myelination described below, presumably reflected in structural and functional connectivity.[28,29]

Diffusion tensor imaging can be used to identify mean diffusivity and fractional anisotropy (FA). With maturation of myelination there are changes in these measures. These measures can be applied to known long white matter tracts and show the myelination (decreased mean diffusability and increased anisotropy) of the known major matter tracts[29–34] (Figure 18.4). The main maturation of these tracts occurs before the age of two, after which there are modest changes,

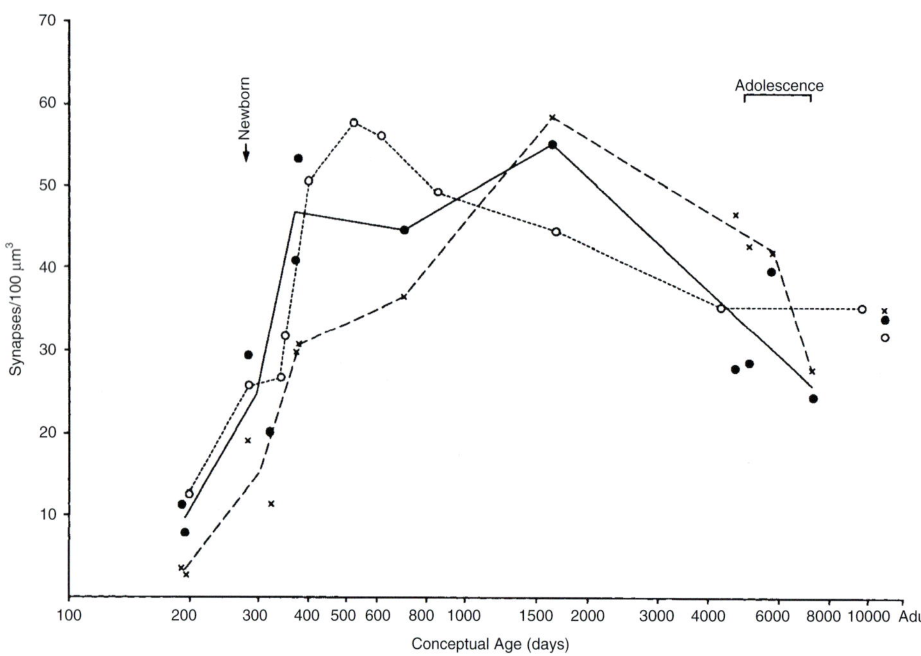

Figure 18.1. Synaptic density in auditory (filled circles), calcrine (open circles) and prefrontal cortex (xs) across ages. From Huttenlocher and Dabholkar et al., *J Comp Neurol.* 1997. Copyright © 1997 Wiley-Liss, Inc.

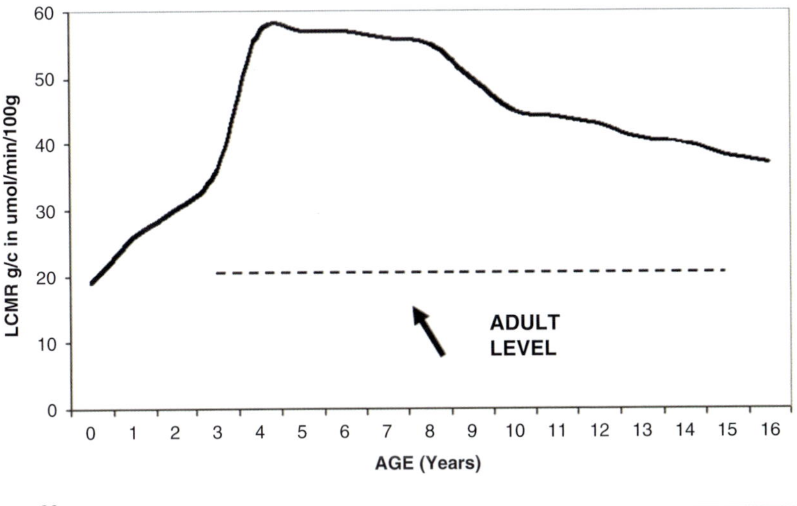

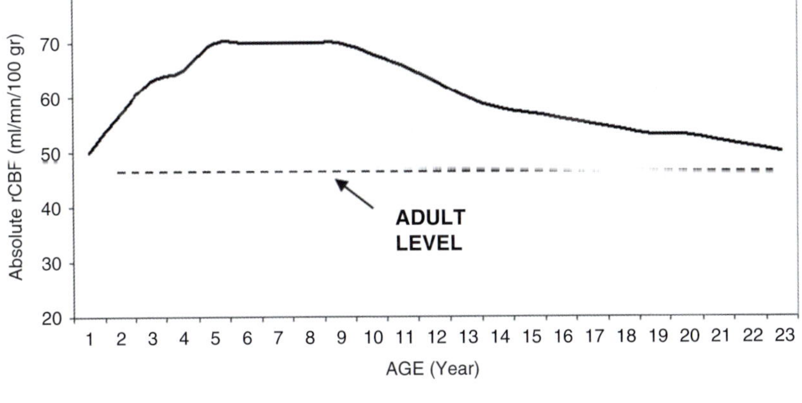

Figure 18.2. A. LCMRGlc across ages ascertained by [18]FDG-PET B. rCBF across ages ascertained by Xenon SPECT. Adapted from Chugani et al., *Ann Neurol.* 1987. Copyright © 1987 American Neurological Association; Chiron et al., *J Nucl Med.* 1992. Originally published in Chiron et al. Changes in regional cerebral blood flow during brain maturation in children and adolescents. *J Nucl Med.* 1992;**33**:696–703. © by the Society of Nuclear Medicine and Molecular Imaging, Inc.

although full association cortex maturation—e.g., in frontal lobes—is not fully mature until the mid-20s (when automobile insurance rates decline).[2] As with previous modalities, sensory motor tracts are the first to fully myelinate. The fibers that sustain language mature in the first few years of life. Developmental milestones generally reflect the myelination of these systems, for example, fixing and following in the first month of life, reaching at age four months, and walking at age 12 months. Language processing and expression begins during infancy though main language skills are solidly present by 4–5 years—receptive language followed by expressive language.[35–37]

Imaging of infants is particularly challenging as the lack of myelination and high water content makes visualization of gray-white matter boundaries problematic. There are known and well-delineated changes in T1- and T2-weighted images over the first two years of life. Maturation follows myelination, which occurs in a medial to lateral and caudal to rostral pattern. During these early years, myelination is marked by increased T1 W white matter signal as T1 shortens. In contrast, T2 shortens with maturation of myelination reflected in reduced white matter signal. For example, at 12 months the T2 W signal in frontal subcortical white matter is the same as gray matter.[4] Therefore, it may be very difficult to identify malformations of cortical development between 3 and 18 months and they may not be well visualized until after 24 months.[38] The optimal time to identify malformations of cortical development is less than 3 months and after 24 months. FLAIR imaging and 3D T1 weighted images are the least helpful in this age range (<2 years); rather, T2-weighted sequences in three different axes (or 3D) are the best. Thus, any study that seeks to identify changes over time as a cause or consequence of epilepsy must view observed changes and findings in the context of normal changes in signal that reflect

195

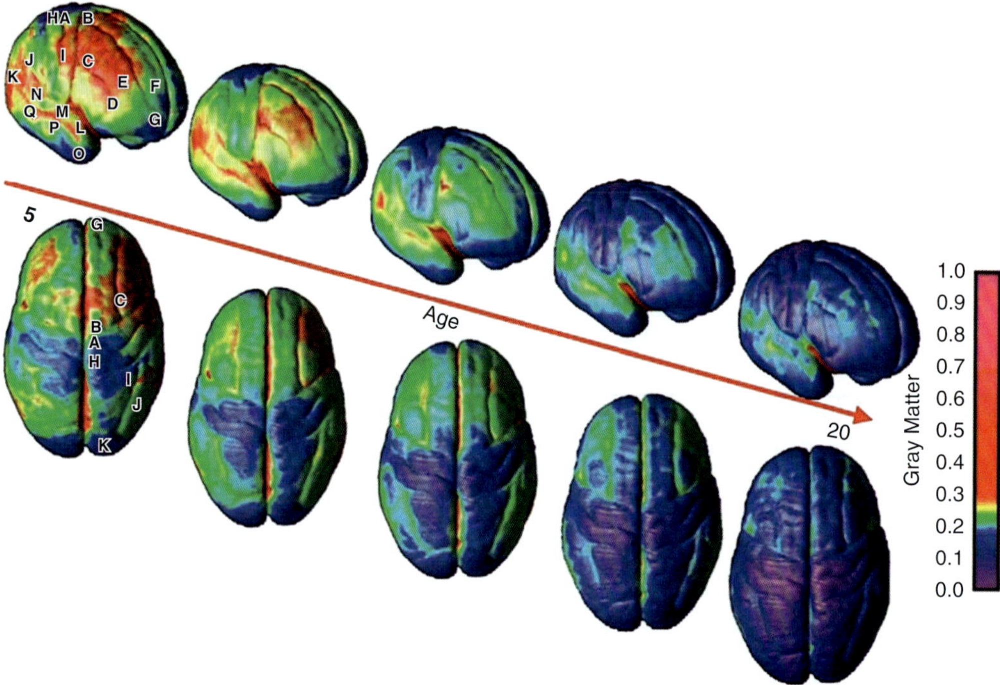

Figure 18.3. Changes in regional cortical thickness with age. From Gogtay et al., *PNAS* 2004. Copyright (2004) National Academy of Sciences, USA.

white matter maturation, especially if they are voxel based.

Malformations of cortical development are the most common finding and etiology in children with new-onset seizures and epilepsy in infants (those older than one month but younger than 2–3 years).[39,40] Identification of etiology has prognostic features. The identification of structural abnormalities is associated with high likelihood of continued seizures and increases the risk of developmental arrest associated with epileptic encephalopathy (and may be surgically remediable).

Mesial temporal sclerosis (MTS) is mostly associated with injury before age 5 years. The principal historical cause is febrile seizures (mostly but not always complex or prolonged), followed by encephalitis, meningitis, and other brain injury. Cross-sectional and longitudinal studies and imaging and anatomic studies find evidence of two hits: typically an initial insult often, but not always, accompanied by status epilepticus, and then progressive atrophy presumably reflecting the effect of continued seizures or interictal activity. MTS can be identified in infants but is more commonly found in older patients.[41–48] Imaging

children with febrile status may identify those with acute increased signal and acute increased size of the hippocampal formation that may be a marker of risk for developing MTS.[45,46] The fundamental findings of MTS such as atrophy and increased signal are the same in children and adults. The FEBSTAT study emphasizes the need not only for control subject imaging but for longitudinal imaging as well. There is evidence that early T2 signal abnormalities, especially those in CA1 and CA3, may place children at risk to develop MTS and mesial temporal lobe epilepsy. More important, a number of children have fallen off their growth curves for normal hippocampal formation development, suggesting the presence of more subtle but perhaps equally important findings.[46] Other data suggest the presence of clear MTS and focal cortical dysplasia indicate poor likelihood of long-term seizure control.[49]

18.5 Functional Imaging: Task Based

Functional MRI (fMRI) using the BOLD image technique has been used to identify and map cognitive networks in adults and children.

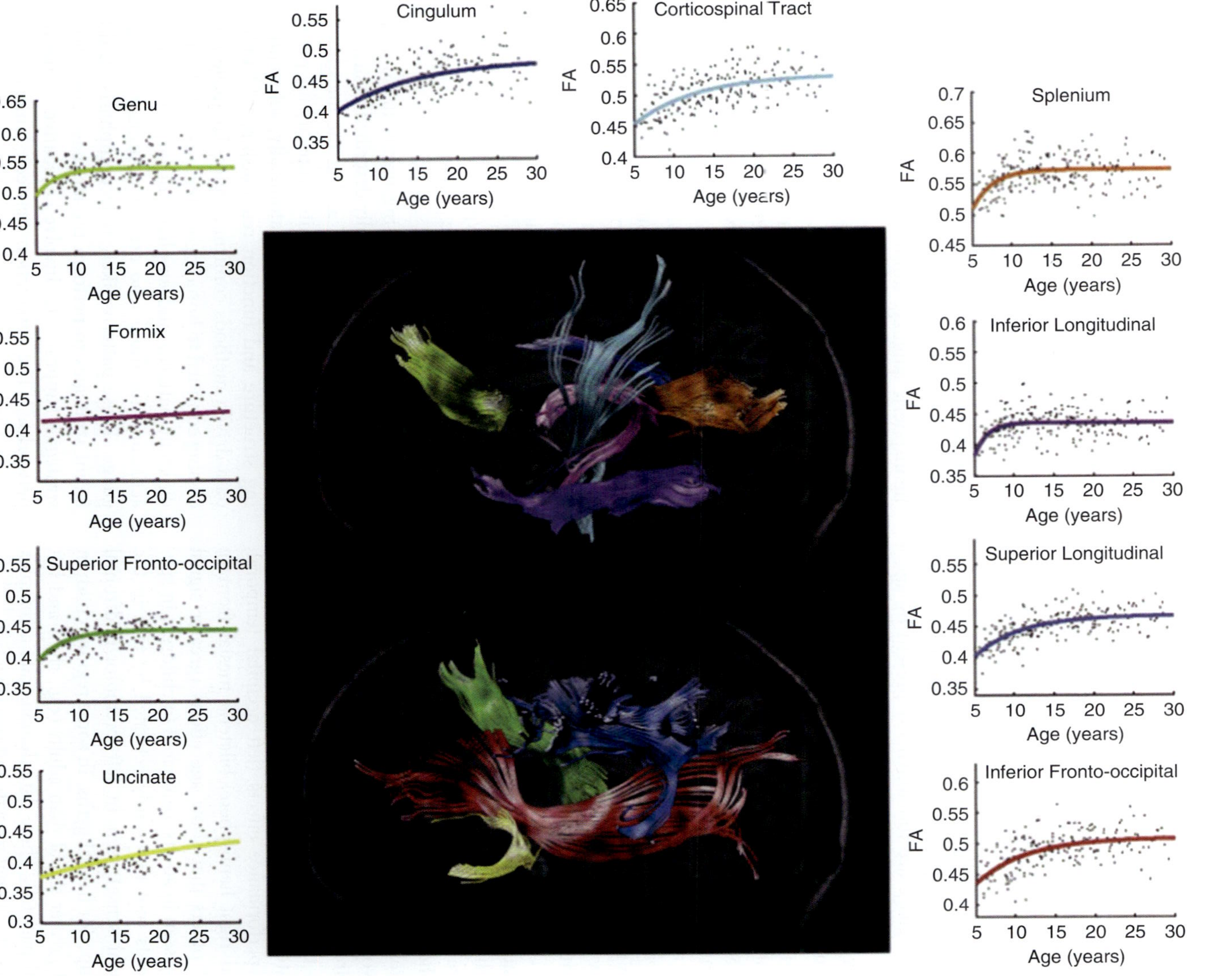

Figure 18.4. Fractional anisotropy in white matter tracts across ages (*n* = 202). From Lebel et al., *NeuroImage* 2008. Reprinted from Lebel C, Walker L, Leemans A, Phillips L, Beaulieu C. *NeuroImage*. 2008;**40**(3):1044–55. Copyright (2008), with permission from Elsevier.

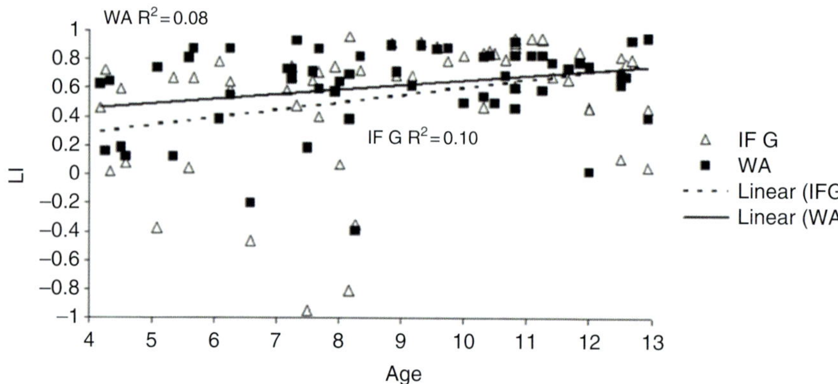

Figure 18.5. Changes in regional laterality indices across ages for Broca's and Wernicke's areas during an auditory word definition task compared to reverse speech. From Berl et al., *Hum Brain Mapp.* 2012. Copyright © 2012 Wiley Periodicals, Inc.

18.5.1 Language

Studies of language fMRI show that the networks that sustain language are fundamentally present by age 4 years.[9–13,18,50–54] They are regionally distinct and lateralized, representing a diffuse and distributed network for language processing that spans both hemispheres. The temporal lobe "receptive" areas in typically developing children reach adult levels of laterality by age 4 to 7 years seem to be firmly established by age 7 years. The frontal "expressive" language areas are also clearly present and lateralized by age 4 years, but do not achieve adult levels of laterality until age 10 years (determined both by laterality index [the strength of laterality of activation] and proportion of those who meet criteria for atypical language dominance)[18,50,53] (Figure 18.5). Younger children have greater variability in measures or activation laterality (evidence of immaturity) than do adults, and most "atypical" laterality of language dominance is due to activation in the right hemisphere, which appears to diminish with time and is not related to effort or task difficulty.[53] Studies of single word processing (several forms of verbal fluency) in children 7 through the late 20s find (1) some brain regions that are unchanged with age, mainly in the distributed language network; (2) other regions that exhibit performance effects (left occipital temporal, scattered right hemisphere, bilateral mesial frontal); and (3) others that show age effects (greater signal intensity seen in adults, compared to children, in specific voxels (not necessarily activated by task in the traditional sense) seen in left premotor/midfrontal gyrus/superior frontal gyrus. In some regions children showed greater signal than adults (e.g., right and left occipital-parietal). These data demonstrate a complex interplay of regional growth and specialization.[19,20]

Epilepsy populations show a high incidence of atypical language dominance, but the expression of hemispheric language dominance needs to be appreciated in the context of normal development. Atypical language BOLD activation laterality is associated with brain injury before age 6 years, either onset of epilepsy or other evidence of brain injury.[55–59] Imaging children at onset of epilepsy observes these findings are present at the onset of epilepsy rather than as a cause of epilepsy (the sole exception may be Rasmussen's encephalitis, where resection of dominant hemisphere after age 8 years typically results in an aphasia that exhibits a gradual albeit incomplete recovery).[60] The anterior regions are more amenable to establishing atypical representation until later childhood and are associated with atypical handedness. Also, frontal systems appear to move along with posterior systems but not vice versa. Thus, atypical language reflects persistence of an immature system of language representation.[53,58,59,61–63]

Atypical language processing occurs in homologues of the left regions classically described as sustaining language functions[54,62–65] and at the margins of language areas. There is a limited capacity for plasticity, which is constrained anatomically, substantiating structure and function relationships.[64,65] Other data-driven methods confirm atypical language processing confined to homologous regions, but also find different patterns and emphasis of left hemispheric activation for those who are left dominant for language. These patterns likely reflect emphasis on different strategies to perform the language task, those strategies appear to involve executive function networks.[54] The interaction between language and working memory networks provides a transition to the developmental expression of working memory and executive function networks.

18.5.2 Executive Function/Working Memory

Executive functions, and working memory in particular, are a set of skills required for goal-directed behavior. Working memory is the ability to keep information in mind. These skills are critical for learning and are associated with higher order cognitive skills. The neural network that underlies these skills involves several regions that span the frontal lobe, parietal lobe, and cerebellum. In addition, lateral prefrontal cortex (PFC) is a lynchpin region found across animal and human studies (for review, see D'Esposito[66]); however, the PFC is not fully developed until late adulthood. As such, several studies have found differences between children and adults on working memory paradigms using fMRI.[17,67–69] In general, adults engage a wider range of regions than children. Although relatively small ($n < 35$ children), fMRI studies of typically developing children suggest that the developmental differences in working memory networks likely reflect different strategies.[68,70,71] In one study, authors propose a developmental shift from the early maturing premotor/striatal/parietal/cerebellar networks in children to ventral prefrontal/inferior temporal networks in adults.[70] Other studies noted that children do not engage as many regions in response to increasing working memory load[68,69] and the cerebellum in particular is not as active in children.[68] In a longitudinal study, authors propose that the hippocampus is recruited in younger adolescents but not in young adults.[71] These differences may be related to methodological variances among the studies including the tasks used and ages included.

Complementary to functional studies, structural MR studies document white matter developmental changes in children. General findings are that (1) large changes occur in children under age 3 years but continue at a steady rate into early adulthood; (2) trajectories of change differ by region; and (3) changes correlate with cognitive development.[72–76] Studies with children show modest correlations between diffusion parameters in the superior longitudinal fasciculus (SLF) and working memory skills in children.[77,78] Ongoing and future studies are examining different diffusion acquisitions and metrics that may be more sensitive to the relationship between white matter structure and neuropsychological performance.

Overall findings in patients with epilepsy show that (1) there are decreases in FA and increases in MD often, but not consistently, in frontal white matter and corpus callosum; (2) differences extend beyond the location of seizure focus; and (3) parameters are often associated with neuropsychological performance. However, limitations of current studies include small sample sizes and heterogeneous patients groups across and within studies. In addition, very few studies confirm findings of other studies because regions that are significantly different are variable, which are in part due to differences in analytic approach (some target specific a priori regions of interest (ROIs) vs. whole-brain analysis, TBSS vs. extracting mean FA from an ROI, etc.). A brief summary of findings from diffusion studies in epilepsy populations follows.

18.5.3 General Studies Examining Differences between Controls and Epilepsy Groups

Twenty children imaged 5–6 years after initial diagnosis with idiopathic (mixed focal and generalized) epilepsy, 9 of whom still had active epilepsy, were compared to 29 controls using FSL tract-based spatial statistics. The investigation observed reduced FA and increased MD, for the active group only mainly in the fornix, internal capsule; corpus callosum, superior corona radiata, and superior fronto-occipital fasciculus.[79] In another study, 30 children with FLE, assessed with MRTrix, showed reduced FA in the cognitively impaired epilepsy group in posterior regions.[80] Fifteen children with Sturge-Weber Syndrome showed decreased FA in ipsilateral prefrontal white matter that also correlated with cognition.[81] Eight children with temporal lobe epilepsy showed lower FA in splenium of corpus callosum and internal capsule in addition to correlation with age of onset.[82] In eight children with temporal-lobe epilepsy (TLE), diffusivity but not FA differences was observed in temporal lobe and cingulate gyrus.[83] Mixed white matter abnormalities were identified in all of 24 patients with focal epilepsy and were concordant for some patients with magnetoencephalography (MEG) findings.[84] In the largest study to date, encompassing 40 and 45 children with nonlesional epilepsy, standard FA and MD measurements, as well as graph theoretical analyses, demonstrated white matter differences in frontal, temporal, parietal, and occipital regions, in both FLE and TLE groups, regardless of seizure focus laterality. FA in several right hemisphere regions was associated with neuropsychological performance and right temporal FA correlated with seizure onset.[85] The graph theory analysis found more areas of abnormality in the FLE group but no correlation with IQ or seizure onset.[86]

18.5.4 Studies Related to Language and/or Executive Functioning

In a study of 12 intractable epilepsy patients the FA was decreased in the SLF II, which connects parietal and PFC and is involved in attention and other frontal lobe functions.[87] A study of 19 children with idiopathic epilepsy described reduced FA in the posterior corpus callosum and cingulum. A study of 23 children with intractable focal epilepsy being considered for surgery found reduced FA in several white matter tracts including inferior fronto-occipital (IFO), inferior longitudinal fasciculus (ILF), and the SLF.[88] In 33 patients with epilepsy and focal cortical dysplasia, diffusion parameters (FA and apparent diffusion coefficient) of left language-related tracts (uncinate, arcuate fasciculus, IFO fasciculus) predicted language impairment.[89] Some studies have attempted to use diffusion to determine language dominance by examining the arcuate fasciculus in particular with mixed results in small sample sizes ($n = 6$ and 13, respectively).[90,91]

18.5.5 Memory

Studies of memory functioning in typically developing children using neuroimaging techniques are sparse. Verbal memory in typically developing children (age 7–19) is related to left hippocampal and basal ganglia activation, which declines as age increases.[92] Overall, mesial temporal lobe activation may decrease across typical development, while prefrontal cortex activity increases.[93,94] Groups of typically developing children across the age span (8-year-olds, 10- to 11-year-olds, and 14-year-olds to young adults) demonstrate functional changes across development with the hippocampus and parahippocampal gyrus becoming increasingly more specialized for detailed recollection.[95] Functional developmental changes in regions of the hippocampus have also been found; episodic recall is associated with activity in the posterior hippocampus in TD children, and in the anterior hippocampus in TD adults.[96] Indirect high frequency (HF) activation in children performing a language word definition task are more bilaterally represented in typically developing children

and children with left focal epilepsy than in adult controls and patients with left focal epilepsy where HF activation generally follow language dominance.[97] These data suggest that material specificity may not be established until adolescence and may explain better postoperative memory performance in children compared to adults.[98,99] In sum, across development mesial temporal lobe activation may decrease, while PFC activity increases and the hippocampus demonstrates functional developmental changes across the longitudinal axis (posterior to anterior).

18.6 Functional and Structural Connectivity

Independent component analysis and principal component analysis of resting-state data reveal several distinct distributed networks representing identifiable cognitive systems. The networks include those involved in executive function and cognitive control.[100] Functional connectivity examines the strength of connections based on signal fluctuations between cortical regions. These connections may be adjacent (small world) or more distant (large world). Recent developmental studies of resting-state networks reveal that functional brain development is characterized by a reconfiguration of the balance between long-range and short-range cortical connectivity. Studies with infants show the presence of motor and sensory networks but less reliable identification of frontal-parietal networks.[101–103] Between 7 and 9 years, "small worldness" of the whole-brain network is manifested, but the strength of connectivity of the midline frontal-parietal network remains weak but increases with age thereafter.[104–107] Application of graph-theoretical methods to analysis of large-scale cortical networks demonstrates that individual network nodes are incorporated into different networks with increasing age from childhood to adulthood, suggesting reorganization rather than strengthening of the same networks with development.[104,108] Further, correlations among nodes closer in distance (segregation) weaken with age, whereas those among nodes further away strengthen with age (integration), and the children have stronger and more subcortical-cortical connections while cortical-cortical connections are more prominent in young adults.[109] As regards intrahemispheric and interhemispheric connectivity during childhood development, there are data to suggest intrahemispheric connectivity develops before intrahemispheric connectivity[97,101,110] and anterior to posterior connections develop last.[106]

Most studies of connectivity show that structural and functional connectivity is less complex and strong in epilepsy populations compared to controls in focal epilepsy[111–118] and generalized epilepsy.[118–122] The graph metrics are simpler, which may be common to many developmental and other brain-based disorders. Epilepsy, like most central nervous system and developmental disorders, is associated with simpler, less complex network topology (in other words simpler organizational patterns). What remains unclear is if this is an active process or effectively represents a developmental dysmaturity. The observations may reflect age of insult or epilepsy onset rather than effects of disease on a mature system.

Data for language systems in connectivity in language networks show that there is greater inter- than intrahemispheric connectivity with homologous regions that reverses with age. As children age, intrahemispheric connectivity becomes stronger.[97] Also, there are age-related changes in connectivity in the dorsal stream (subserved by the arcuate fasciculus and superior longitudinal fasciculus) and the ventral stream (extreme capsule, uncinate, and inferior longitudinal fasciculus). Stronger connectivity measures are associated with better language abilities. Children with focal epilepsy exhibit particular vulnerability to maturation of the ventral stream. Furthermore, for children with epilepsy, decreased connectivity in the ventral stream is related to poorer performance on language measures.[123]

There are few studies of EEG fMRI activity and connectivity in children. Studies in childhood absence show a distributed network that is regionally specific involving thalamus and some cortical regions. In addition, the time course can be delineated and relate to the spike or after going slow wave. Studies in adults reveal the remote effects of spike activity of BOLD response.[124] Most pediatric studies have been conducted in children with generalized epilepsy, e.g., absence, or in young adults with a history of a childhood epilepsy syndrome. These findings vary in activation and deactivation based in spike or slow wave activity.[125,126] Studies of spike foci in patients with focal epilepsy and rapid generalization and some patients with Lennox-Gastaut syndrome, or with patients with frontal lobe epilepsy with rapid secondarily generalization that share some features of Lennox-Gastaut syndrome, indicate that frontal foci show connectivity to the broadly defined attention network.[127–129] These data suggest that some epileptiform activity, depending on location, taps into the distributed networks into which that region is linked that relate to the basic cognitive systems identified in early adult connectivity work. They highlight the effect of epilepsy on distributed networks, and may account for some of the cognitive impairments in patients with epilepsy.

18.7 Conclusions

Brain maturation may be imaged in different modalities. As synaptic connections are formed and pruned, there is regional growth and shrinkage of the cortical ribbon linked to the maturation of the long matter tracts that connect adjacent and far-reaching brain areas. The changes in synaptic density may be assessed with measures of blood flow and cerebral metabolism. Finally, the numerous networks that sustain essential brain and cognitive functions may be assessed by fMRI tasks that show their maturation and perturbations in response to epilepsy. Connectivity analysis and graph theory provide additional means to assess the local and remote functional and white matter connections. These factors need to be considered when evaluating the effect of epilepsy or its cause on brain structure and function at the time of onset and over time.

References

1. Huttenlocher PR, Dabholkar AS. Regional differences in synaptogenesis in human cerebral cortex. *J Comp Neurol.* 1997;**387**(2):167–78.

2. Yakovlev P, Lecours A. The myelogenetic cycles of regional maturation of the brain. In: Dekaban AS, ed. *Regional Development of the Brain in Early Life.* Oxford: Blackwell; 1967:3–70.

3. Yerys BE, Jankowski KF, Shook D, et al. The fMRI success rate of children and adolescents: typical development, epilepsy, attention deficit/hyperactivity disorder, and autism spectrum disorders. *Hum Brain Mapp.* 2009;**30**(10):3426–35.

4. Almli CR, Rivkin MJ, McKinstry RC. The NIH MRI study of normal brain development (objective-2): newborns, infants, toddlers, and preschoolers. *NeuroImage.* 2007;**35**(1):308–25.

5. Shaw P, Kabani NJ, Lerch JP, et al. Neurodevelopmental trajectories of the human cerebral cortex. *J Neurosci.* 2008;**28**(14):3586–94.

6. Wierenga LM, Langen M, Oranje B, Durston S. Unique developmental trajectories of cortical thickness and surface area. *NeuroImage*. 2014;87:120–6.

7. Gogtay N, Giedd JN, Lusk L, et al. Dynamic mapping of human cortical development during childhood through early adulthood. *Proc Natl Acad Sci USA*. 2004;**101**(21):8174–9.

8. Raznahan A, Lerch JP, Lee N, et al. Patterns of coordinated anatomical change in human cortical development: a longitudinal neuroimaging study of maturational coupling. *Neuron*. 2011;**72**(5):873–84.

9. Gaillard WD, Balsamo LM, Ibrahim Z, Sachs BC, Xu B. fMRI identifies regional specialization of neural networks for reading in young children. *Neurology*. 2003;**60**(1):94–100.

10. Gaillard WD, Pugliese M, Grandin CB, et al. Cortical localization of reading in normal children: an fMRI language study. *Neurology*. 2001;**57**(1):47–54.

11. Gaillard WD, Balsamo L, Xu B, et al. Language dominance in partial epilepsy patients identified with an fMRI reading task. *Neurology*. 2002;**59**(2):256–65.

12. Ahmad Z, Balsamo LM, Sachs BC, Xu B, Gaillard WD. Auditory comprehension of language in young children. *Neurology*. 2003;**60**(10):1598–605.

13. Balsamo LM, Xu B, Grandin CB, et al. A functional magnetic resonance imaging study of left hemisphere language dominance in children. *Arch Neurol*. 2002;**59**(7):1168–74.

14. Berl MM, Mayo J, Parks EN, et al. Regional differences in the developmental trajectory of lateralization of the language network. *Hum Brain Mapp*. 2014;**35**:270–84.

15. Casey BJ, Cohen JD, Jezzard P, et al. Activation of prefrontal cortex in children during a nonspatial working memory task with functional MRI. *NeuroImage*. 1995;**2**(3):221–9.

16. Casey BJ, Trainor RJ, Orendi JL, et al. A developmental functional MRI study of prefrontal activation during performance of a go-no-go task. *J Cogn Neurosci*. 1997;**9**(6):835–47.

17. Thomas KM, King SW, Franzen PL, et al. A developmental functional MRI study of spatial working memory. *NeuroImage*. 1999;**10**:327–38.

18. Holland SK, Plante E, Weber Byars A, Strawsburg RH, Schmithorst VJ, Ball WS. Normal fMRI brain activation patterns in children performing a verb generation task. *NeuroImage*. 2001;**14**(4):837–43.

19. Schlaggar BL, Brown TT, Lugar HM, Visscher KM, Miezin FM, Petersen SE. Functional neuroanatomical differences between adults and school-age children in the processing of single words. *Science*. 2002;**296**(5572):1476–9.

20. Brown TT, Lugar HM, Coalson RS, Miezin FM, Petersen SE, Schlaggar BL. Developmental changes in human cerebral functional organization for word generation. *Cereb Cortex*. 2005;**15**(3):275–90.

21. Souweidane MM, Kim KH, McDowall R, et al. Brain mapping in sedated infants and young children with passive-functional magnetic resonance imaging. *Pediatr Neurosurg*. 1999;**30**(2):86–92.

22. Ogg RJ, Laningham FH, Clarke D, et al. Passive range of motion functional magnetic resonance imaging localizing sensorimotor cortex in sedated children. *J Neurosurg Pediatr*. 2009;**4**(4):317–22.

23. Bernal B, Grossman S, Gonzalez R, Altman N. FMRI under sedation: what is the best choice in children? *J Clin Med Res*. 2012;**4**(6):363–70.

24. Li W, Wait SD, Ogg RJ, et al. Functional magnetic resonance imaging of the visual cortex performed in children under sedation to assist in presurgical planning. *J Neurosurg Pediatr*. 2013;**11**(5):543–6.

25. Chugani HT, Phelps ME, Mazziotta JC. Positron emission tomography study of human brain functional development. *Ann Neurol*. 1987;**22**(4):487–97.

26. Van Bogaert P, Wikler D, Damhaut P, Szliwowski HB, Goldman S. Regional changes in glucose metabolism during brain development from the age of 6 years. *NeuroImage*. 1998;**8**(1):62–8.

27. Chiron C, Raynaud C, Maziere B, et al. Changes in regional cerebral blood flow during brain maturation in children and adolescents. *J Nucl Med*. 1992;**33**(5):696–703.

28. Sowell ER, Peterson BS, Thompson PM, Welcome SE, Henkenius AL, Toga AW. Mapping cortical change across the human life span. *Nat Neurosci*. 2003;**6**(3):309–15.

29. Lebel C, Walker L, Leemans A, Phillips L, Beaulieu C. Microstructural maturation of the human brain from childhood to adulthood. *NeuroImage*. 2008;**40**(3):1044–55.

30. Qiu A, Mori S, Miller MI. Diffusion tensor imaging for understanding brain development in early life. *Annu Rev Psychol*. 2015;66:853–76.

31. Geng X, Gouttard S, Sharma A, et al. Quantitative tract-based white matter development from birth to age 2 years. *NeuroImage*. 2012;**61**(3):542–57.

32. Nie J, Li G, Shen D. Development of cortical anatomical properties from early childhood to early adulthood. *NeuroImage*. 2013;76:216–24.

33. Kochunov P, Glahn DC, Lancaster J, et al. Fractional anisotropy of cerebral white matter and thickness of cortical gray matter across the lifespan. *NeuroImage*. 2011;**58**(1):41–9.

34. Wu M, Lu LH, Lowes A, et al. Development of superficial white matter and its structural interplay with cortical gray matter in children and adolescents. *Hum Brain Mapp*. 2014;**35**(6):2806–16.

35. Su P, Kuan C-C, Kaga K, Sano M, Mima K. Myelination progression in language-correlated regions in brain of normal children determined by quantitative MRI assessment. *Int J Pediatr Otorhinolaryngol*. 2008;**72**(12):1751–63.

36. Brauer J, Anwander A, Friederici AD. Neuroanatomical prerequisites for language functions in the maturing brain. *Cereb Cortex*. 2011;**21**(2):459–66.

37. Lebel C, Beaulieu C. Lateralization of the arcuate fasciculus from childhood to adulthood and its relation to cognitive abilities in children. *Hum Brain Mapp*. 2009;**30**(11):3563–73.

38. Gaillard WD, Chiron C, Helen Cross J, et al. Guidelines for imaging infants and children with recent-onset epilepsy. *Epilepsia*. 2009;**50**(9):2147–53.

39. Hsieh DT, Chang T, Tsuchida TN, et al. New-onset afebrile seizures in infants: role of neuroimaging. *Neurology*. 2010;**74**(2):150–6.

40. Eltze CM, Chong WK, Cox T, et al. A population-based study of newly diagnosed epilepsy in infants. *Epilepsia*. 2013;**54**(3):437–45.

41. Theodore WH, Bhatia S, Hatta J, et al. Hippocampal atrophy, epilepsy duration, and febrile seizures in patients with partial seizures. *Neurology*. 1999;**52**(1): 132–6.

42. Theodore WH, Gaillard WD, De Carli C, Bhatia S, Hatta J. Hippocampal volume and glucose metabolism in temporal lobe epileptic foci. *Epilepsia*. 2001;**42**(1): 130–2.

43. Theodore WH, Kelley K, Toczek MT, Gaillard WD. Epilepsy duration, febrile seizures, and cerebral glucose metabolism. *Epilepsia*. 2004;**45**(3):276–9.

44. Mathern GW, Adelson PD, Cahan LD, Leite JP. Hippocampal neuron damage in human epilepsy: Meyer's hypothesis revisited. *Prog Brain Res*. 2002;**135**: 237–51.

45. Shinnar S, Bello JA, Chan S, et al. MRI abnormalities following febrile status epilepticus in children: the FEBSTAT study. *Neurology*. 2012;**79**(9):871–7.

46. Lewis DV, Shinnar S, Hesdorffer DC, et al. Hippocampal sclerosis after febrile status epilepticus: the FEBSTAT study. *Ann Neurol*. 2014;**75**(2):178–85.

47. Wainwright MS, Martin PL, Morse RP, et al. Human herpesvirus 6 limbic encephalitis after stem cell transplantation. *Ann Neurol*. 2001;**50**(5):612–9.

48. Kadom N, Tsuchida T, Gaillard WD. Hippocampal sclerosis in children younger than 2 years. *Pediatr Radiol*. 2011;**41**(10):1239–45.

49. Spooner CG, Berkovic SF, Mitchell LA, Wrennall JA, Harvey AS. New-onset temporal lobe epilepsy in children: lesion on MRI predicts poor seizure outcome. *Neurology*. 2006;**67**(12):2147–53.

50. Gaillard WD, Sachs BC, Whitnah JR, et al. Developmental aspects of language processing: fMRI of verbal fluency in children and adults. *Hum Brain Mapp*. 2003;**18**(3):176–85.

51. Balsamo LM, Xu B, Gaillard WD. Language lateralization and the role of the fusiform gyrus in semantic processing in young children. *NeuroImage*. 2006;**31**(3):1306–14.

52. Berl MM, Duke ES, Mayo J, et al. Functional anatomy of listening and reading comprehension during development. *Brain Lang*. 2010;**114**(2):115–25.

53. Berl MM, Mayo J, Parks EN, et al. Regional differences in the developmental trajectory of lateralization of the language network. *Hum Brain Mapp*. 2014;**35**:270–84.

54. You X, Adjouadi M, Guillen MR, et al. Sub-patterns of language network reorganization in pediatric localization related epilepsy: a multisite study. *Hum Brain Mapp*. 2011;**32**(5):784–99.

55. Rasmussen T, Milner B. The role of early left-brain injury in determining lateralization of cerebral speech functions. *Ann N Y Acad Sci*. 1977;**299**:355–69.

56. Springer JA, Binder JR, Hammeke TA, et al. Language dominance in neurologically normal and epilepsy subjects: a functional MRI study. *Brain J Neurol*. 1999;**122**(pt 11):2033–46.

57. Woermann FG, Jokeit H, Luerding R, et al. Language lateralization by Wada test and fMRI in 100 patients with epilepsy. *Neurology*. 2003;**61**(5):699–701.

58. Gaillard WD, Berl MM, Moore EN, et al. Atypical language in lesional and nonlesional complex partial epilepsy. *Neurology*. 2007;**69**(18):1761–71.

59. Berl MM, Zimmaro LA, Khan OI, et al. Characterization of atypical language activation patterns in focal epilepsy: language activation patterns. *Ann Neurol*. 2013;**75**:33–42.

60. Hertz-Pannier L, Chiron C, Jambaque I, et al. Late plasticity for language in a child's non-dominant hemisphere. *Brain*. 2002;**125**:361–72.

61. Duke ES, Tesfaye M, Berl MM, et al. The effect of seizure focus on regional language processing areas. *Epilepsia*. 2012;**53**(6):1044–50.

62. Staudt M, Grodd W, Niemann G, Wildgruber D, Erb M, Krageloh-Mann I. Early left periventricular brain lesions induce right hemispheric organization of speech. *Neurology*. 2001;**57**(1):122–5.

63. Staudt M, Lidzba K, Grodd W, Wildgruber D, Erb M, Krageloh-Mann I. Right-hemispheric organization of language following early left-sided brain lesions:

functional MRI topography. *NeuroImage*. 2002;**16**(4): 954–67.

64. Rosenberger LR, Zeck J, Berl MM, et al. Interhemispheric and intrahemispheric language reorganization in complex partial epilepsy. *Neurology*. 2009;**72**(21):1830–6.

65. Mbwana J, Berl MM, Ritzl EK, et al. Limitations to plasticity of language network reorganization in localization related epilepsy. *Brain*. 2009;**132**(pt 2): 347–56.

66. D'Esposito M. From cognitive to neural models of working memory. *Philos Trans R Soc B Biol Sci*. 2007;**362**(1481):761–72.

67. Klingberg T, Forssberg H, Westerberg H. Increased brain activity in frontal and parietal cortex underlies the development of visuospatial working memory capacity during childhood. *J Cogn Neurosci*. 2002;**14** (1):1–10.

68. O'Hare ED, Lu LH, Houston SM, Bookheimer SY, Sowell ER. Neurodevelopmental changes in verbal working memory load-dependency: an fMRI investigation. *NeuroImage*. 2008;**42**(4):1678–85.

69. Thomason ME, Race E, Burrows B, Whitfield-Gabrieli S, Glover GH, Gabrieli JD. Development of spatial and verbal working memory capacity in the human brain. *J Cogn Neurosci*. 2009;**21**(2):316–32.

70. Ciesielski KT, Lesnik PG, Savoy RL, Grant EP, Ahlfors SP. Developmental neural networks in children performing a categorical N-back task. *NeuroImage*. 2006;**33**(3):980–90.

71. Finn AS, Sheridan MA, Kam CLH, Hinshaw S, D'Esposito M. Longitudinal evidence for functional specialization of the neural circuit supporting working memory in the human brain. *J Neurosci*. 2010;**30**(33): 11062–7.

72. Dean DC, O'Muircheartaigh J, Dirks H, et al. Characterizing longitudinal white matter development during early childhood. *Brain Struct Funct*. 2015;**220** (4):1921–33.

73. Krogsrud SK, Fjell AM, Tamnes CK, et al. Changes in white matter microstructure in the developing brain— a longitudinal diffusion tensor imaging study of children from 4 to 11 years of age. *NeuroImage*. 2016;**124**:473–86.

74. Lebel C, Walker L, Leemans A, Phillips L, Beaulieu C. Microstructural maturation of the human brain from childhood to adulthood. *NeuroImage*. 2008;**40**(3): 1044–55.

75. Peters BD, Ikuta T, DeRosse P, et al. Age-related differences in white matter tract microstructure are associated with cognitive performance from childhood to adulthood. *Biol Psychiatry*. 2014;**75**(3):248–56.

76. Schmithorst VJ, Wilke M, Dardzinski BJ, Holland SK. Cognitive functions correlate with white matter architecture in a normal pediatric population: a diffusion tensor MRI study. *Hum Brain Mapp*. 2005;**26**(2):139–47.

77. Østby Y, Tamnes CK, Fjell AM, Walhovd KB. Morphometry and connectivity of the fronto-parietal verbal working memory network in development. *Neuropsychologia*. 2011;**49**(14):3854–62.

78. Sala-Llonch R, Palacios EM, Junqué C, Bargalló N, Vendrell P. Functional networks and structural connectivity of visuospatial and visuoperceptual working memory. *Front Hum Neurosci*. 2015;**9**:340.

79. Amarreh I, Dabbs K, Jackson DC, et al. Cerebral white matter integrity in children with active versus remitted epilepsy 5 years after diagnosis. *Epilepsy Res*. 2013;**107** (3):263–71.

80. Braakman HMH, Vaessen MJ, Jansen JFA, et al. Pediatric frontal lobe epilepsy: white matter abnormalities and cognitive impairment. *Acta Neurol Scand*. 2014;**129**(4):252–62.

81. Alkonyi B, Govindan RM, Chugani HT, Behen ME, Jeong J-W, Juhász C. Focal white matter abnormalities related to neurocognitive dysfunction: an objective diffusion tensor imaging study of children with Sturge-Weber syndrome. *Pediatr Res*. 2011;**69**(1):74–9.

82. Meng L, Xiang J, Kotecha R, et al. White matter abnormalities in children and adolescents with temporal lobe epilepsy. *Magn Reson Imaging*. 2010;**28** (9):1290–8.

83. Nilsson D, Go C, Rutka JT, et al. Bilateral diffusion tensor abnormalities of temporal lobe and cingulate gyrus white matter in children with temporal lobe epilepsy. *Epilepsy Res*. 2008;**81**(2–3):128–35.

84. Widjaja E, Geibprasert S, Otsubo H, Snead OC, Mahmoodabadi SZ. Diffusion tensor imaging assessment of the epileptogenic zone in children with localization-related epilepsy. *Am J Neuroradiol*. 2011;**32**(10):1789–94.

85. Widjaja E, Kis A, Go C, Raybaud C, Snead OC, Smith ML. Abnormal white matter on diffusion tensor imaging in children with new-onset seizures. *Epilepsy Res*. 2013;**104**(1–2):105–11.

86. Widjaja E, Zamyadi M, Raybaud C, Snead OC, Doesburg SM, Smith ML. Disrupted global and regional structural networks and subnetworks in children with localization-related epilepsy. *Am J Neuroradiol*. 2015; **36**:1362–8.

87. Holt RL, Provenzale JM, Veerapandiyan A, et al. Structural connectivity of the frontal lobe in children with drug-resistant partial epilepsy. *Epilepsy Behav*. 2011;**21**(1):65–70.

88. Kim H, Harrison A, Kankirawatana P, et al. Major white matter fiber changes in medically intractable neocortical epilepsy in children: a diffusion tensor imaging study. *Epilepsy Res.* 2013;**103**:211–20.

89. Paldino MJ, Hedges K, Zhang W. Independent contribution of individual white matter pathways to language function in pediatric epilepsy patients. *NeuroImage Clin.* 2014;**6**:327–32.

90. Saporta ASD, Kumar A, Govindan RM, Sundaram SK, Chugani HT. Arcuate fasciculus and speech in congenital bilateral perisylvian syndrome. *Pediatr Neurol.* 2011;**44**(4):270–4.

91. Tiwari VN, Jeong J-W, Asano E, Rothermel R, Juhasz C, Chugani HT. A sensitive diffusion tensor imaging quantification method to detect language laterality in children correlation with the Wada test. *J Child Neurol.* 2011;**26**(12):1516–21.

92. Maril A, Davis PE, Koo JJ, et al. Developmental fMRI study of episodic verbal memory encoding in children. *Neurology.* 2010;**75**(23):2110–6.

93. Menon V, Boyett-Anderson JM, Reiss AL. Maturation of medial temporal lobe response and connectivity during memory encoding. *Brain Res Cogn Brain Res.* 2005;**25**(1):379–85.

94. Ofen N, Kao Y-C, Sokol-Hessner P, Kim H, Whitfield-Gabrieli S, Gabrieli JDE. Development of the declarative memory system in the human brain. *Nat Neurosci.* 2007;**10**(9):1198–205.

95. Ghetti S, DeMaster DM, Yonelinas AP, Bunge SA. Developmental differences in medial temporal lobe function during memory encoding. *J Neurosci.* 2010;**30**(28):9548–56.

96. Demaster DM, Ghetti S. Developmental differences in hippocampal and cortical contributions to episodic retrieval. *Cortex.* 2013;**49**(6):1482–93.

97. Sepeta LN, Croft LJ, Zimmaro LA, et al. Reduced language connectivity in pediatric epilepsy. *Epilepsia.* 2015;**56**(2):273–82.

98. Smith ML, Elliott IM, Lach L. Cognitive, psychosocial, and family function one year after pediatric epilepsy surgery. *Epilepsia.* 2004;**45**(6):650–60.

99. Smith ML, Olds J, Snyder T, Elliott I, Lach L, Whiting S. A follow-up study of cognitive function in young adults who had resective epilepsy surgery in childhood. *Epilepsy Behav.* 2014;**32**:79–83.

100. Buckner RL, Krienen FM. The evolution of distributed association networks in the human brain. *Trends Cogn Sci.* 2013;**17**(12):648–65.

101. Fransson P, Skiold B, Horsch S, et al. Resting-state networks in the infant brain. *Proc Natl Acad Sci USA.* 2007;**104**(39):15531–6.

102. Gao W, Zhu H, Giovanello KS, et al. Evidence on the emergence of the brain's default network from 2-week-old to 2-year-old healthy pediatric subjects. *Proc Natl Acad Sci USA.* 2009;**106**(16):6790–5.

103. Smyser CD, Inder TE, Shimony JS, et al. Longitudinal analysis of neural network development in preterm infants. *Cereb Cortex.* 2010;**20**(12):2852–62.

104. Fair DA, Dosenbach NUF, Church JA, et al. Development of distinct control networks through segregation and integration. *Proc Natl Acad Sci USA.* 2007;**104**(33):13507–12.

105. Fair DA, Cohen AL, Dosenbach NUF, et al. The maturing architecture of the brain's default network. *Proc Natl Acad Sci USA.* 2008;**105**(10):4028–32.

106. Kelly AMC, Di Martino A, Uddin LQ, et al. Development of anterior cingulate functional connectivity from late childhood to early adulthood. *Cereb Cortex.* 2009;**19**(3):640–57.

107. Supekar K, Musen M, Menon V. Development of large-scale functional brain networks in children. *PLOS Biol.* 2009;**7**(7):e1000157.

108. Fair DA, Cohen AL, Power JD, et al. Functional brain networks develop from a "local to distributed" organization. *PLOS Comput Biol.* 2009;**5**(5).

109. Uddin LQ, Supekar K, Menon V. Typical and atypical development of functional human brain networks: insights from resting-state FMRI. *Front Syst Neurosci.* 2010;**4**:21.

110. Liu Y, Liang M, Zhou Y, et al. Disrupted small-world networks in schizophrenia. *Brain.* 2008;**131**(4):945–61.

111. Widjaja E, Zamyadi M, Raybaud C, Snead OC, Smith ML. Abnormal functional network connectivity among resting-state networks in children with frontal lobe epilepsy. *AJNR Am J Neuroradiol.* 2013;**34**(12):2386–92.

112. Widjaja E, Zamyadi M, Raybaud C, Snead OC, Smith ML. Impaired default mode network on resting-state FMRI in children with medically refractory epilepsy. *AJNR Am J Neuroradiol.* 2013;**34**(3):552–7.

113. Vaessen MJ, Braakman HMH, Heerink JS, et al. Abnormal modular organization of functional networks in cognitively impaired children with frontal lobe epilepsy. *Cereb Cortex.* 2013;**23**(8):1997–2006.

114. Vaessen MJ, Jansen JFA, Braakman HMH, et al. Functional and structural network impairment in childhood frontal lobe epilepsy. *PLOS ONE.* 2014;**9**(3):e90068.

115. Braakman HMH, Vaessen MJ, Jansen JFA, et al. Frontal lobe connectivity and cognitive impairment

in pediatric frontal lobe epilepsy. *Epilepsia*. 2013;**54**(3):446–54.

116. Ibrahim GM, Cassel D, Morgan BR, et al. Resilience of developing brain networks to interictal epileptiform discharges is associated with cognitive outcome. *Brain J Neurol*. 2014;**137**(pt 10):2690–702.

117. Ibrahim GM, Morgan BR, Lee W, et al. Impaired development of intrinsic connectivity networks in children with medically intractable localization-related epilepsy. *Hum Brain Mapp*. 2014;**35**(11):5686–700.

118. Luo C, Yang T, Tu S, et al. Altered intrinsic functional connectivity of the salience network in childhood absence epilepsy. *J Neurol Sci*. 2014;**339**(1–2):189–95.

119. Killory BD, Bai X, Negishi M, et al. Impaired attention and network connectivity in childhood absence epilepsy. *NeuroImage*. 2011;**56**(4):2209–17.

120. Masterton RA, Carney PW, Jackson GD. Cortical and thalamic resting-state functional connectivity is altered in childhood absence epilepsy. *Epilepsy Res*. 2012;**99**(3):327–34.

121. Maneshi M, Moeller F, Fahoum F, Gotman J, Grova C. Resting-state connectivity of the sustained attention network correlates with disease duration in idiopathic generalized epilepsy. *PLOS ONE*. 2012;7(12):e50359.

122. Xue K, Luo C, Zhang D, et al. Diffusion tensor tractography reveals disrupted structural connectivity in childhood absence epilepsy. *Epilepsy Res*. 2014;**108**(1):125–38.

123. Croft LJ, Baldeweg T, Sepeta L, Zimmaro L, Berl MM, Gaillard WD. Vulnerability of the ventral language network in children with focal epilepsy. *Brain J Neurol*. 2014;**137**(pt 8):2245–57.

124. Kobayashi E, Bagshaw AP, Grova C, Dubeau F, Gotman J. Negative BOLD responses to epileptic spikes. *Hum Brain Mapp*. 2006;**27**(6):488–97.

125. Bai X, Vestal M, Berman R, et al. Dynamic time course of typical childhood absence seizures: EEG, behavior, and functional magnetic resonance imaging. *J Neurosci*. 2010;**30**(17):5884–93.

126. Berman R, Negishi M, Vestal M, et al. Simultaneous EEG, fMRI, and behavior in typical childhood absence seizures. *Epilepsia*. 2010;**51**(10):2011–22.

127. Pillay N, Archer JS, Badawy RAB, Flanagan DF, Berkovic SF, Jackson G. Networks underlying paroxysmal fast activity and slow spike and wave in Lennox-Gastaut syndrome. *Neurology*. 2013;**81**(7):665–73.

128. Archer JS, Warren AEL, Stagnitti MR, Masterton RAJ, Abbott DF, Jackson GD. Lennox-Gastaut syndrome and phenotype: secondary network epilepsies. *Epilepsia*. 2014;**55**(8):1245–54.

129. Pedersen M, Curwood EK, Archer JS, Abbott DF, Jackson GD. Brain regions with abnormal network properties in severe epilepsy of Lennox-Gastaut phenotype: multivariate analysis of task-free fMRI. *Epilepsia*. 2015;**56**(11):1767–73.

Imaging Comorbidities in Epilepsy: Depression

William H. Theodore

19.1 Introduction

A wide range of comorbidities affects patients with epilepsy, including psychiatric, neuropsychological, social, and systemic disorders. Depression is one of the most prominent and disabling. Patients with depression as well as epilepsy have lower income, QOLIE-89 scores, as well as marriage and employment rates. Some studies suggest a relation with poor response to antiepileptic drugs or even surgical outcome.[1]

Estimates of depression prevalence in epilepsy vary widely, from about 10–40%, depending on whether data come from community surveys or clinical samples.[2] Rates are usually highest in patients being evaluated for drug-resistant epilepsy, either because of the severity of the seizure disorder or thorough investigation. Several studies based on community health surveys have been performed. In the Canadian Community Health Survey and the US Centers for Disease Control Health Styles Survey, the risk for depression in epilepsy patients at any time point was about twice that of the overall population.[3,4] A recent review suggested that the lifetime risk for experiencing any psychiatric disorder, of which depression and anxiety are the most common, is 30–35%.[1] Patients often describe prominent anxiety, as a well as an intermittent, waxing and waning course described as a "dysthymic" or "interictal dysphoric" disorder.[5,6] Patients may report depression, irritability, or less commonly elation, hours to days before a seizure. Postictal depression and anxiety can last up to 14 days.[7,8]

Epidemiological studies suggest a potential bidirectional relationship: a history of major depression and attempted suicide independently increase the risk for unprovoked seizures.[9–11]

The etiology of depression in patients with epilepsy most likely is multifactorial, including both social and biological factors[1] (Table 19.1). Imaging studies have the potential to help elucidate the pathophysiology of some of these conditions, and suggest treatment approaches.

19.2 MRI and PET Glucose Metabolism in Primary Major Depressive Disorders

In patients with primary major depressive disorders (MDD) but not epilepsy, MRI studies have shown

Table 19.1 Possible Causes of Depression in People with Epilepsy

A. Seizure-related

Age of onset and duration

Seizure frequency, severity

Temporal lobe epilepsy (TLE)

Left focus localization

B. Neurobiological

Serotonin

Norepinephrine

Dopamine

GABA-glutamate

C. Sociopsychological

Unemployment

Low education

Social isolation

Stigma

Chronic disease

Lack of control

D. Antiepileptic drugs (AEDs)

Barbiturates

Topiramate

Levetiracetam

Zonisamide

Vigabatrin

Others?

atrophy in orbitomedial prefrontal cortex, as well as hippocampus and other limbic regions. Reduced hippocampal volume is the most consistent funding, and may be restricted to subfields, particularly cornu ammonis and dentate gyrus.[12] The patients do not have increased signal consistent with the focal gliosis found in epilepsy, however. There is some evidence for progressive atrophy, possibly related to stress-induced increased glucocorticoid release and inhibited hippocampal neo-neurogenesis.[13] Positron emission tomography (PET) studies of cerebral metabolism using [^{18}F]-2-deoxy glucose showed decreased cerebral metabolic rate for glucose (CMRglc) in dorsomedial and anterolateral prefrontal cortex and cingulate gyrus.[14]

19.2.1 MRI, Depression, and Epilepsy

Structural Imaging studies have shown a number of possible findings that may be associated with depression in people with epilepsy. Patients with mesial temporal sclerosis (MTS) have been reported to have higher depression rating scores, either without a laterality effect,[15] with right MTS,[16] or left MTS.[17] In contrast, however, several investigators reported no relation of depression to MTS.[18,19] One study reported an increased incidence of depression in nonlesional compared with lesional focal epilepsy.[20] Two others reported that depressed patients with a right temporal focus, with or without MTS, had greater left hippocampal volume reduction than patients without depression.[17,21] In contrast, in a study of 28 patients, those with TLE and depression had larger hippocampal volumes than those with TLE alone.[22]

One study found positive relations between bilateral amygdala volume and depression.[23] In contrast, patients with the "dysphoric" disorder of epilepsy, including emotional instability, dysphoria, irritability, and aggression, had significantly reduced amygdalar volumes, compared to patients with epilepsy alone and those with depression.[24]

Several new MRI-based approaches have been used to examine relationships between structural and functional networks and depression in epilepsy. TLE patients were compared with healthy controls using voxel-based morphometry (VBM).[25] Patients received both the Structured Clinical Interview for DSM-IV and the Beck Depression Inventory (BDI). Those with depression had more areas of gray matter volume loss, compared with control values, than patients with TLE alone. In a study using quantitative,

surface-based MRI analysis, increasing severity of depression was associated with orbitofrontal thinning in controls, but thickening in TLE patients.[26] Right and left mesial TLE patients and controls were investigated using resting-state blood-oxygen-level-dependent functional MRI for functional connectivity.[27] Right TLE patients showed a relation between high depression and anxiety scores and normal hippocampal to amygdala functional connectivity values, while left TLE patients showed the reverse relationship. A study comparing patients with TLE to healthy controls compared BDI-II scores with functional connectivity (FC) between medial temporal and prefrontal regions as well as diffusion tensor imaging for uncinate fasciculus fractional anisotropy as well as amygdalar and hippocampal mean diffusivity to predict depressive symptoms.[28] Higher depression rating scores were associated with stronger connectivity of hippocampus ipsilateral to the seizure focus with anterior prefrontal cortex (this measure was the strongest predictor of depressive symptoms), lower and bilateral uncinate fasciculus fractional anisotropy. Patients with left temporal foci had higher ipsilateral hippocampal mean diffusivity, while those with right temporal foci had lower contralateral uncinate fasciculus fractional anisotropy.

Seventeen TLE patients with a lifetime affective disorder diagnosis, 31 without psychiatric history, and 30 healthy controls performed a visuospatial working memory paradigm to investigate the default mode network.[29] TLE patients with lifetime affective disorder had significantly greater deactivation in subgenual anterior cingulate cortex than either patients with TLE but not depression or healthy controls. The result was not affected by current psychiatric drug treatment or the severity of current depression or anxiety measured by the BDI. In this study there were no significant between group differences in gray matter volume or connectivity.

Overall these disparate data do suggest that depression in people with epilepsy is associated with deficits in cortical structure and connectivity, particularly affecting limbic regions. However, many of the studies are small, and methodology differs. Other factors such as AED response, that may themselves be related to depression, could have affected the results of many studies.[30] A large body of work has shown that in patients with TLE, hippocampal volume ipsilateral to the seizure focus is inversely associated with epilepsy duration, and may be affected by a history of complex or prolonged febrile seizures. Association studies cannot examine cause-effect relationships. The complexity of interpreting these results is increased even more by data, for example, showing that an "unhealthy western diet" is associated with lower left hippocampal volume.[31]

Studies in patients with major depressive disorders (MDD) but not epilepsy have shown results paralleling some findings in epilepsy but also illustrating the difficulties of interpretation. A review of task-based and resting-state fMRI studies of patients younger than 25 years old found increased activity in anterior cingulate, ventromedial and orbitofrontal cortex, and amygdala most consistently and commonly reported in studies involving emotional and reward processing, and affective cognition, as well as resting-state connectivity.[32] Meta-analysis by another group found increased activation in some of the same regions, but more task-specific changes including hypoactivation in caudate, hyperactivation in thalamus and parahippocampal gyrus, hypoactivity in posterior insula, and hyperactivity in dorsolateral prefrontal and superior temporal cortex.[33] A second meta-analysis found frontoparietal hypoconnectivity and default mode network hyperconnectivity.[34] Some of these differences could have been due not only to variation in the studies included, although there was substantial overlap, but also to the vagaries of the analytic approaches. It is likely that in patients with epilepsy only large multicenter studies that will have the power to account for a variety of factors including seizure frequency, epilepsy duration, and AED treatment, will lead to firmer conclusions.

19.2.2 Depression, Imaging, and Epilepsy Surgery

The effect of temporal lobectomy on depression in epilepsy is variable and not always predictable.[18] In a retrospective study, 30 patients with left MTS had gray matter volumes measured using voxel-based morphometry (VBM).[35] Development of postoperative depression, not necessarily in the immediate perioperative period, was associated with reduced preoperative gray matter volume in orbitofrontal cortices, ipsilateral cingulate gyrus and thalamus. The results were not affected by outcome for seizure control. In another study, patients with mesial and lateral temporal lobe resections were compared.[36] In the mesial but not lateral temporal group postoperative depression was associated

with significantly smaller contralateral hippocampal volume. Postoperative BDI scores correlated significantly with hippocampal and amygdala resection extent in 35 patients.[37]

19.3 Cerebral Metabolism in Depression and Epilepsy

Using [^{18}F]fluorodeoxyglucose PET to measure cerebral glucose metabolism, patients with reduced mesial temporal CMRglc ipsilateral to their seizure focus had higher BDI scores, and a study using 1H-magnetic resonance spectroscopy found a relation between the extent of the temporal lobe abnormality and depression rating scores.[38] Several studies found that bilateral or unilateral frontal hypometabolism was associated with depression in epilepsy.[39,40] These results are interesting in light of the more recent data suggesting alterations in frontal lobe cortical thickness and connectivity in patients with TLE and depression.[26] In each of these studies, patients with drug-resistant epilepsy being considered for surgery were evaluated. These patients may be more likely to have both depression and extensive PET hypometabolism. A study of 11 patients using the N-methyl d-aspartate receptor agonist [^{18}F]GE-179 VT found that patients with focal epilepsy overall had increased binding, but that there was reduced receptor availability in patients taking antidepressants, suggesting increased exogenous occupancy that might be related to drug effects.[41]

19.3.1 Serotonin and Epilepsy

The most interesting approaches to PET imaging of depression in epilepsy used serotonin (5HT) 1A receptor ligands. The serotonin system includes at least 16 receptor subtypes, that are involved in modulation of a wide variety of physiologic processes including vascular and nonvascular smooth muscle contractility, platelet aggregation, appetite, wakefulness, sleep, mood, and anxiety.[42] Activation of the G-protein coupled 5HT1A receptors in particular has antidepressant and anxiolytic effects. Serotoninergic neurons have cell bodies in the midbrain raphe and project widely to neocortical limbic regions; 5HT1A receptors are particularly abundant in the latter. Their activation leads to increased potassium and reduced calcium conductance, increased norepinephrine and decreased glutamate release, and CA1 membrane hyperpolarization. 5HT1A receptor activation has antiseizure effects in several epilepsy models.

19.3.2 5HT1A PET Imaging in Primary Major Depressive Disorders

Previous work in patients with primary MDDs but not epilepsy had shown reduced 5HT1A receptor binding in several brain regions, including lateral and mesial temporal and parieto-occipital cortex, insula, anterior and posterior cingulate in both treated with selective serotonin reuptake inhibitors (SSRIs) and untreated patients.[43] The reductions are present when patients are in remission, suggesting a trait rather than state effect.[44] Some monkeys have a naturally occurring "depressive disorder"; these animals show reduced 5HT1A binding compared to nondepressed monkeys in similar regions to those found in humans.[45]

In contrast, other investigators reported increased binding in the same regions in MDD patients who had never been exposed to SSRIs, and no significant abnormalities in those who did have previous treatment, even if they were not taking the drugs at the time of study.[46] A study using [11C] WAY-100635 that compared patients before and after SSRI treatment showed no difference between the two states. However, before treatment, patients who proved later to be SSRI nonresponders had higher bilateral orbitofrontal binding than those who did respond.[47]

19.3.3 5HT1A Imaging in Epilepsy

Patients with mesial and lateral temporal foci have reduced 5HT1A receptor using [18F]-Trans-4-Fluoro-N-(2-[4-(2-methoxyphenyl)piperazin-1-yl]ethyl)-N-(2-pyridyl)cyclohexanecarboxamide ([18F]-FCWAY).[48–51] These reductions are not due simply to tissue volume loss, as shown by studies that performed partial volume correction.[51] 5HT1A receptor imaging may help predict outcome after temporal lobectomy.[52] Two studies investigated the relation of scales measuring concurrent depressive symptoms to 5HT1A receptor binding or number. [18F]-FCWAY ligand free fraction corrected volume of distribution (V/f1) in hippocampus ipsilateral to the seizure focus was inversely correlated with the BDI, with a nonsignificant trend for contralateral hippocampus[53] (Figure 19.1). Patients with a BDI score more than 20, suggesting moderate to severe depression, had significantly lower [18F]-FCWAY binding than those with lower BDI (Figure 19.2). The presence of MTS, focus laterality, or gender did not have a significant effect on the BDI. Similar results had been

found in ipsilateral insula using [11C]WAY and the Montgomery-Asberg scale.[49] In a study relating 5HT1A receptor binding to hippocampal volume, BDI, and neuropsychological test scores, there was a significant effect of the interaction of left hippocampal [18F]FCWAY V/f1 and left hippocampal volume on delayed auditory memory, but not of either alone.[54] Focus laterality and BDI did not affect memory in this study, but BDI did correlate inversely with 5HT1A receptor binding as in the previous reports.

In contrast, a study using a different ligand, [18F]-MPPF, in 24 patients with drug-resistant TLE and MRI evidence of hippocampal sclerosis, BDI scores ranging from 0 to 34, and no prior antidepressant exposure, found that total BDI score and symptoms of psychomotor anhedonia and negative cognition correlated positively with [18F]MPPF binding in raphe nuclei and insula contralateral to the seizure focus.[55] Somatic symptoms correlated positively with [18F]MPPF binding in the hippocampus and parahippocampus ipsilateral to seizure onset, left midcingulate gyrus and bilateral inferior dorsolateral frontal cortex. There may be several potential explanations for the difference between this study and the previous ones that showed relatively decreased rather than increased binding in patients with epilepsy and depression. The authors suggested that there might be differences in the patient population, including uniform presence of MTS, and no prior exposure to antidepressant drugs (which would not have distinguished the patients from those in several previous studies). However, the most likely reason for the discrepancy is the relative difference in binding affinity between the PET ligands used. [18F]MPPF has much lower affinity for 5HT1A receptors than [18F]-FCWAY, and is very close to the binding potential of serotonin itself. Reduced synaptic 5HT availability, which presumably is present in patients with depression, could lead to relatively increased receptor occupancy by the exogenous ligand since more binding sites would be available, when depressed epilepsy patients are compared with nondepressed patients.[55] [18F]FCWAY would not be affected by this mechanism, since it binds more avidly that 5HT itself. In fact, the studies using the two ligands complement each other, confirming the role of 5HT1A receptors in depression and epilepsy, while showing both reduced receptor availability ([18F]FCWAY), and also reduced synaptic 5HT availability. Moreover, they suggest widespread and partly bilateral abnormalities,

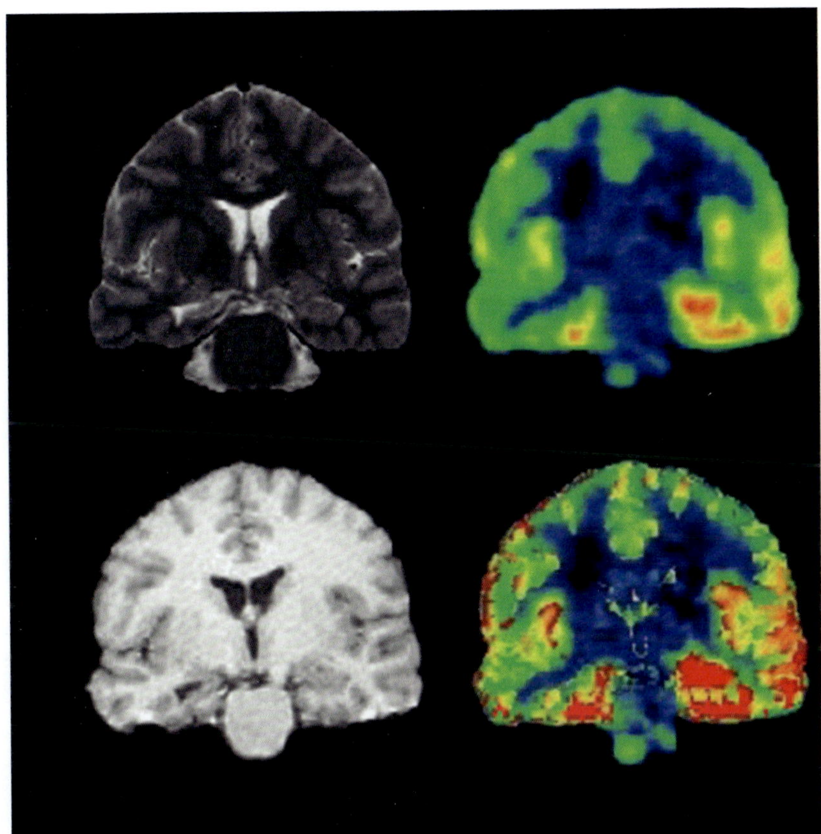

Figure 19.1. Original and partial volume corrected (PVC) [^{18}F]-FCWAY PET scans showing reduced binding in a patient with right temporal mesial temporal sclerosis and the effect of partial volume correction.

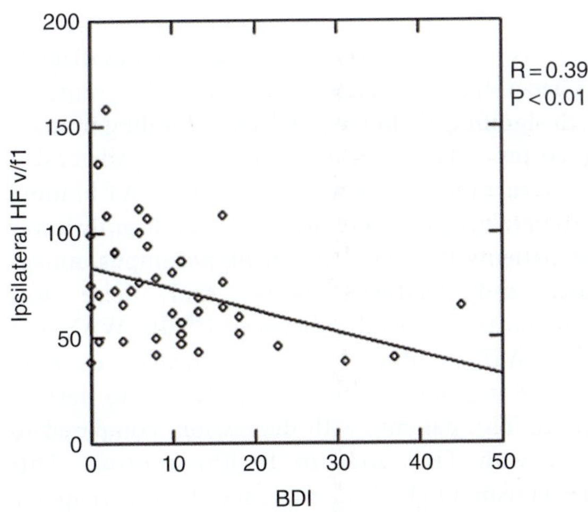

Figure 19.2. Correlation of Beck Depression Inventory with 5HT1A receptor binding measured with [^{18}F]-FCWAY in hippocampus ipsilateral to the seizure onset zone ($R = .38$, $p < .02$). Reproduced from Theodore et al. 2007.[53]

involving not only mesial temporal structures, but cingulate cortex and insula as well.

Some studies in MDD suggested an effect of SSRIs on 5HT1A receptor binding.[46,47] Others however did not find an effect, possibly because of the relatively low occupancy of the receptor sites.[43,56] Some antiepileptic drugs (AEDs) may increase 5HT synthesis and release or block reuptake. However, no AED effects on [^{18}F]-FCWAY binding were found in patients taking these drugs, once correction was made for the increase in FCWAY plasma free fraction, probably due to protein binding interactions with AEDs.[57]

In addition to studies using state-related depression scales such as the BDI, epilepsy patients may present with a comorbid MDD diagnosis based on the Structured Clinical Interview for DSM-IV, which identifies an underlying depressive illness even if patients are euthymic at the time of study.[58] This study, using [^{18}F]FCWAY, found relatively reduced binding in TLE patients with MDD, extending into nonlesional limbic brain areas outside the epileptic focus. After

Bonferroni correction, the depressed group had lower values in anterior cingulate, hippocampus, and medial and superior temporal lobe (Figure 19.3). Seizure focus side was associated with ipsilateral decreases in [18F]-FCWAY V/f1 in hippocampus, parahippocampal gyrus, left amygdala, left fusiform gyrus, and right superior temporal lobe. The only significant group (depressed versus nondepressed) by focus (right versus left) interaction was lower [18F]FCWAY V/f1 raphe values in depressed patients with right temporal foci. However, this did not survive correction for multiple comparisons. Nevertheless, it is an interesting finding, as the [18F]MPPF study also found effects in raphe, where the 5HT cell bodies, as well as autoreceptors are located.[55] The presence of comorbid anxiety, or MTS, did not affect the results. When only patients with concurrent depression were included, only the hippocampus remain significantly reduced after correction for multiple comparisons.

19.4 Serotonin Transport

The serotonin transporter (5HTT), responsible for reuptake of synaptically released ligand, has been implicated in the pathophysiology of depression. "Short" alleles of the encoding gene are associated with reduced transcriptional activity and 5HT reuptake. In patients with one or two short alleles, possible associations with depression, increased response to stress, fear, and "suicidality" in relation to stressful life events, in comparison to individuals homozygous for the long allele have been described.[59]

Some data suggest a role for 5HTT in epilepsy including a higher frequency of some alleles in TLE and MTS compared with healthy controls, and possibly differing AED response. Altered 5HT reuptake and thus synaptic availability might affect modulation of hippocampal excitability. However, not all studies have found a relation between 5HTT allelic variants and TLE.[60]

Several PET ligands have been used to image the 5HTT transporter in patients with depression. Some reported increased availability in patients with MDD and bipolar disease compared to controls, in thalamus, insula, prefrontal cortex, and cingulate gyrus.[61,62] However, others reported decreased availability in amygdala, hippocampus, thalamus, putamen, anterior cingulate, and midbrain.[63,64] A recent review found meta-analysis indicates a trend toward reduced serotonin transporter availability in patients with MDD.

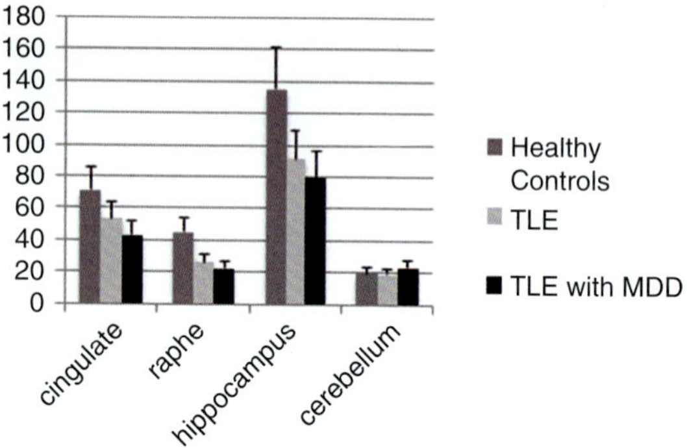

Figure 19.3. Free-fraction corrected 5HT1A serotonin receptor volume of distribution measured with [18F]FCWAY PET in healthy controls, patients with TLE alone, and TLE with depression, showing the additive effects of depression.

The authors suggested that inconsistencies in results among studies might be due to symptom heterogeneity and might therefore be important for stratification of patients into clinical subsets.[65] Another implication is that much larger studies would be needed to elucidate fully patterns of altered transport (or indeed 5HT receptor binding, given the variability in results discussed above). Unfortunately, PET studies are expensive.

Thirteen patients with TLE who had been evaluated with the BDI as well as the SCID had imaging with the 5-HTT transporter ligand [11C]-3-amino-4-(2-dimethylaminomethylphenylsulfanyl)-benzonitrile ([11C]DASB) as well as [18F]FCWAY.[66] Regional [11C]DASB binding did not differ between patients and controls. However, depression diagnosis had a significant effect on [11C]DASB asymmetry, with significantly lower [11C]DASB binding in insular cortex and a trend for fusiform gyrus ipsilateral to the seizure focus (Figure 19.4). [18F]FCWAY binding ipsilateral to the seizure focus was significantly lower for patients than controls in hippocampus, amygdala, and fusiform gyrus. And there was a significant correlation between [18F]FCWAY and [11C]DASB binding. These data suggest relatively reduced transporter activity ipsilateral to seizure foci in TLE patients with depression, compared to those with TLE alone or healthy controls. This mechanism might lead to reduced 5HT reuptake and thus increased synaptic availability. Since [11C]DASB and [18F]FCWAY binding were correlated, reduced transport might represent a compensatory mechanism for 5HT1A receptor loss.

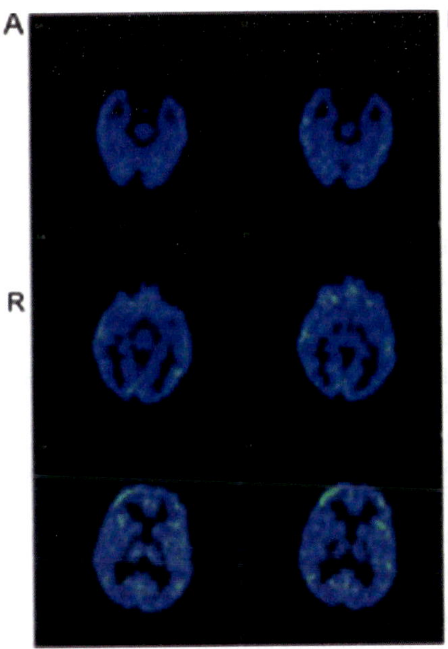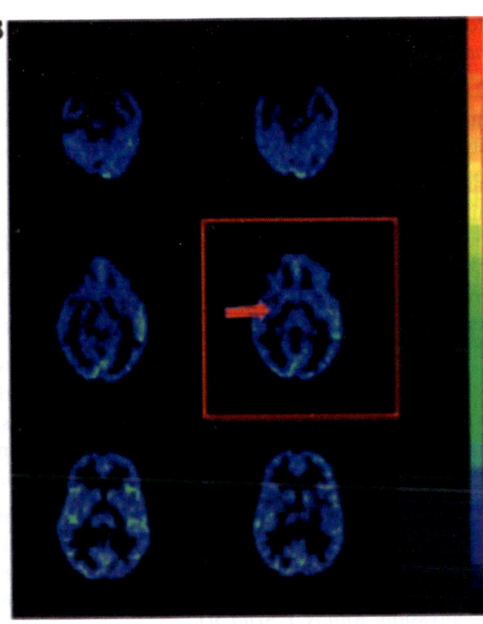

Figure 19.4. (A) [^{11}C]-DASB PET in a patient with no history of depression shows symmetrical insula binding. (B) [^{11}C]DASB PET in a patient with a history of depression shows relatively reduced right insula binding. Reproduced from Martinez et al. 2013.[66]

Taken together, studies of both 5HT1A receptors and 5-HTT suggest a strong role of 5HT1A receptors in TLE patients with depression, and support the use of SSRIs in treating them. Moreover, a review of US Food and Drug Administration data from antidepressant drug trials showed that patients had fewer seizures during treatment than placebo arms, suggesting that the drugs may have antiseizure effects.[67]

19.5 Conclusions

Several lines of evidence, including structural and functional MRI studies, as well as PET studies of 5HT1A receptors and 5HT transport, suggest a role for alterations in limbic structures, as well as serotonergic neurotransmission, in the pathophysiology of depression in patients with epilepsy. At present, most of the data are cross-sectional. It is uncertain, for example, if patients with epilepsy are more likely to become depressed due to limbic dysfunction, or whether some exogenous factors, such as seizure frequency and epilepsy duration, in addition to focus localization, influence both clinical symptoms and anatomo-physiological correlates of depression in epilepsy.

It is encouraging that there are strong parallels between imaging results in patients with depression in epilepsy and MDD alone. These observations support the epidemiologic data showing reciprocal relations between the two disorders. The PET 5HT1A and 5HTT imaging data strongly support the use of SSRIs in treatment. Emerging data on the role of altered FC may lead to new therapeutic approaches involving direct modulation of neuronal activity.

References

1. Kanner A. Psychiatric comorbidities through the life of the seizure disorder: a complex relation with a not so complex solution. *Epilepsy Currents*. 2014;14:323–8.

2. Fiest KM, Dykeman J, Patten SB et al. Depression in epilepsy: a systematic review and meta-analysis. *Neurology*. 2013;80:1–10.

3. Fuller-Thomson E, Brennenstuhl S. The association between depression and epilepsy in a nationally representative sample *Epilepsia*. 2009;50(5): 1051–8.

4. Kobau, R, Gilliam F, Thurman DJ. Prevalence of self-reported epilepsy or seizure disorder and its associations with self-reported depression and anxiety: results from the 2004 Healthstyles Survey. *Epilepsia*. 2006;47:1915–21.

5. Kanner AM, Schachter SC, Barry JJ, et al. Depression and epilepsy, pain and psychogenic non-epileptic seizures: clinical and therapeutic perspectives. *Epilepsy Behav*. 2012;24:169–81.

6. Mula M, Jauch R, Cavanna A, et al. Interictal dysphoric disorder and periictal dysphoric symptoms in patients with epilepsy. *Epilepsia*. 2010;51:1139–45.

7. Kanner AM, Soto A, Gross-Kanner H. Prevalence and clinical characteristics of postictal psychiatric symptoms in partial epilepsy *Neurology*. 2004;62:708–13.

64. Oquendo MA, Hastings RS, Huang YY, et al. Brain serotonin transporter binding in depressed patients with bipolar disorder using positron emission tomography. *Arch Gen Psychiatry*. 2007;**64**:201–8.

65. Spies M, Knudsen GM, Lanzenberger R, et al. The serotonin transporter in psychiatric disorders: insights from PET imaging. *Lancet Psychiatry*. 2015;**2**: 743–55.

66. Martinez A, Finegersh A, Cannon DM et al. The 5-HT1A receptor and 5-HT transporter in temporal lobe epilepsy. *Neurology*. 2013;**80**:1465–71.

67. Alper K, Schwartz KA, Kolts RL, et al. Seizure incidence in psychopharmacological clinical trials: an analysis of Food and Drug Administration (FDA) summary basis of approval reports. *Biol Psychiatry*. 2007;**62**:345–54.

Chapter 20

Tracking Epilepsy Disease Progression with Neuroimaging

Boris C. Bernhardt, Ana Carolina Coan, Lorenzo Caciagli, Andrea Bernasconi, and Neda Bernasconi

20.1 Introduction

Epilepsy is one of the most prevalent neurological disorders, affecting ~1% of the general population. Since the early hypothesis that "seizures beget seizures" by William Gowers,[1] there has been a considerable body of animal and human research suggesting that several forms of epilepsy may be progressive.[2,3] Epilepsy progression may be defined as a cumulative impact of the disease over time. Domains affected by progression may range from clinical manifestations, such as progressively worse seizure control and seizure severity, to cumulatively altered EEG patterns and more extensive brain abnormalities, cognitive decline, together with increased challenges for adequate socio-affective functioning and quality of life.

Neuroimaging has become a key component of the workup in individual patients with epilepsy, as it allows for the in vivo localization of epileptogenic lesions.[4–9] Neuroimaging is especially important in patients with pharmacoresistant focal epilepsies for whom surgery is, overall, the most effective treatment to arrest seizures.[10,11] Particularly magnetic resonance imaging (MRI) has been central to unveiling structural and functional manifestations of the disease. Quantitative imaging studies have furthermore been pivotal in characterizing whole-brain phenotypes,[12–19] providing meaningful complements to genetic studies and experimental approaches.

Throughout the last decades, individual- and group-level neuroimaging assessments have addressed progression in different epilepsy syndromes. In the current chapter, we aim to overview neuroimaging markers and study designs that have been used in the evaluation of disease progression. We evaluate the evidence put forward by previous imaging studies assessing disease progression, particularly those in temporal lobe epilepsy (TLE) and idiopathic generalized epilepsy (IGE). In TLE, multiple studies have suggested cumulative brain atrophy in patients with longer disease duration, particularly in the mesiotemporal lobe, but also in thalamic and neocortical regions.[3,20,21] These findings have been complemented by diffusion MRI work suggesting cumulative derangement in white matter architecture and microstructure, together with functional and metabolic studies emphasizing cumulative reorganization in interregional networks.[22,23] Although progressive brain alterations have been less extensively studied in IGE than in TLE, several reports have also suggested progressive changes at the level of the thalamus and neocortical regions.[24,25] Collectively, findings largely support the concept that the disease course in several epileptic populations may be progressive. A critical evaluation of the literature, however, also reveals that the overwhelming majority of previous assessments were either cross-sectional designs with heterogeneous age-control procedures or single-cohort longitudinal studies restricted to small patient groups. We believe that evidence is still partial to determine which specific patient subgroups present with a progressive disease course. Moreover, our understanding on which mechanisms contribute to progression needs to be further refined.[2,26]

20.2 Tracking Progression with Neuroimaging Studies

Neuroimaging modalities have been highly useful in the study of epilepsy, particularly MRI, positron emission tomography (PET), single-photon emission computed tomography (SPECT), and magnetoencephalography (MEG).[27,28] As these techniques probe complementary aspects of the nervous system, with variable trade-offs in terms of costs, spatial resolution, and temporal resolution, the promise rests on their combined application to further consolidate our understanding of pathological substrates, pathophysiological mechanisms, and disease progression in epileptic disorders.

Undoubtedly, MRI has played a pivotal role in the investigation and management of epilepsy, particularly the drug-resistant forms.[27,29] Indeed, since its introduction to research and the clinical practice in the late 1980s, ongoing advances in MRI acquisition and analysis have revolutionized diagnostic and therapeutic approaches.[28] Given its unmatched spatial resolution, noninvasiveness, and whole-brain coverage, MRI has been the most important modality to reveal lesions co-occurring with the site of seizure origin.[29] The existence of different MRI sequences provides flexibility to model multiple aspects of brain organization and pathology, ranging from alteration in morphology (through the analysis of volumetry or cortical thickness based on T1-weighted images) to gliosis (through analysis on FLAIR/T2 intensity), myelination and iron deposition (based on quantitative T1 mapping and T2*/SWI analysis), fiber architecture and tissue microstructure (inferred from diffusion MRI parameters), local function (derived from fMRI derivatives), and interregional connectivity (based on functional MRI connectivity, structural covariance, or diffusion tractography). In addition to being of high utility in defining the surgical target,[6,30–32] recent work has also proposed several MRI markers to serve as prognostic biomarkers in the prediction of long-term surgical outcome.[14,33–36]

Given its noninvasiveness, MRI can provide markers that track disease progression in individual patients over time. So far, however, the overwhelming majority of studies on progressive brain changes have employed cross-sectional designs, which related MRI metrics to disease duration, estimates of seizure frequency, a history of generalized seizures, or age. Statistical methods are largely based on generalized linear models, including straightforward regressions (e.g., between duration of epilepsy and hippocampal volume) or group comparisons (e.g., comparisons of hippocampal volume between new-onset and long-duration patients). Cross-sectional designs are cost-effective and logistically less complex than longitudinal ones. As they do not carry the burden of repeated assessments over time, they may also allow the inclusion of a broader range of measures to explore variable interrelationships that may generate novel hypotheses about disease progression. Nevertheless, cross-sectional studies suffer from the crucial confound of mixing between- and within-subject effects.[37,38] In other words, these designs compare patient groups with short duration to those with long duration (or those with few seizures to those with frequent seizures); yet, results have been frequently interpreted to signify progressive disease trajectory of individual patients. In the latter case, differences across age groups in lifestyle, fitness, but also available treatment, together with other cohort effects, might have influenced findings.

Despite their high costs and logistic challenges, longitudinal assessments are undeniably the more appropriate design to test hypotheses regarding within-subject trajectories, and to infer causality.[39] Within the broad category, *single-cohort* variants follow an individual cohort over time. If carried out over long intervals, such studies may become costly, patients' follow-up may become increasingly difficult, and test-retest reliability may be compromised by updates in neuroimaging hardware and analysis techniques. *Multicohort* longitudinal designs overcome some of these limitations, and combine longitudinal and cross-sectional elements. They follow several patient and control cohorts. Further stratification is possible, for example, with respect to levels of drug-control (e.g., controlled vs. pharmacoresistant), duration (e.g., early onset vs. chronic long standing, or according to an interval scale), age, gender, and variables that may interact with disease trajectories, such as initial precipitating events or genetic markup. In effect, *stratified multicohort* designs theoretically provide the most accurate estimation of disease trajectories, control for age and cohort effects, and adequately model interindividual variability. Longitudinal designs can be analyzed using statistical methods that model both within- and between-subject effects on disease trajectories. A prominent design choice is *multilevel* modeling, with mixed-effects models being a prominent subcategory. These models estimate fixed effects of a given variable of interest, such as duration or time from baseline, on a dependent variable, such as hippocampal volume or cortical thickness, while also taking into account the within-participant dependence of observations. These techniques can flexibly model uneven sampling intervals, missing data, and an imbalanced number of samples across individuals. By including all data possible, they may increase statistical sensitivity compared to the more widely used repeated-measures ANOVAs in the case of imbalanced or incomplete data. Multilevel models can be used to infer population-level trajectories, to evaluate between-group differences of trajectories, and to examine individual variability in progressive disease course. Notably, and despite their increased costs and challenges to follow a given individual, longitudinal studies have the benefit of requiring drastically smaller samples than cross-sectional studies to capture subtle effects.[40] In a power analysis by Steen and colleagues, for example, detecting a 5% difference in brain volume in a 2-sample cross-sectional study required a sample size of 73 per group, while a similar-sample longitudinal study demanded just 5 individuals per group of patients.[40]

Yet, and as we describe in the following section, longitudinal studies have been surprisingly scarce in the epilepsy neuroimaging literature and stratified multicohort studies are so far nonexistent. One particular challenge is the group of patients with drug-resistant focal epilepsy, who are recommended to undergo surgery once a lesion is detected. This implies that well-characterized patients are operated more readily, while surgery may be delayed for many years in more challenging candidates (with a possibly more severe disease course). Moreover, given that drug-resistant patients are treated with variable drugs and doses throughout the course of their disease, it has been almost impossible to properly evaluate the role of antiepileptic drugs in disease progression. Future prospective studies are, therefore, recommended to devise meaningful strategies that account for such selective attrition and therapy-related confounding effects.

20.2.1 Studies in Temporal Lobe Epilepsy

TLE is the most common drug-resistant epilepsy in adults and is commonly associated with hippocampal sclerosis, the marked cell loss, and/or gliosis seen on histological specimens.[41] Quantitative MRI studies have been pivotal in revealing hippocampal pathology in vivo; indeed, hippocampal volume loss measured on T1-weighted images has been shown to correlate with the degree of neuronal loss[42] while increased T2 signal is thought to index reactive astrogliosis.[43]

In the mid-1990s, several cross-sectional studies in TLE brought forward the first evidence that the extent of ipsilateral hippocampal damage as measured on MRI may increase as a function of disease duration,[44] seizure frequency estimates,[45–47] and a history of generalized seizures.[47] In a study by Kalviainen and colleagues, hippocampal volume was reported to correlate negatively with both the total number of partial or generalized seizures in patients with left TLE.[47] These data were complemented by the finding of prolonged T2 relaxation times in patients with more frequent seizures and a longer disease duration.[47]

Subsequent cross-sectional assessments by multiple groups have reproduced these findings, mainly in drug-resistant cohorts.[48–52] For example, Theodore et al. demonstrated in 35 unilateral TLE patients with a history of secondary generalization that duration, but not age at seizure onset, correlated with the degree of ipsilateral hippocampal atrophy.[53] Several studies measured hippocampal volumes using automated techniques, and were able to demonstrate results with similar effect sizes.[48,54,55] Notably, while most assessments reported evidence for progression in the ipsilateral hippocampus, some studies have also suggested cumulative atrophy in the contralateral hippocampus.[49,56]

It is notable that there has been no standardized methodology in these studies to control for confounds of normal aging. Indeed, approaches range from the complete omission of age control procedures to the reporting of no age effect in controls, corrections for age or age at seizure onset, or statistical interaction analysis of aging between patients and controls.

Longitudinal imaging comparisons between hippocampal volume changes in patients relative to controls may provide a rather direct control for aging, but these analyses are so far virtually inexistent. In addition to several interesting case studies,[57–59] only few single-cohort designs have suggested within-patient disease progression. Analyzing serial MRI data in 24 patients recruited from a first seizure clinic, Briellmann and colleagues observed ipsilateral hippocampal volume decrease of 9% over a period of 3.5 years.[60] Atrophy rate was modulated by the number of generalized seizures between both scans, suggesting a harmful effect of even a few generalized seizures on the brain. In a longitudinal study evaluating a small sample of seizure-free patients and those with continuing seizures over a similar follow-up period, Fuerst and colleagues observed progressive ipsilateral atrophy in the latter but not former subgroup, also suggesting a link between seizures and disease progression.[61]

Cross-sectional volumetric studies have suggested that progressive changes are not limited to the hippocampus, but may also involve adjacent mesiotemporal regions, such as the entorhinal cortex and amygdala (Figure 20.1A).[62] Cumulatively increased structural compromise in mesiotemporal

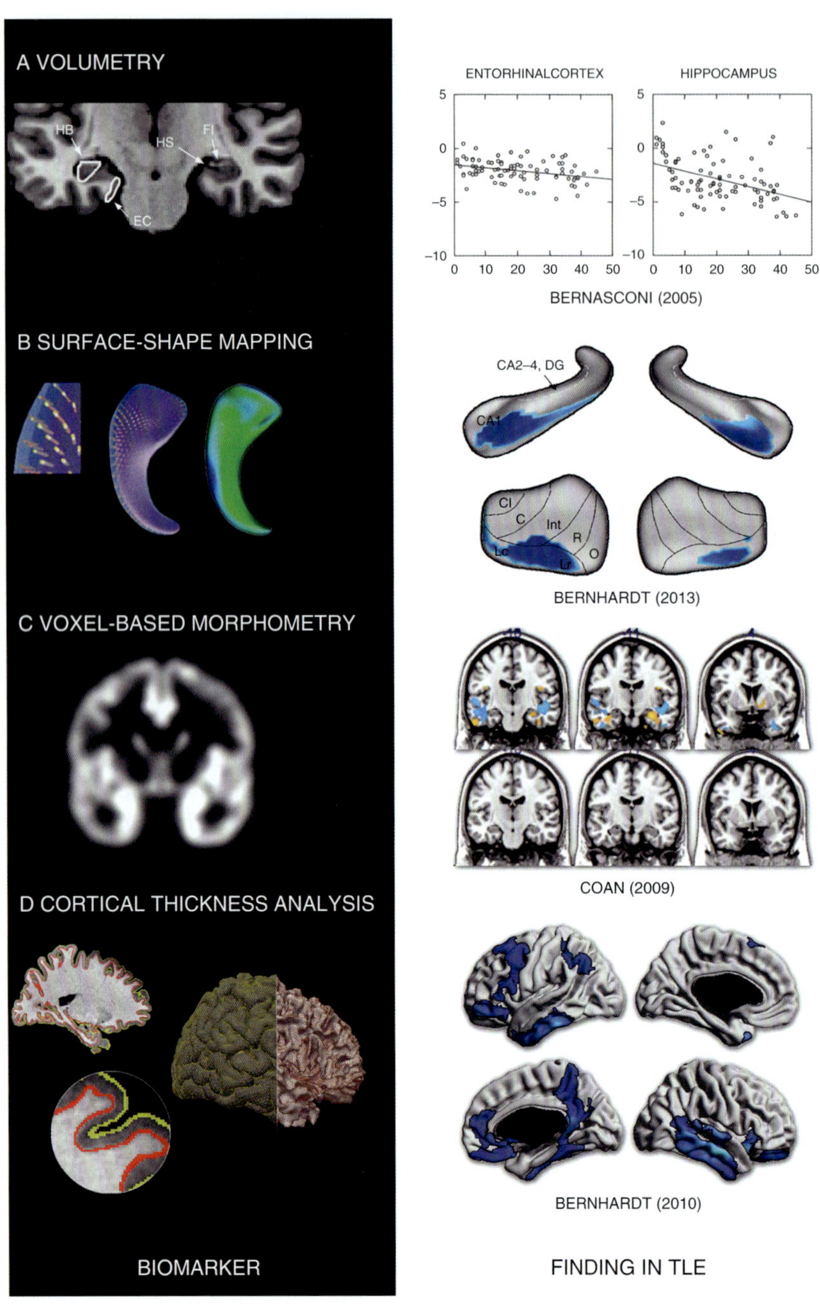

Figure 20.1. Imaging biomarkers (left column) and findings in temporal lobe epilepsy (TLE, right column). Examples are shown for mesiotemporal volumetry (A), hippocampal surface-shape mapping (B), voxel-based morphometry (C), and cortical thickness analysis (D). The selected findings in TLE collectively suggest more marked atrophy in patients with a longer duration of epilepsy. Adapted from Bernhardt et al.,[12] Coan et al.,[20] Bernasconi et al.,[62] and Bernhardt et al.[63] with permission.

networks may extend to both hemispheres, as evidenced by a recent surface-shape analysis.[63] In this study, a high similarity was observed between cross-sectional findings in 134 patients and longitudinal analysis in a subgroup of 31 (Figure 20.1B).[63] Notably, surface-based mapping of local atrophy has further refined the pattern of progressive changes in the hippocampus itself, by consistently pointing to high effect sizes particularly in the CA1 subregion.[63,64] Given that progressive atrophy preferentially colocalized with areas displaying marked neuronal loss on histology,[41,65] these results emphasize the ability of advanced structural MRI processing to unveil lesional tissue and progressive structural damage otherwise not detected on visual evaluation or global volumetry.[6]

Considering the thalamus, data derived from volumetry, MR spectroscopy, voxel-based morphology, as well as surface-shape mapping have collectively yielded findings that suggest progressive damage and neuronal dysfunction in TLE,[66–69] in line with the concept that the thalamus plays an important role in the pathophysiological network of this condition.[70,71]

Cross-sectional assessments of the neocortex based on voxel-based morphometry,[67,72] whole-brain volumetric techniques,[48] and MRI-based cortical thickness measurements[15,21,73–75] have shown widespread and multilobar atrophy increasing with longer disease duration. Again, the few longitudinal studies performed to date have been rather sensitive despite the modest samples studied, and could consistently reveal increased atrophy over time across all lobes (Figure 20.1C–D).[12,20,21] These studies could furthermore suggest a modulation of progressive trajectories by several clinical variables, including seizure frequency,[20,21] focus lateralization,[20] and duration at baseline.[21]

As in the case of cross-sectional designs, different approaches have been used to account for potential aging effects in these longitudinal studies. In a series of studies,[21,76] the absence of a longitudinal control group was compensated by cross-sectionally comparing age effects on structural markers between patients and controls. This analysis indicated more marked age-related cortical thinning in the former group, suggesting that progressive atrophy in patients is likely due to not typical aging but rather additional disease-specific effects. In the study of Coan et al.,[20] the finding of marked gray matter loss between both scans in the patient sample visually contrasted with that of no significant gray matter changes in controls.

A recent systematic review and meta-analysis of MRI volumetric studies of the hippocampus indicated overall moderate effect sizes supportive of more severe ipsilateral atrophy in patients with longer epilepsy duration and more frequent seizures.[77] Additional synthesis of whole-brain morphometric studies emphasized that changes often extend to extratemporal and subcortical regions, collectively supporting that TLE is likely progressive. Notably, quantitative synthesis of study design variability also indicated that previous work was mainly based on cross-sectional inference and effects of chronological aging were rather inconsistently addressed, emphasizing the need for future studies with longitudinal study designs and more rigorous age control procedures.[77]

Data from other neuroimaging modalities, although performed less frequently, also support the concept of TLE being a progressive disorder. In a cross-sectional diffusion tractography assessment, Keller and colleagues reported a correlation between epilepsy duration and decreased white matter fiber anisotropy (a marker of ordered fiber arrangement, axonal membranes, and myelination)[78] in the ipsilateral temporal lobe, bilateral thalamus, and posterior corpus callosum.[23] The hypothesis that progressive white matter pathology may be preferentially located in temporolimbic networks affected by seizure spread was also supported by the finding of cumulatively decreased anisotropy in the uncinate fasciculus and cingulum.[75] The parallel analysis of cortical thickness in these patients furthermore revealed progressive cortical thinning in the ipsilateral parietal and contralateral frontal lobe.[75]

Findings in the structural domain have been complemented by reports of progressively altered intrinsic functional connectivity based on task-free ("resting-state") functional MRI. Zhang et al.,[79] for example, observed that the degree of functional connectivity decreases between mesiotemporal regions and parietal midline cortices correlated with epilepsy duration. Duration effects on interhemispheric hippocampal functional connectivity have been reported as well.[80] More recent graph-theoretical analyses could show more marked functional connectome changes in patients with a longer disease duration,[81] a finding in accordance to a longitudinal analysis of networks derived from cortico-cortical structural covariance patterns.[13]

Assessing [^{18}F]-fluorodeoxyglucose (^{18}FDG)-PET data in children with new-onset partial seizures (most of them having a temporal seizure focus), Gaillard, Theodore, and colleagues suggested decreased glucose metabolism may be less common in children than in studies of adults with chronic epilepsy.[82] In a subsequent report based on 91 patients with drug-resistant TLE, the same group demonstrated a correlation between ipsilateral hippocampal hypometabolism and disease duration;[83] moreover, longitudinal PET in children revealed a relation between seizure frequency and the degree of hypometabolism.[84]

20.2.2 Studies in Idiopathic Generalized Epilepsy

IGE refers to a group of epilepsy syndromes characterized by generalized spike and slow-wave discharges on EEG, occurring in runs of 2.5–4 Hz, with normal background activity.[85–87] Although their exact pathophysiological substrate remains unknown, abundant studies in animal models and human patients have suggested a major role of cortico-thalamic networks in the generation and maintenance of generalized epileptic activity.[88–91]

As in TLE, the majority of previous neuroimaging studies in IGE has addressed progressive changes using cross-sectional designs. Using MR spectroscopy, Bernasconi and colleagues demonstrated a correlation between decreased thalamic NAA/Cr ratio, a marker for neuronal loss and dysfunction, and duration of epilepsy (Figure 20.2A).[24] Thalamic duration effects have also been reported using voxel-based morphometric techniques[92,93] and volumetric assessments.[25] In a sample of 50 IGE, a combined voxel-based morphometric and surface-shape analysis localized duration effects on atrophy primarily in anterior-medial and posterior-dorsal aspects of bilateral thalami.[94]

Several studies have also suggested progressive alterations in neocortical regions. In frontal cortices, there are cross-sectional data supporting progressive gray matter loss based on voxel-based morphometric and cortical thickness assessments in cohorts with juvenile myoclonic epilepsy[95,96] and with generalized tonic-clonic seizures only.[25,93] In the latter IGE subgroup, a previous study showed duration-related atrophy in both thalamic as well as neocortical regions, suggesting progressive compromise of the thalamocortical system (Figure 20.2B).[25] Furthermore, the study could observe faster cortical thinning in pharmacoresistant patients relative to those with well-controlled seizures.[25]

Cross-sectional diffusion MRI analyses have complemented other structural neuroimaging assessments and provided broad support to the notion of abnormal subcortico-cortical connectivity in IGE.[22,23,97,98] Moreover, in samples with juvenile myoclonic epilepsy, patients with more frequent generalized seizures show reduced fiber anisotropy in thalamocortical tracts alone[22] or in concert with other subcortical bundles.[99] Progressive changes in thalamocortical connectivity have independently been suggested using resting-state functional connectivity analysis in a heterogeneous IGE population.[100] Last, recent graph theoretical analysis has shown effects of duration on several parameters relating to network topology.[101,102]

Longitudinal studies in adult IGE populations are so far nonexistent. In a series of cross-sectional and longitudinal imaging studies on new-onset pediatric cases with IGE, differences in brain structure were observed at baseline between patients and controls,[103,104] indicative of preexisting brain anomalies, together with abnormal trajectories early in the course of the disease, possibly suggesting combinations of alterations in developmental trajectories and progressive changes.[104–106]

20.3 Conclusions

There is abundant neuroimaging data concluding that TLE, being among the most common and widely studied epilepsies, is likely progressive. While relatively consistent findings suggestive of disease progression have been reported at the level of the ipsilateral hippocampus, cumulative atrophy in TLE has been shown to also involve mesiotemporal, thalamic, and neocortical regions—not necessarily restricted to the hemisphere ipsilateral to the focus. Overall, these data together with recent meta-analytical synthesis[77] support the concept that TLE is a progressive disorder with a system-level impact on brain networks.[107,108] While progression has been less frequently studied in IGE, there is repeated support for disease progression in several patient subgroups (e.g., juvenile myoclonic epilepsy and patients with generalized tonic-clonic seizures), likely involving thalamic and/or prefrontal and frontocentral networks.

It is noteworthy that most findings have been derived from cross-sectional studies with a relatively heterogeneous control for aging, and that some cross-sectional studies did not support evidence for progressive changes on variable neuroimaging measures in both TLE[109] and IGE.[110–112] Carefully evaluating the few longitudinal studies published to date, a crucial shortcoming is the lack of statistical comparisons in population slopes between patients and controls. It, thus, still remains to be tested whether longitudinally measured progressive changes in TLE and IGE are indeed different from aging, and to which extent.

Future longitudinal studies should help to clarify mechanisms underlying disease progression. In animal models, single seizures have been shown to cause apoptotic changes or intermediate forms of neuronal

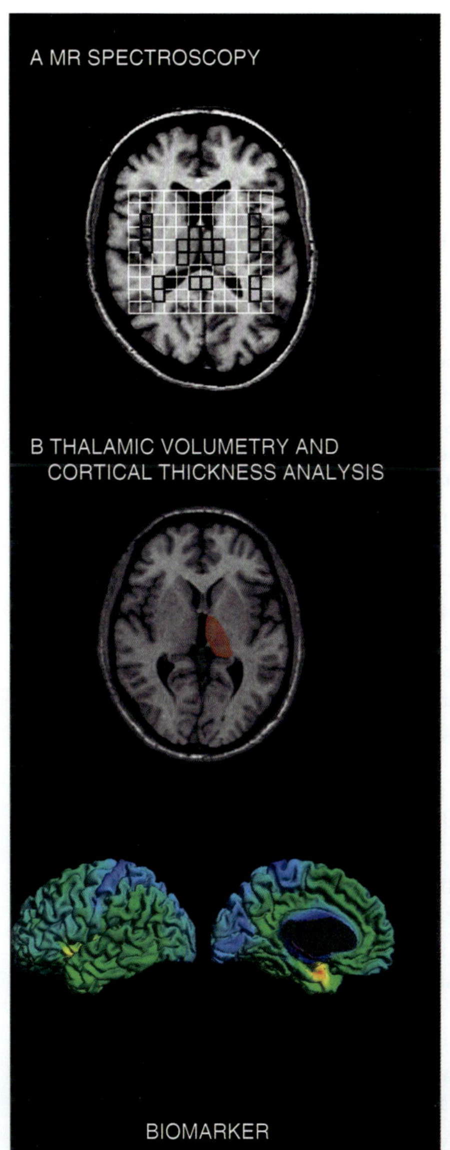

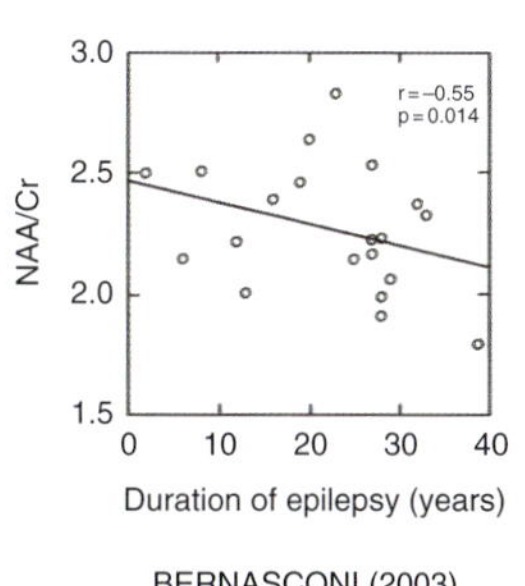

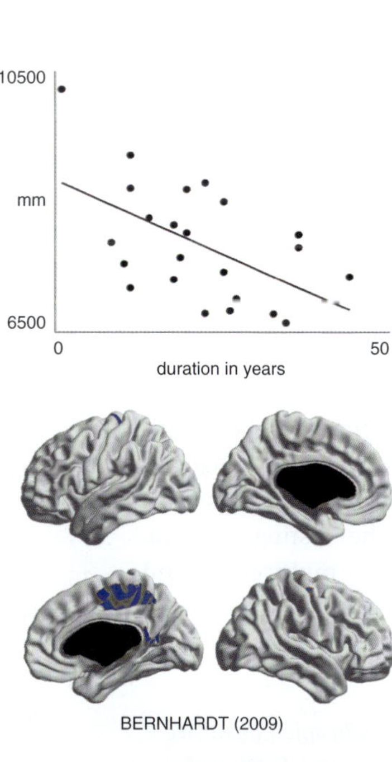

Figure 20.2. Imaging biomarkers (left column) and findings in idiopathic generalized epilepsy, IGE (right column). Examples are shown for thalamic spectroscopy (A) and a combined thalamic volumetric and cortical thickness analysis (B). The selected findings in IGE suggest more marked compromise of the thalamocortical system in patients with a longer duration of epilepsy. Adapted from Bernasconi et al.[24] and Bernhardt et al.[25] with permission.

death.[113,114] Similar findings have been observed after experimental[115] and human status epilepticus.[116] In animals, seizures can increase the susceptibility of network synchronization, lowering the threshold to generate new seizures.[117,118] Seizures have been suggested to damage the brain, possibly through the up-regulation of neuronal-axonal excitability markers,[119] particularly glutamate.[120] Furthermore, disruptions in cortical GABAergic circuits have been described, potentially contributing to genesis or maintenance of seizure activity.[121]

Effects of antiepileptic drugs on brain networks are largely unknown. Phenytoin[122] has been suggested to induce cerebellar atrophy, and valproic acid therapy has been associated with pseudoatrophy[123] and cortical thinning.[124] On the other hand, the same drugs have been attributed neuroprotective effects promoting neurogenesis.[125,126] Given that drug-resistant patients often have a history of multiple and combined drug trials, this cohort may pose practical challenges on the implementation of prospective studies with well-controlled and standardized medical treatment.

Progression may be not only heterogeneous across different epilepsy syndromes but also variable within a specific syndrome. Considering TLE, clinical observations and research findings suggest that some patients may have a more severe disease course and accelerated decline compared to others. Genetic studies may provide important data that could explain interindividual differences in susceptibility for seizure-related atrophy. Moreover, it remains to be evaluated how additive effects of challenges in psychosocial functioning, lifestyle choices, and comorbid depression can be distinguished from the effects of epilepsy and seizures.

In sum, further longitudinal studies that follow patient and control cohorts using advanced neuroimaging are needed to provide high-level evidence for disease progression and identify its underlying factors. Future studies should statistically compare trajectories and identify patient subgroups based on patterns of progression. As the necessity for treatment in drug-resistant cohorts precludes tracking of within-subject change over time, structured and accelerated designs that systematically enroll patients at different time points in their disease course, from new onset to chronic long standing, are recommended.[127] In the ideal case, such an investigation should include both drug-resistant and well-controlled patients and closely monitor medication dose, seizure counts, and psychosocial functioning throughout the testing interval. Such a rich design clearly demands multicentric and multidisciplinary efforts.

References

1. Gowers WR. *Epilepsy and Other Chronic Convulsive Disorders: Their Causes, Symptoms and Treatment.* London: J&A Churchill; 1881.

2. Sutula TP, Hagen J, Pitkanen A. Do epileptic seizures damage the brain? *Curr Opin Neurol.* 2003;**16**: 189–95.

3. Cascino GD. Temporal lobe epilepsy is a progressive neurologic disorder: time means neurons! *Neurology.* 2009;**72**:1718–9.

4. Bernasconi A, Antel SB, Collins DL, et al. Texture analysis and morphological processing of magnetic resonance imaging assist detection of focal cortical dysplasia in extra-temporal partial epilepsy. *Ann Neurol.* 2001;**49**:770–5.

5. Colliot O, Bernasconi N, Khalili N, et al. Individual voxel-based analysis of gray matter in focal cortical dysplasia. *NeuroImage.* 2006;**29**:162–71.

6. Bernasconi A, Bernasconi N, Bernhardt BC, et al. Advances in MRI for "cryptogenic" epilepsies. *Nat Rev Neurol.* 2011;**7**:99–108.

7. Hong S, Kim H, Bernasconi N, et al. Automated detection of cortical dysplasia type II in MRI-negative epilepsy. *Neurology.* 2014;**83**:48–55.

8. Huppertz HJ, Grimm C, Fauser S, et al. Enhanced visualization of blurred gray-white matter junctions in focal cortical dysplasia by voxel-based 3D MRI analysis. *Epilepsy Res.* 2005;**67**:35–50.

9. Wagner J, Weber B, Urbach H, et al. Morphometric MRI analysis improves detection of focal cortical dysplasia type II. *Brain.* 2011;**134**:2844–54.

10. Wiebe S, Blume WT, Girvin JP, et al. A randomized, controlled trial of surgery for temporal-lobe epilepsy. *N Engl J Med.* 2001;**345**:311–8.

11. Engel J Jr, McDermott MP, Wiebe S, et al. Early surgical therapy for drug-resistant temporal lobe epilepsy: a randomized trial. *JAMA.* 2012;**307**:922–30.

12. Bernhardt BC, Bernasconi N, Concha L, et al. Cortical thickness analysis in temporal lobe epilepsy: reproducibility and relation to outcome. *Neurology.* 2010;**74**:1776–84.

13. Bernhardt BC, Chen Z, He Y, et al. Graph-theoretical analysis reveals disrupted small-world organization of cortical thickness correlation networks in temporal lobe epilepsy. *Cereb Cortex.* 2011;**21**:2147–57.

14. Hong S, Bernhardt BC, Schrader DV, et al. Whole-brain MRI phenotyping of dysplasia-related frontal lobe epilepsy. *Neurology.* 2016;**86**:643–50.

15. McDonald CR, Hagler DJ Jr, Ahmadi ME, et al. Regional neocortical thinning in mesial temporal lobe epilepsy. *Epilepsia.* 2008;**49**:794–803.

16. Keller SS, Roberts N. Voxel-based morphometry of temporal lobe epilepsy: An introduction and review of the literature. *Epilepsia.* 2008;**49**:741–57.

17. Koepp MJ, Woermann FG. Imaging structure and function in refractory focal epilepsy. *Lancet Neurol.* 2005;**4**:42–53.

18. Woermann FG, Free SL, Koepp MJ, et al. Voxel-by-voxel comparison of automatically segmented cerebral gray matter—a rater-independent comparison of structural MRI in patients with epilepsy. *NeuroImage.* 1999;**10**:373–84.

19. Woermann FG, Free SL, Koepp MJ, et al. Abnormal cerebral structure in juvenile myoclonic epilepsy demonstrated with voxel-based analysis of MRI. *Brain.* 1999;**122**:2101–8.

20. Coan AC, Appenzeller S, Bonilha L, et al. Seizure frequency and lateralization affect progression of atrophy in temporal lobe epilepsy. *Neurology.* 2009;**73**: 834–42.

21. Bernhardt BC, Worsley KJ, Kim H, et al. Longitudinal and cross-sectional analysis of atrophy in pharmacoresistant temporal lobe epilepsy. *Neurology.* 2009;**72**:1747–54.

22. Deppe M, Kellinghaus C, Duning T, et al. Nerve fiber impairment of anterior thalamocortical circuitry in juvenile myoclonic epilepsy. *Neurology.* 2008;**71**:1981–5.

23. Keller SS, Schoene-Bake JC, Gerdes JS, et al. Concomitant fractional anisotropy and volumetric abnormalities in temporal lobe epilepsy: cross-sectional evidence for progressive neurologic injury. *PLOS ONE.* 2012;**7**:e46791.

24. Bernasconi A, Bernasconi N, Natsume J, et al. Magnetic resonance spectroscopy and imaging of the thalamus in idiopathic generalized epilepsy. *Brain.* 2003;**126**:2447–54.

25. Bernhardt BC, Rozen DA, Worsley KJ, et al. Thalamo-cortical network pathology in idiopathic generalized epilepsy: insights from MRI-based morphometric correlation analysis. *NeuroImage.* 2009;**46**:373–81.

26. Coan AC, Cendes F. Epilepsy as progressive disorders: what is the evidence that can guide our clinical decisions and how can neuroimaging help? *Epilepsy Behav.* 2013;**26**:313–21.

27. Duncan J. The current status of neuroimaging for epilepsy. *Curr Opin Neurol.* 2009;**22**:179–84.

28. Kuzniecky RI, Knowlton RC. Neuroimaging of epilepsy. *Semin Neurol.* 2002;**22**:279–88.

29. Bernasconi N, Bernasconi A. Epilepsy: imaging the epileptic brain–time for new standards. *Nat Rev Neurol.* 2014;**10**:133–4.

30. Duncan JS. Imaging in the surgical treatment of epilepsy. *Nat Rev Neurol.* 2010;**6**:537–50.

31. Hong SJ, Kim H, Schrader D, et al. Automated detection of cortical dysplasia type II in MRI-negative epilepsy. *Neurology.* 2014;**83**:48–55.

32. Kim H, Bernhardt BC, Kulaga-Yoskovitz J, et al. Multivariate hippocampal subfield analysis of local MRI intensity and volume: application to temporal lobe epilepsy. *Med Image Comput Comput Assist Interv.* 2014;**17**:170–8.

33. Jack CR Jr, Sharbrough FW, Cascino GD, et al. Magnetic resonance imaging-based hippocampal volumetry: correlation with outcome after temporal lobectomy. *Ann Neurol.* 1992;**31**:138–46.

34. Bernhardt BC, Hong SJ, Bernasconi A, et al. Magnetic resonance imaging pattern learning in temporal lobe epilepsy: classification and prognostics. *Ann Neurol.* 2015;**77**:436–46.

35. Bonilha L, Keller SS. Quantitative MRI in refractory temporal lobe epilepsy: relationship with surgical outcomes. *Quant Imaging Med Surg.* 2015;**5**:204–24.

36. Keller SS, Richardson MP, Schoene-Bake JC, et al. Thalamotemporal alteration and postoperative seizures in temporal lobe epilepsy. *Ann Neurol.* 2015;**77**:760–74.

37. Kraemer HC, Yesavage JA, Taylor JL, et al. How can we learn about developmental processes from cross-sectional studies, or can we? *Am J Psychiatry.* 2000;**157**:163–71.

38. Di Martino A, Fair DA, Kelly C, et al. Unraveling the miswired connectome: a developmental perspective. *Neuron.* 2014;**83**:1335–53.

39. Mills KL, Tamnes CK. Methods and considerations for longitudinal structural brain imaging analysis across development. *Dev Cogn Neurosci.* 2014;**9**:172–90.

40. Steen RG, Hamer RM, Lieberman JA. Measuring brain volume by MR imaging: impact of measurement precision and natural variation on sample size requirements. *AJNR Am J Neuroradiol.* 2007;**28**:1119–25.

41. Blumcke I, Thom M, Aronica E, et al. International consensus classification of hippocampal sclerosis in temporal lobe epilepsy: a task force report from the ILAE Commission on Diagnostic Methods. *Epilepsia.* 2013;**54**:1315–29.

42. Cascino GD, Jack CR Jr, Parisi JE, et al. Magnetic resonance imaging-based volume studies in temporal lobe epilepsy: pathological correlations. *Ann Neurol.* 1991;**30**:31–6.

43. Briellmann RS, Kalnins RM, Berkovic SF, et al. Hippocampal pathology in refractory temporal lobe epilepsy: T2-weighted signal change reflects dentate gliosis. *Neurology.* 2002;**58**:265–71.

44. Spencer SS, McCarthy G, Spencer DD. Diagnosis of medial temporal lobe seizure onset: relative specificity and sensitivity of quantitative MRI. *Neurology.* 1993;**43**:2117–24.

45. Van Paesschen W, Connelly A, King MD, et al. The spectrum of hippocampal sclerosis: a quantitative magnetic resonance imaging study. *Ann Neurol.* 1997;**41**:41–51.

46. Salmenpera T, Kalviainen R, Partanen K, et al. Hippocampal damage caused by seizures in temporal lobe epilepsy. *Lancet.* 1998;**351**:35.

47. Kalviainen R, Salmenpera T, Partanen K, et al. Recurrent seizures may cause hippocampal damage in temporal lobe epilepsy. *Neurology.* 1998;**50**:1377–82.

48. Seidenberg M, Kelly KG, Parrish J, et al. Ipsilateral and contralateral MRI volumetric abnormalities in chronic unilateral temporal lobe epilepsy and their clinical correlates. *Epilepsia.* 2005;**46**:420–30.

49. Jokeit H, Ebner A. Long term effects of refractory temporal lobe epilepsy on cognitive abilities: a cross

sectional study. *J Neurol Neurosurg Psychiatry.* 1999;67:44–50.

50. Fuerst D, Shah J, Kupsky WJ, et al. Volumetric MRI, pathological, and neuropsychological progression in hippocampal sclerosis. *Neurology.* 2001;57:184–8.

51. Marques CM, Caboclo LO, da Silva TI, et al. Cognitive decline in temporal lobe epilepsy due to unilateral hippocampal sclerosis. *Epilepsy Behav.* 2007;10:477–85.

52. Tasch E, Cendes F, Li LM, et al. Neuroimaging evidence of progressive neuronal loss and dysfunction in temporal lobe epilepsy. *Ann Neurol.* 1999;45:568–76.

53. Theodore WH, Bhatia S, Hatta J, et al. Hippocampal atrophy, epilepsy duration, and febrile seizures in patients with partial seizures. *Neurology.* 1999;52:132–6.

54. Alhusaini S, Doherty CP, Scanlon C, et al. A cross-sectional MRI study of brain regional atrophy and clinical characteristics of temporal lobe epilepsy with hippocampal sclerosis. *Epilepsy Res.* 2012;99:156–66.

55. Pulsipher DT, Seidenberg M, Morton JJ, et al. MRI volume loss of subcortical structures in unilateral temporal lobe epilepsy. *Epilepsy Behav.* 2007;11:442–9.

56. Pacagnella D, Lopes TM, Morita ME, et al. Memory impairment is not necessarily related to seizure frequency in mesial temporal lobe epilepsy with hippocampal sclerosis. *Epilepsia.* 2014;55:1197–204.

57. Worrell GA, Sencakova D, Jack CR, et al. Rapidly progressive hippocampal atrophy: evidence for a seizure-induced mechanism. *Neurology.* 2002;58:1553–6.

58. Jackson GD, Chambers BR, Berkovic SF. Hippocampal sclerosis: development in adult life. *Dev Neurosci.* 1999;21:207–14.

59. O'Brien TJ, So EL, Meyer FB, et al. Progressive hippocampal atrophy in chronic intractable temporal lobe epilepsy. *Ann Neurol.* 1999;45:526–9.

60. Briellmann RS, Berkovic SF, Syngeniotis A, et al. Seizure-associated hippocampal volume loss: a longitudinal magnetic resonance study of temporal lobe epilepsy. *Ann Neurol.* 2002;51:641–4.

61. Fuerst D, Shah J, Shah A, et al. Hippocampal sclerosis is a progressive disorder: a longitudinal volumetric MRI study. *Ann Neurol.* 2003;53:413–6.

62. Bernasconi N, Natsume J, Bernasconi A. Progression in temporal lobe epilepsy: differential atrophy in mesial temporal structures. *Neurology.* 2005;65:223–8.

63. Bernhardt BC, Kim H, Bernasconi N. Patterns of subregional mesiotemporal disease progression in temporal lobe epilepsy. *Neurology.* 2013;81:1840–7.

64. Maccotta L, Moseley ED, Benzinger TL, et al. Beyond the CA1 subfield: Local hippocampal shape changes in MRI-negative temporal lobe epilepsy. *Epilepsia.* 2015;56:780–8.

65. Yilmazer-Hanke DM, Wolf HK, Schramm J, et al. Subregional pathology of the amygdala complex and entorhinal region in surgical specimens from patients with pharmacoresistant temporal lobe epilepsy. *J Neuropathol Exp Neurol.* 2000;59:907–20.

66. Bernhardt BC, Bernasconi N, Kim H, et al. Mapping thalamocortical network pathology in temporal lobe epilepsy. *Neurology.* 2012;78:129–36.

67. Bonilha L, Rorden C, Appenzeller S, et al. Gray matter atrophy associated with duration of temporal lobe epilepsy. *NeuroImage.* 2006;32:1070–9.

68. Natsume J, Bernasconi N, Andermann F, et al. MRI volumetry of the thalamus in temporal, extratemporal, and idiopathic generalized epilepsy. *Neurology.* 2003;60:1296–300.

69. Bernasconi A, Tasch E, Cendes F, et al. Proton magnetic resonance spectroscopic imaging suggests progressive neuronal damage in human temporal lobe epilepsy. *Prog Brain Res.* 2002;135:297–304.

70. Bertram EH, Mangan PS, Zhang D, et al. The midline thalamus: alterations and a potential role in limbic epilepsy. *Epilepsia.* 2001;42:967–78.

71. Sinjab B, Martinian L, Sisodiya SM, et al. Regional thalamic neuropathology in patients with hippocampal sclerosis and epilepsy: a postmortem study. *Epilepsia.* 2013;54:2125–33.

72. Keller SS, Mackay CE, Barrick TR, et al. Voxel-based morphometric comparison of hippocampal and extrahippocampal abnormalities in patients with left and right hippocampal atrophy. *NeuroImage.* 2002;16:23–31.

73. Bernhardt BC, Worsley KJ, Besson P, et al. Mapping limbic network organization in temporal lobe epilepsy using morphometric correlations: insights on the relation between mesiotemporal connectivity and cortical atrophy. *NeuroImage.* 2008;42:515–24.

74. Lin JJ, Salamon N, Lee AD, et al. Reduced neocortical thickness and complexity mapped in mesial temporal lobe epilepsy with hippocampal sclerosis. *Cereb Cortex.* 2007;17:2007–18.

75. Kemmotsu N, Girard HM, Bernhardt BC, et al. MRI analysis in temporal lobe epilepsy: cortical thinning and white matter disruptions are related to side of seizure onset. *Epilepsia.* 2011;52:2257–66.

76. Bernasconi N, Bernhardt BC. Temporal lobe epilepsy is a progressive disorder. *Nat Rev Neurol.* 2010;6:1.

77. Caciagli L, Bernasconi A, Wiebe S, et al. Time is brain? A meta-analysis on progressive atrophy in intractable temporal lobe epilepsy *Neurology.* 2017;89:506–16.

78. Concha L, Livy DJ, Beaulieu C, et al. In vivo diffusion tensor imaging and histopathology of the fimbria-fornix in temporal lobe epilepsy. *J Neurosci.* 2010;**30**:996–1002.

79. Zhang Z, Lu G, Zhong Y, et al. Altered spontaneous neuronal activity of the default-mode network in mesial temporal lobe epilepsy. *Brain Res.* 2010;**1323**: 152–60.

80. Morgan VL, Rogers BP, Sonmezturk HH, et al. Cross hippocampal influence in mesial temporal lobe epilepsy measured with high temporal resolution functional magnetic resonance imaging. *Epilepsia.* 2011;**52**:1741–9.

81. Wang J, Qiu S, Xu Y, et al. Graph theoretical analysis reveals disrupted topological properties of whole brain functional networks in temporal lobe epilepsy. *Clin Neurophysiol.* 2014;**125**:1744–56.

82. Gaillard WD, Kopylev L, Weinstein S, et al. Low incidence of abnormal (18)FDG-PET in children with new-onset partial epilepsy: a prospective study. *Neurology.* 2002;**58**:717–22.

83. Theodore WH, Kelley K, Toczek MT, et al. Epilepsy duration, febrile seizures, and cerebral glucose metabolism. *Epilepsia.* 2004;**45**:276–9.

84. Gaillard WD, Weinstein S, Conry J, et al. Prognosis of children with partial epilepsy: MRI and serial 18FDG-PET. *Neurology.* 2007;**68**:655–9.

85. Nordli DR Jr. Idiopathic generalized epilepsies recognized by the International League Against Epilepsy. *Epilepsia.* 2005;**46**(suppl 9):48–56.

86. ILAE. Commission on Classification and Terminology of the International League Against Epilepsy: proposal for classification of epilepsies and epileptic syndromes. *Epilepsia.* 1989;**30**:389–99.

87. Andermann F, Berkovic SF. Idiopathic generalized epilepsy with generalized and other seizures in adolescence. *Epilepsia.* 2001;**42**:317–20.

88. Gloor P. Generalized epilepsy with spike-and-wave discharge: a reinterpretation of its electrographic and clinical manifestations. The 1977 William G. Lennox Lecture, American Epilepsy Society. *Epilepsia.* 1979;**20**: 571–88.

89. Blumenfeld H. From molecules to networks: cortical/ subcortical interactions in the pathophysiology of idiopathic generalized epilepsy. *Epilepsia.* 2003;**44** (suppl 2):7–15.

90. Castro-Alamancos MA, Calcagnotto ME. Presynaptic long-term potentiation in corticothalamic synapses. *J Neurosci.* 1999;**19**:9090–7.

91. Meeren HK, Pijn JP, Van Luijtelaar EL, et al. Cortical focus drives widespread corticothalamic networks during spontaneous absence seizures in rats. *J Neurosci.* 2002;**22**:1480–95.

92. Kim JH, Lee JK, Koh SB, et al. Regional grey matter abnormalities in juvenile myoclonic epilepsy: a voxel-based morphometry study. *NeuroImage.* 2007;**37**:1132–7.

93. Huang W, Lu G, Zhang Z, et al. Gray-matter volume reduction in the thalamus and frontal lobe in epileptic patients with generalized tonic-clonic seizures. *J Neuroradiol.* 2011;**38**:298–303.

94. Kim JH, Kim JB, Seo WK, et al. Volumetric and shape analysis of thalamus in idiopathic generalized epilepsy. *J Neurol.* 2013;**260**:1846–54.

95. Tae WS, Hong SB, Joo EY, et al. Structural brain abnormalities in juvenile myoclonic epilepsy patients: volumetry and voxel-based morphometry. *Korean J Radiol.* 2006;7:162–72.

96. Tae WS, Kim SH, Joo EY, et al. Cortical thickness abnormality in juvenile myoclonic epilepsy. *J Neurol.* 2008;**255**:561–6.

97. O'Muircheartaigh J, Vollmar C, Barker GJ, et al. Abnormal thalamocortical structural and functional connectivity in juvenile myoclonic epilepsy. *Brain.* 2012;**135**:3635–44.

98. Koepp MJ, Woermann F, Savic I, et al. Juvenile myoclonic epilepsy—neuroimaging findings. *Epilepsy Behav.* 2013;**28**(suppl 1):S40–4.

99. Kim JH, Suh SI, Park SY, et al. Microstructural white matter abnormality and frontal cognitive dysfunctions in juvenile myoclonic epilepsy. *Epilepsia.* 2012;**53**:1371–8.

100. Kim JB, Suh SI, Seo WK, et al. Altered thalamocortical functional connectivity in idiopathic generalized epilepsy. *Epilepsia.* 2014;**55**:592–600.

101. Xue K, Luo C, Zhang D, et al. Diffusion tensor tractography reveals disrupted structural connectivity in childhood absence epilepsy. *Epilepsy Res.* 2014;**108**: 125–38.

102. Zhang Z, Liao W, Chen H, et al. Altered functional-structural coupling of large-scale brain networks in idiopathic generalized epilepsy. *Brain.* 2011;**134**:2912–28.

103. Bonilha L, Tabesh A, Dabbs K, et al. Neurodevelopmental alterations of large-scale structural networks in children with new-onset epilepsy. *Hum Brain Mapp.* 2014;**35**: 3661–72.

104. Tosun D, Dabbs K, Caplan R, et al. Deformation-based morphometry of prospective neurodevelopmental changes in new onset paediatric epilepsy. *Brain.* 2011;**134**:1003–14.

105. Pulsipher DT, Dabbs K, Tuchsherer V, et al. Thalamofrontal neurodevelopment in new-onset pediatric idiopathic generalized epilepsy. *Neurology.* 2011;**76**:28–33.

106. Lin JJ, Dabbs K, Riley JD, et al. Neurodevelopment in new-onset juvenile myoclonic epilepsy over the first 2 years. *Ann Neurol.* 2014;**76**:660–8.

107. Bernhardt BC, Hong S, Bernasconi A, et al. Imaging structural and functional brain networks in temporal lobe epilepsy. *Front Hum Neurosci.* 2013;**7**:624.

108. Caciagli L, Bernhardt BC, Hong SJ, et al. Functional network alterations and their structural substrate in drug-resistant epilepsy. *Front Neurosci.* 2014;**8**:411.

109. Dabbs K, Becker T, Jones J, et al. Brain structure and aging in chronic temporal lobe epilepsy. *Epilepsia.* 2012;**53**:1033–43.

110. Woermann FG, Sisodiya SM, Free S, et al. Quantitative MRI in patients with idiopathic generalized epilepsy: evidence of widespread cerebral structural changes. *Brain.* 1998;**121**:1661–7.

111. Savic I, Lekvall A, Greitz D, et al. MR spectroscopy shows reduced frontal lobe concentrations of N-acetyl aspartate in patients with juvenile myoclonic epilepsy. *Epilepsia.* 2000;**41**:290–6.

112. Ronan L, Alhusaini S, Scanlon C, et al. Widespread cortical morphologic changes in juvenile myoclonic epilepsy: evidence from structural MRI. *Epilepsia.* 2012;**53**:651–8.

113. Bengzon J, Kokaia Z, Elmer E, et al. Apoptosis and proliferation of dentate gyrus neurons after single and intermittent limbic seizures. *Proc Natl Acad Sci USA.* 1997;**94**:10432–7.

114. Bengzon J, Mohapel P, Ekdahl CT, et al. Neuronal apoptosis after brief and prolonged seizures. *Prog Brain Res.* 2002;**135**:111–9.

115. Pitkanen A, Nissinen J, Nairismagi J, et al. Progression of neuronal damage after status epilepticus and during spontaneous seizures in a rat model of temporal lobe epilepsy. *Prog Brain Res.* 2002;**135**:67–83.

116. Pohlmann-Eden B, Gass A, Peters CNA, et al. Evolution of MRI changes and development of bilateral hippocampal sclerosis during long lasting generalized status epilepticus. *J Neurol Neurosurgery.* 2004;**75**:898–900.

117. Sutula TP. Mechanisms of epilepsy progression: current theories and perspectives from neuroplasticity in adulthood and development. *Epilepsy Res.* 2004;**60**:161–71.

118. Bertram E. The relevance of kindling for human epilepsy. *Epilepsia.* 2007;**48**(suppl 2):65–74.

119. Isokawa M, Levesque M, Fried I, et al. Glutamate currents in morphologically identified human dentate granule cells in temporal lobe epilepsy. *J Neurophysiol.* 1997;**77**:3355–69.

120. Zilles K, Qu MS, Kohling R, et al. Ionotropic glutamate and GABA receptors in human epileptic neocortical tissue: quantitative in vitro receptor autoradiography. *Neuroscience.* 1999;**94**:1051–61.

121. Ragozzino D, Palma E, Di Angelantonio S, et al. Rundown of GABA type A receptors is a dysfunction associated with human drug-resistant mesial temporal lobe epilepsy. *Proc Natl Acad Sci USA.* 2005;**102**:15219–23.

122. McLain LW Jr, Martin JT, Allen JH. Cerebellar degeneration due to chronic phenytoin therapy. *Ann Neurol.* 1980;**7**:18–23.

123. Papazian O, Canizales E, Alfonso I, et al. Reversible dementia and apparent brain atrophy during valproate therapy. *Ann Neurol.* 1995;**38**:687–91.

124. Pardoe HR, Berg AT, Jackson GD. Sodium valproate use is associated with reduced parietal lobe thickness and brain volume. *Neurology.* 2013;**80**:1895–900.

125. Hao Y, Creson T, Zhang L, et al. Mood stabilizer valproate promotes ERK pathway-dependent cortical neuronal growth and neurogenesis. *J Neurosci.* 2004;**24**:6590–9.

126. Magarinos AM, McEwen BS, Flugge G, et al. Chronic psychosocial stress causes apical dendritic atrophy of hippocampal CA3 pyramidal neurons in subordinate tree shrews. *J Neurosci.* 1996;**16**:3534–40.

127. Thompson WK, Hallmayer J, O'Hara R, et al. Design considerations for characterizing psychiatric trajectories across the lifespan: application to effects of APOE-epsilon4 on cerebral cortical thickness in Alzheimer's disease. *Am J Psychiatry.* 2011;**168**:894–903.

Imaging Biomarkers to Study Cognition in Epilepsy

21

Silvia B. Bonelli and John S. Duncan

21.1 Introduction

Cognitive impairment is a frequent comorbidity in epilepsy and has a major impact on quality of life in these patients. Cognitive deficits either can result from the underlying disease as a consequence of seizures or interictal epileptic activity, or can be caused by adverse effects of antiepileptic drugs (AEDs).[1]

In temporal lobe epilepsy (TLE), as the most common type of focal epilepsy, typically material-specific memory impairment and naming difficulties have been reported.[2] Even in some patients with epilepsy of recent onset[3] cognitive impairment can be observed, which supports the hypothesis that cognitive problems cannot fully be explained by adverse effects of medication. In general, patients who are refractory to AED treatment are at higher risk of suffering from cognitive impairment.

In patients with medically refractory TLE, anterior temporal lobe resection (ATLR) is an effective and safe treatment option, leading to seizure remission in up to 60–70% of these patients.[4,5] Unfortunately this procedure may be complicated by a decline in language and memory abilities. Therefore an important emphasis during presurgical evaluation is to identify the epileptic brain tissue that has to be removed for the patient to become seizure-free, and to predict the cognitive cost of this. Neuropsychological deficits must be avoided, which requires accurate localization of the brain areas that are responsible for motor, language, and memory functions (eloquent cortex).

In recent years, epilepsy surgery has been carried out earlier in the course of the disease, and so the potential benefits must be carefully weighed against the potential risks of decline. In order to successfully identify eloquent brain areas and guide resection, a range of imaging techniques such as functional MRI (fMRI) and diffusion tensor imaging (DTI) have been employed.

In particular fMRI has proven a valid and reliable tool to investigate cognitive functions during presurgical assessment of patients with focal epilepsy, and

has increasingly replaced invasive techniques such as the intracarotid amytal test (IAT) as it is noninvasive, cheaper, and repeatable. Compared to standard neuropsychological assessment, it has the potential to provide additional information about the lateralization and localization of language and memory function and specifically allows evaluation of functional reorganization processes (over time). There is an increasing interest in its possible role for identification of the eloquent cortex and prediction of postoperative cognitive changes.

21.2 Methodological Considerations and Limitations of Cognitive fMRI

In the past, invasive methods have been routinely used to identify the eloquent cortex: the IAT was widely used to lateralize function, and electrocortical stimulation mapping (ESM) is still the gold standard to localize eloquent cortex. In recent years, however, noninvasive methods, predominantly fMRI, have been established for this purpose.[6,7] Obvious advantages of noninvasive methods are patient safety and tolerability. In addition they can be applied in healthy volunteers for research purposes allowing systematic comparison of different study groups. There are, however, some methodological aspects that need to be considered.

Functional MRI is an activation-based method. The rationale behind this is that if a certain cognitive function is used, relevant brain areas will be activated. Blood flow will increase in these areas to compensate for the higher demand of oxygen. In fMRI, this is displayed by the BOLD contrast, which represents the regional changes in blood flow over time. To be able to get a reasonable temporal resolution echo planar imaging (EPI), a fast MRI sequence, is used. Functional

sides, respectively, based on the number of activated voxels for the whole hemisphere or using regions of interest (ROIs) targeted to known language areas. As these methods are very much dependent on the various thresholds chosen, more recent methods are based on calculation of an overall weighted bootstrapped lateralization index, taking thresholds into account.[13,14]

21.3.2 Language fMRI in Epilepsy

Numerous different language paradigms have been applied to group studies, which may partly explain some differences in the results between studies.[7,15] Many studies comparing language fMRI with the classic invasive methods in particular the IAT showed that fMRI is a valid method for identifying the language-dominant hemisphere,[15–17] with an overall concordance of 90% between the two techniques. Functional MRI seems more likely to elicit bilateral language representation compared to the IAT; however, the meaning of this finding for language function after surgery is not fully understood.[17] Concordance between fMRI and the IAT is the highest for right TLE patients with left language dominance and for frontal language areas, and the lowest for left TLE patients with left language dominance.[18] Atypical language dominance on fMRI[17] and interhemispheric language dissociation[19] are correlated with IAT/fMRI discordance. This suggests that fMRI may be more sensitive than the IAT or cortical stimulation to map the whole network involved in language processing, but is less specific at an individual level as compared to cortical stimulation.

From a clinical perspective in patients with focal epilepsy it is advisable to use a combination of expressive and receptive fMRI tasks (e.g., verbal fluency, reading comprehension, auditory comprehension) to reduce interrater variability and help to evaluate language laterality.[20] Of members of the healthy population, 90% have left-sided language representation, while the remaining 10% show atypical (i.e., bilateral, right-hemispheric) language representation. Patients with epilepsy have a significantly higher incidence of atypical language representation (16% bilateral, 6% right-lateralized, 78% left-lateralized), which is additionally influenced by variables such as earlier age of epilepsy onset and the underlying pathology.[21] It has been speculated that an early insult to the originally language-dominant hemisphere may lead to reorganization of language function to homotopic regions in the contralateral hemisphere. At the same time it is not obligatory that lateralization for different language functions (i.e., receptive and expressive) within one patient is the same ("crossed" language lateralization).[22,23] Temporal lobe foci have wide-ranging effects on the distributed language system, and TLE patients are more likely to have atypical language representation in Wernicke's area compared with a frontal focus, while the effects of a frontal lobe focus appear restricted to anterior rather than posterior language processing areas.[24] There is an increase and posterior shift of language related activation in the right inferior frontal gyrus after left hemisphere injury. However, activations in the left inferior frontal gyrus remain in the same location.[25] There is also evidence that the hippocampus itself may be an important factor for establishing language dominance as patients with left hippocampal sclerosis compared to patients with left frontal and lateral temporal lesions had a higher incidence of atypical language representation.[26] Language areas activated for abstract and concrete words are also different for TLE patients, with or without HS, suggesting that HS is associated with altered functional organization of cortical networks involved in lexical and semantic processing.[27] Significant correlations between fMRI activation in the left hippocampus during basic language tasks and out of scanner naming performance have been shown in controls and patients with right TLE, while in left TLE patients naming function was reallocated to the left middle frontal gyrus, supporting the hypothesis that naming ability depends on the integrity of the hippocampus and connecting frontotemporal networks.[28,29] This also supports observations that surgical removal of the dominant hippocampus in patients with left TLE may cause language alterations, mainly naming deficits. In pediatric patients with lesions in close proximity to Broca's area, expressive language function could be demonstrated in the perilesional cortex.[30] In summary, apart from age at disease onset, other factors such as an underlying structural lesion, neuroanatomical substrates, as well as the underlying pathology of the lesion seem to contribute to the reorganization process of language functions, which needs to be taken into account for the clinical interpretation of such data.[31]

In patients with left TLE, Janszky and colleagues found that atypical language representation was associated with increased interictal epileptic discharges,[32] which suggests that fMRI might also be suitable for investigation of dynamic changes in cognitive networks.[33] AEDs can affect cognitive processing.[34] Topiramate has an effect on activation in the basal ganglia, anterior cingulate, and posterior visual cortex and can cause reduced deactivation of the default mode network-related areas during a language task suggesting interference in cognitive processing by topiramate.[35,36]

There is increasing evidence that pre- and post-operatively reorganization of language function can occur to the contralateral but also within the ipsilateral hemisphere.[37] Atypical language lateralization to the right hemisphere may shift back to the left hemisphere in seizure-free patients after left selective amygdalo-hippocampectomy.[38] In a large longitudinal study it has been shown that four months following ATLR, left TLE patients demonstrated greater bilateral fMRI activation during a verbal fluency task compared to preoperatively, suggesting an early post-operative reorganization of language function to the contralateral hemisphere, with evidence of multiple systems supporting language function. Right TLE patients on the contrary showed no postoperative change in language activation patterns.[29] These findings were corroborated by functional connectivity analysis showing greater postoperative than preoperative connectivity to the contralateral frontal lobe in patients with left TLE, while there was no postoperative change in functional connectivity in right TLE.[29]

In terms of efficiency of this postoperative reorganization, left TLE patients without significant naming decline after left temporal lobe surgery showed a significant correlation between postoperative language fMRI activation and postoperative naming performance in the posterior remnant of the left hippocampus, while patients with a significant naming decline demonstrated a correlation in homotopic language areas (MFG) of the contralateral hemisphere,[29] emphasizing the importance of ipsilateral reorganization even within the damaged hippocampus for maintaining naming function. Similarly, Wong and colleagues showed that after ATLR left TLE patients showed reduced activation in the left MFG and right IFG, whereas no difference was observed in the right TLE patients.[39]

Language fMRI may, however, be useful in patients other than those who are undergoing presurgical evaluation. In patients with generalized epilepsy, fMRI showed that language function was impaired as represented by reduced suppression of the default mode network, an inadequate suppression of activation in the left anterior temporal lobe and the posterior cingulate cortex, and an aberrant activation in the right hippocampal formation.[40] In primary reading epilepsy, a study combining EEG and fMRI showed specific regions that were involved in seizure generation during reading.[41] Language fMRI has also been used to study language networks in different epilepsy syndromes, showing that patients suffering from benign epilepsy were less likely to have left lateralized language representation compared with healthy controls.[26] Children with benign epilepsy with centrotemporal spikes (BECTS) and other epilepsy types also showed bihemispheric language networks, which may represent a compensatory response for ongoing epileptic activity in the brain.[42,43] Also, anterior language networks are affected more in BECTS, resulting in language difficulties for functions dependent on the integrity of anterior language regions, e.g., sentence production.[44] Brain activation patterns can be assessed repeatedly during language development in certain epilepsies, which may provide prognostic information regarding potential language achievement in relation to seizures and help finding new rehabilitation strategies.[41,45]

21.3.3 Localization of Language Function and Prediction of Postoperative Deficits

Aphasia is rare after epilepsy surgery, but more subtle language decline such as word finding and naming difficulties has been reported in up to 50% of patients after ATLR of the language-dominant hemisphere.[46] There is evidence that the risk for postoperative naming decline increases with both age of seizure onset and the extent of the resection of the lateral temporal neocortex.[47] A recent fMRI study investigated the impact of focal epilepsy on the developing language system. In 21 children with focal epilepsy and left language dominance the authors demonstrated decreased activation of the ventral language network (compared to healthy controls), which was associated with poorer language outcome. The authors concluded that childhood-onset epilepsy preferentially alters the maturation of the ventral language system, and that this was related to poorer language ability.[48] Using a semantic decision paradigm Sabsevitz and colleagues demonstrated that greater left- than right-sided activation in particular in the temporal structures was associated with a significantly higher risk for postoperative naming deficits.[49] In addition, in left

TLE, a predictive value of fMRI language lateralization for verbal memory decline[50,51] could be shown, emphasizing a link between the inferior frontal cortex and the hippocampus. Using a simple verbal fluency task in left TLE patients preoperative left middle frontal activation was predictive of clinically significant postoperative naming decline, with high sensitivity and relatively lower specificity.[29] However, prediction models for specific language tasks on an individual level are not yet sufficient to be applied in clinical routine. Future studies will need to investigate whether specific language paradigms, particularly auditory and visual naming paradigms, will allow more accurate prediction in individual patients.

Compared to lateralization, accurate localization of language areas with fMRI is not yet established. First of all, test-retest series have shown that the localization of areas that were activated during a specific language fMRI task was less reliable than lateralization.[12] Furthermore, ESM studies showed only imperfect overlap with activation clusters of fMRI: in some cases electric stimulation of fMRI activated brain areas did not result in language disturbances,[52] while in others crucial areas were not displayed during fMRI.[53] The differences may be related to either the applied language paradigms or the statistical thresholds. Rutten and coworkers developed an fMRI imaging protocol that involved four different language tasks for the intraoperative localization of critical language areas. The authors suggested that retrospectively, regions where no fMRI activation was present during four different language tasks (verbal fluency, word finding, naming, sentence comprehension) could have safely been removed without performing ESM.[54]

To date, language fMRI localization is not suitable for resection decision but may be helpful in planning electrode placement for ESM.[6]

21.4 Memory Functions

Memory impairment is common in patients with epilepsy. Working and long-term memory (autobiographical, verbal, visual memory) may become affected, in a material specific way, based on the site of lesion. Functional MRI can reveal memory networks noninvasively and reliably, and also the effect of surgery on these networks.

21.4.1 Autobiographical Memory

This network including the hippocampus, the medial prefrontal cortex, temporal poles, the retro-splenial, and lateral parietal cortex showed reduced activation in patients with left HS and transient epileptic amnesia.[55,56] The connectivity of a sclerosed left hippocampus was also reduced, whereas connections between extrahippocampal nodes were increased.[55,56] In patients with transient epileptic amnesia, there was reduced activation of the right hemisphere, more specifically of the posterior parahippocampal gyrus, the temporoparietal junction, and the cerebellum, for midlife and recent memories.[55,56] In addition, there was reduced effective connectivity between the right posterior parahippocampal gyrus and the right middle temporal gyrus.[56] These findings suggest that there is functional reorganization of the neural network supporting autobiographical memory retrieval in patients with TLE and transient epileptic amnesia.[55,56]

21.4.2 Material Specificity and Episodic Memory

A cognitive process that enables the explicit recollection of unique events and the context in which they occurred,[57] including the transformation of an experience into an enduring memory trace (memory encoding) and the subsequent recollection of this event at a later time (memory retrieval), is defined as episodic memory. Several brain regions including the medial temporal lobe (MTL) and prefrontal cortices are involved in these processes. The MTL consists of the hippocampus, the amygdala, and the parahippocampal regions and is strongly associated with memory function. TLE and ATLRs are both associated with reduced memory function. Initial lesion studies have provided evidence that the hippocampi play a crucial role in memory functioning.[58] While bilateral injury to these areas leads to a characteristic amnesic syndrome[59] patients with unilateral TLE often present with memory impairment, which is specific to certain materials (e.g., verbal and visual). After temporal lobe surgery to the language-dominant hemisphere more often verbal memory decline can be observed,[60] while temporal lobe surgery in the nondominant hemisphere is more likely to result in visuospatial memory decline.[61] Sustained anterograde amnesic syndrome after unilateral ATLR, however, is rare. Most of these patients have subsequently been found to have evidence of contralateral hippocampal pathology, on postoperative EEG,[62] postmortem pathological findings,[63] or postoperative volumetric MRI.[64]

Many fMRI studies have demonstrated material-specific lateralization of memory function in prefrontal but also medial-temporal regions.[65–68] For clinical purposes, paradigms are usually applied that show bilateral MTL activation in healthy controls.[69,70] Previous fMRI studies in patients reported reduced activation in the temporal lobe ipsilateral to the seizure onset.[66,67,71,72] These results were comparable with the Wada test.[67] Most of these studies employed blocked analyses and showed more posterior hippocampal activations. More recent studies using event-related analyses showed material-specific lateralization of memory function in more anterior hippocampal regions during successful memory encoding[68] and therefore in an area that is most likely to be resected during standard ATLR (Figure 21.2). The reduced activation within the affected temporal lobe and increased contralateral MTL activation during memory fMRI have provided further evidence of reorganization of memory function in TLE.[70,72] It is a matter of debate if reorganization toward the healthy hemisphere is effective, and whether it may be protective for memory decline after surgery. By correlation of fMRI activation and performance on standard neuropsychological memory tests, it has been shown that higher MTL activation ipsilateral to the pathology was associated with better memory performance while contralateral, compensatory activation correlated with poorer performance.[70] Functional reorganization of networks involving extratemporal and temporal structures for verbal, visual, and non-material-specific memory encoding suggests compensatory mechanisms to mitigate the failure of the sclerosed hippocampus.[73–75] In one study that investigated patients with left and right TLE, higher MTL activation was observed in the healthy temporal lobe. However, better verbal memory was still related to left MTL activation.[76] This explains why good verbal memory is a risk factor for postoperative decline.

21.4.3 Prediction of Postoperative Memory Changes

In two-thirds of patients with medically refractory TLE, an ATLR leads to seizure freedom, but may be complicated by memory impairment, typically verbal memory decline following left, and visual memory decline following right ATLR. Investigation of patients' ability to sustain memory is critical for planning an ATLR as memory decline is not an inevitable consequence of temporal lobe surgery. One of the ultimate goals of clinical neuroimaging is to accurately predict likelihood and severity of postoperative memory decline in order to make an informed decision regarding surgical treatment.

Two different models of hippocampal function have been proposed to explain memory deficits following unilateral ATLR, the hippocampal reserve model, and the functional adequacy theory.[77] According to the hippocampal reserve model, postoperative memory decline depends on the capacity or reserve of the contralateral hippocampus to support memory following surgery, while the functional adequacy model suggests that it is the capacity of the hippocampus that is to be resected that determines whether changes in memory function will be observed.

Over recent years many studies have focused on the identification of prognostic indicators for risk of memory loss after ATLR. The severity of hippocampal sclerosis on MRI turned out to be an important predictor, being inversely correlated with a decline in verbal memory following left ATLR, with less severe HS increasing the risk of memory decline.[78] Another recognized prognostic factor for memory decline after ATLR was preoperative performance on neuropsychological tests, with higher preoperative scores indicating a greater risk for postoperative decline.[79,80]

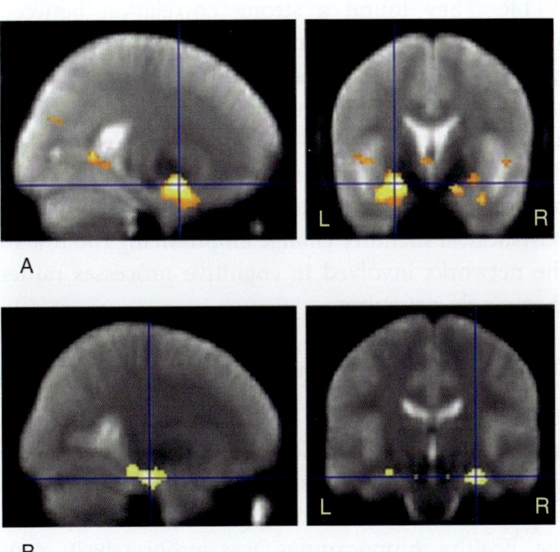

Figure 21.2. Material specificity of verbal and visual memory in healthy controls. Main effects of verbal and visual memory fMRI. (A) Word encoding—left hippocampal activation. (B) Face encoding—right hippocampal activation. Threshold $p < .01$, uncorrected. Significant regions are superimposed onto an averaged normalized mean EP image from 30 healthy controls. Adapted from Bonelli et al.[65]

Language lateralization assessed by the IAT or more recently language fMRI has been found helpful to predict memory outcome.[81,82] These risk factors reflect the functional integrity of the resected temporal lobe and suggest that patients with residual memory function in the pathological hippocampus are at greater risk of memory impairment after ATLR (supporting the functional adequacy theory of the ipsilateral MTL, rather than the functional reserve of the contralateral MTL).

Other epilepsy-related factors such as age of epilepsy onset and duration of epilepsy[83] have also been identified as useful predictors of postoperative outcome.

Memory fMRI has also shown to be a potential predictor of postoperative memory decline after ATLR.

Several small studies have investigated the predictive value of fMRI for verbal memory decline.[82,84–86] Only a few fMRI studies have investigated visual memory after ATLR.[71,86,87] In patients with left HS, greater verbal memory encoding activity in the left hippocampus prior to surgery predicted the extent of verbal memory decline following left ATLR.[72,84,86] These findings have since been replicated and extended to patients undergoing right ATLR.[65,86] A recent study in patients with left and right TLE demonstrated that preoperative recruitment of more extensive networks, mainly in extratemporal regions, during memory encoding was associated with better memory outcomes.[75] Other groups employed asymmetry indices to account for contralateral hippocampal activation and demonstrated that relatively higher activation in the ipsilateral hippocampus was associated with greater memory decline.[71,87] Using a complex visual scene-encoding paradigm, Binder and colleagues also demonstrated asymmetric MTL activation in TLE patients, which was not predictive for postoperative verbal memory decline.[50] A recent fMRI study combined mediotemporal activation asymmetry during a memory task with preoperative verbal memory scores and the information regarding the laterality of the epileptogenic focus. Greater ipsi- than contralateral fMRI activation was associated with a greater decline of verbal memory function. Using all the variables provided, postoperative verbal memory deficits could be predicted in 90% of all patients in this study.[88] Material-specific activation paradigms are particularly attractive for prediction of material-specific memory deficits. Stronger left anterior hippocampal activation on word encoding was predictive of a more severe postoperative decline in verbal memory after left-sided temporal lobe resection (Figure 21.3), and stronger right anterior hippocampal activation on face-encoding predicted a more marked decline in visual memory function after right-sided resection. In addition, patients with relatively higher ipsi- than contralateral posterior hippocampal activation had the best postoperative verbal or visual memory outcome. These findings suggested that reorganization of function to the posterior ipsilateral hippocampus was associated with preservation of memory after ATLR. In this study, preoperative memory fMRI was also the strongest predictor of postoperative verbal and visual memory decline compared to other epilepsy-related variables. Symmetry of fMRI activation with verbal memory encoding, lateralization of language function, and performance on preoperative neuropsychological tests predicted clinically significant verbal memory decline in all patients who underwent left ATLR, but were less predictive of visual memory decline after right ATLR.[65]

Taking these results further, Sidhu et al. demonstrated that predominantly left-sided activations within the frontal and medial temporal lobes correlated with significant verbal memory decline after left ATLR. They found a strong correlation between a memory LI using a mask in the left frontal and medial temporal lobes and regression-adjusted change in ability to learn a supra-span word list over serial presentations.[89] With this fMRI method they demonstrated involvement of frontal networks in verbal learning processes that appear useful in predicting postsurgical memory change emphasizing the role of the networks involved in cognitive processes rather than single structures.

21.4.4 Working Memory

Working memory is also affected in TLE patients with hippocampal sclerosis. Compared to healthy controls there was reduced right superior parietal lobe activity in patients with TLE and hippocampal activity from the healthy hippocampus was progressively suppressed as the working memory load increased, with maintenance of good performance in patients with TLE.[90] Stronger functional connectivity between the superior parietal lobe (BOLD activation) and the sclerosed hippocampus (BOLD deactivation) was associated with worse performance, suggesting that

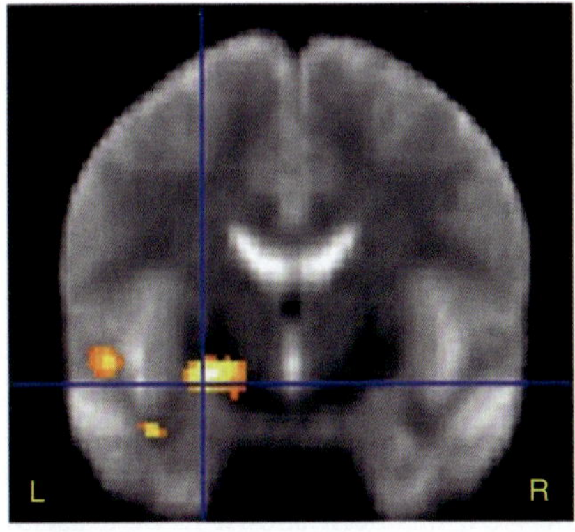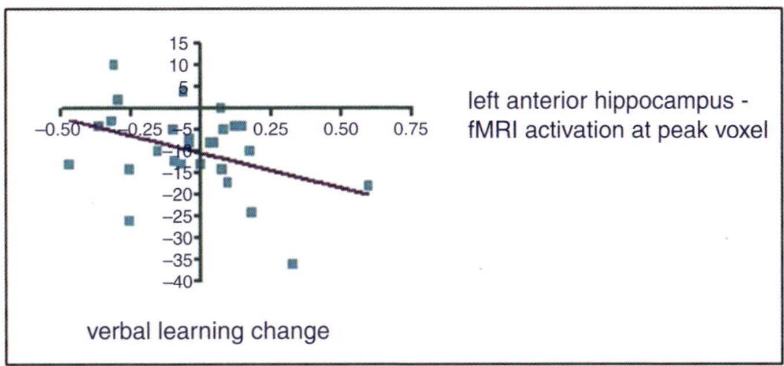

Figure 21.3. Prediction of verbal memory decline using memory fMRI. Left anterior hippocampal activation for encoding words is associated with change in verbal learning scores after left ATLR, characterized by greater verbal memory decline in subjects with greater fMRI activation. Threshold $p < .01$, uncorrected. The correlation at the peak voxel is illustrated on the right. Significant regions are superimposed onto an averaged normalized mean EP image from 30 healthy controls, 15 patients with left and 15 patients with right hippocampal sclerosis. Adapted from Bonelli et al.[65]

the segregation of the task-positive and task-negative functional network was disrupted resulting in working memory dysfunction in TLE.[90,91] Patients with frontal lobe epilepsy (FLE) recruited more widely distributed networks for working memory encoding than did controls; particularly activation of the frontal lobe contralateral to the seizure focus was associated with better performance, suggesting an effective compensatory response of the brain to maintain memory function.[92] Moreover, pediatric patients with FLE showed decreased frontal lobe connectivity, which was associated with cognitive impairment, despite intact fMRI activation patterns for working memory. This decreased frontal lobe connectivity may explain the cognitive problems encountered in children with FLE.[93] In addition, high numbers of secondary generalized seizures may induce functional reorganization of working-memory-related networks, e.g., increased activation and reduced functional connectivity of the prefrontal cortex explaining working memory dysfunction in patients with focal epilepsy.[94,95]

21.4.5 Postoperative Reorganization of Memory Function

Functional MRI may also be used to investigate the extent to which the brain may functionally reorganize following epilepsy surgery. Functional reorganization —both before and after surgery—can occur within the unaffected ipsilateral or contralateral hemisphere, but under which circumstances reorganization becomes effective is still poorly understood.

There is increasing evidence that in TLE patients, memory outcome depends on the extent of the removal of nonlesional functional tissues.[96–99] To date, there are only a few fMRI studies in which memory function was systematically investigated before and after surgery. Cheung and coworkers assessed postoperative memory function using a complex visual scene encoding task. In their study postoperative memory function was significantly associated with functional activation contralateral to the side of resection in nine left and eight right TLE patients; therefore the authors suggested a role for the contralateral medial temporal lobe in supporting postoperative memory.[100] Using a material-specific memory encoding task, Bonelli et al. showed that preoperative memory reorganization to the posterior part of the hippocampus was important to preserve postoperative verbal memory function after left ATLR. Conversely, if reorganization to the ipsilesional posterior remnant of the hippocampus occurred early (within 4 months) after surgery, this additional, early postoperative reorganization worked inefficiently. No significant visual memory reorganization was seen after right ATLR, providing evidence that verbal and visual memory function was affected differently by left and right ATLR.[101] One study examined extratemporal network plasticity after surgery in TLE patients in a ROI in the IFG. Pre- and postoperative signal change within this ROI was compared quantitatively in this word and face classification study. Reduced right frontal activation was shown postoperatively during word encoding in left TLE suggesting a more left lateralized network postoperatively. No differences were seen for visual encoding.[102]

21.5 Functional MRI in the Temporal Lobes—Technical Challenges

Echo planar imaging is the sequence most commonly used for fMRI studies, as these require images to be acquired at high speed. Memory fMRI in the medial temporal lobe is in particular challenging due to limits on resolution and the possibility of geometric distortions and signal drop out caused by susceptibility effects associated with the use of EPI.[103] Inhomogeneities in the magnetic field occur due to the different magnetic properties of bone, tissue, and air, as soon as the head is introduced into the scanner. Brain regions close to borders between sinuses and brain or bone and brain are most affected and therefore most likely to suffer geometric distortions or loss of BOLD signal. The occurrence of material specific memory impairment after ATLR indicates the major relevance of these areas for successful memory function. Intracranial electrophysiological recordings during verbal encoding tasks have shown greater responses in anterior hippocampal and parahippocampal regions for words remembered than those forgotten.[104] At first, these results could not be replicated with functional imaging studies, with many showing encoding related activations in posterior hippocampal and parahippocampal regions, which would be left intact following ATLRs. One possible explanation for this apparent conflict is that anterior temporal regions are subject to signal loss during fMRI sequences, which is most prominent in the inferior frontal and inferior lateral temporal regions.[105] Signal loss leading to sensitivity loss is unrecoverable by image-processing techniques and therefore may be more serious.[9] The anatomical position of the hippocampus that rises from anterior to posterior may explain greater susceptibility-induced signal loss in the anterior (inferior) relative to the posterior (superior) hippocampus, which may have been one reason for the relative lack of anterior hippocampal activation in early fMRI studies of memory. One study demonstrated that there was a differential effect of susceptibility artifacts on the activation in the anterior versus the posterior hippocampus. They showed that the averaged resting voxel intensity in an anterior hippocampal ROI was significantly less than in a posterior hippocampal ROI; intensity decreases were substantial enough to leave many voxels below the threshold at which BOLD effects could be detected.[106] Even more though, in a different study it could be demonstrated that the sensitivity to BOLD changes was proportional to signal intensity at rest so that voxels with a lower baseline signal (such as those in anterior hippocampal regions) would be more difficult to activate than those with higher baseline signals.[107] Shimming, which is a process whereby the static magnetic field is made more homogeneous over the ROI, has been shown to be a useful method to correct some of these artifacts.

Paradigm design may be an alternative explanation for the lack of anterior hippocampal activation seen in many early memory fMRI experiments. Most of the paradigms used were not optimized for detecting subsequent memory effects as, for example, most of these early studies used blocked experimental designs. Studying memory function in general requires more complex paradigms and ways of analysis, as there are several different components involved in memory processing (encoding and retrieval) and as also the nature of the material being encoded or retrieved influences which brain areas are activated.

By using blocked design paradigms we are looking for regions of the brain showing greater activation during task blocks compared with rest blocks, which makes it difficult to separate brain activity due specifically to memory from that due to other cognitive processes involved in the task. To overcome this problem, event-related designs were introduced in recent fMRI memory studies.[9]

In brief, the detection of transient hemodynamic responses to brief stimuli or tasks is defined as event-related fMRI. With this technique, which was derived from techniques used by electrophysiologists to study event-related potentials, trial-based rather than block-based experiments can be carried out. These trial-based designs have a number of methodological advantages, but also some drawbacks: a subject's individual performance on a subsequent cognitive test can be taken into account so that trials can be categorized post hoc according to each subject's performance. When for example studying memory encoding, activations for the different items can be contrasted according to whether they are subsequently remembered or forgotten in a subsequent memory test in each individual. In this way we can identify brain regions with greater activation during encoding of different (material-specific) items that are subsequently remembered compared to items that are subsequently forgotten (subsequent memory effects) representing the neural correlates of memory encoding.[108]

A previous study compared results from a blocked and event-related analysis of memory fMRI of words, pictures, and faces: only the event-related analysis of successfully encoded stimuli showed significant activations in the anterior MTL, whereas simply viewing the different stimuli (using a blocked analysis without taking into account whether items were subsequently remembered or not) revealed predominant activation in the posterior hippocampus.[68] This study provided evidence for a functional dissociation between anterior and posterior hippocampal regions.

In summary, although event-related designs are less powerful than blocked designs at detecting activation and are also more vulnerable to alterations in the hemodynamic response function, they have the big advantage of permitting specifically the detection of subsequent memory effects due to successful encoding.

21.6 Functional Connectivity

The possibility to analyze functional connectivity in patients with epilepsy has significantly improved our understanding of seizure generation and propagation. The analysis of functional connectivity during cognitive tasks allows studying the neuronal networks that subserve these tasks.[109]

During the last few years several studies have demonstrated altered connectivity within the epileptic network but also in networks of cognitive function such as language and memory. In patients with left TLE compared to healthy controls the functional connectivity within the expressive language network was reduced, most likely due to the underlying disease.[110] More specifically, functional connectivity was decreased in the left hemisphere irrespective of the epileptogenic focus[111] and within the prefrontal and frontotemporal networks,[112] which was associated with impaired performance on language assessment.[112] In right TLE, coupling between functional and structural measures for left language lateralization has been shown, while in left TLE decoupling between structural and functional measures suggested more complex language networks.[113]

Other studies evaluating memory function showed that functional connectivity was reduced between the posterior cingulate and the epileptogenic hippocampus and increased between the posterior cingulate and the contralateral hippocampus.[114,115] A recent study used a visual scene encoding task to evaluate memory function in healthy controls compared to patients with TLE. In this study, patients with left TLE demonstrated a significant decrease in functional connectivity to the inferior temporal, occipital, cingulate, and parietal cortices and the thalamus. In addition the authors showed that orbital frontal activity correlated with structural measures of tract coherence in the fornix, which led to the suggestion that this might be the structural correlate of reduced functional connectivity.[116] Vollmar and colleagues reported similar observations of increased connectivity of motor and cognitive networks in patients with juvenile myoclonic epilepsy.[117] However, whether functional connectivity studies may prove clinically useful still needs to be established. A potential application is the use as a clinical predictor of postoperative outcome following surgery, and studies are just beginning to emerge that address this possibility.

21.7 Conclusion

In conclusion, fMRI is a promising, noninvasive tool to study the effects of chronic epilepsy and epilepsy surgery on cognitive functions and factors that are associated with effective functional reorganization/plasticity after surgery.

From a clinical perspective, many of those recent imaging studies increasingly emphasize the need to consider selective as opposed to extended temporal resections (and in particular to spare as much as possible of nonlesional functional tissue) to minimize the risk of cognitive impairment, but at the same time provide the same chance of good seizure control. It is hoped that, in the near future, cognitive fMRI may be able to provide reliable results in individual patients, which when used in concert with other clinical, electrophysiological, and imaging results will not only predict postoperative cognitive decline in single patients but ultimately help prevent them.

Therefore, one of the aims of future work is now to integrate fMRI data with structural MRI into neuronavigation systems, including interventional MRI, and to repeat fMRI studies intraoperatively in order to enable image guided resections.[11] Hurdles to overcome include the different anatomical distortions of T1-weighted MR and EPI fMR images but also physical distortions such as brain shift, which may be as much as 2 cm after surgery.[118]

There is hope that constant improvements in hardware and software and in particular functional MR imaging at higher magnetic fields will lead to improved signal-to-noise ratio. A recent fMRI study compared the presurgical localization of the primary motor hand area on a 3 T and 7 T Siemens scanner with identical investigational procedures and comparable system specific sequences. Results showed significantly higher functional sensitivity of the 7 T system (measured via percentage signal change, mean t-values, number of suprathreshold voxels, and contrast-to-noise ratio), although 7 T data suffered from a significant increase of artifacts such as ghosting and head motion.[119] However, results are promising that ultra-high-field systems provide a clinically relevant increase of functional sensitivity for patients also in the MTL, which might lead to strong and reliable hippocampal activation at an individual level.

Another exciting area is the use of fMRI to study brain network dynamics. Functional connectivity can elicit neuronal networks that contribute to various cognitive tasks. It has been shown that cognitive impairment is often accompanied by reduced functional connectivity. Whether these methods may add to the prediction of postoperative cognitive outcome remains a topic of current research. Together with tractography, which can provide a means to understand the distributed and interrelated structural networks of the brain, functional connectivity can be useful to understand seizure propagation and spread and how this may be related to impaired function. An appealing speculation is that, ultimately, this information, along with data from modalities such as MEG, ESI, and EEG-fMRI, could be used to disconnect seizure foci from the rest of the network.

Finally, fMRI may help to improve the individual prognosis of even subtle cognitive deficits, but also stimulate the development of new therapeutic strategies for cognitive rehabilitation. An exciting possibility for future work could therefore be to evaluate whether it is possible to modify dysfunctional networks prior to surgery, i.e., by applying specific learning strategies to the patients in order to optimize or even improve the chances of a good cognitive outcome following epilepsy surgery.

References

1. Helmstaedter C. Effects of chronic epilepsy on declarative memory systems. *Prog Brain Res.* 2002;**135**:439–53.

2. Helmstaedter C. Neuropsychological aspects of epilepsy surgery. *Epilepsy Behav.* 2004;**5**(suppl 1):S45–55.

3. Witt JA, Helmstaedter C. Should cognition be screened in new-onset epilepsies? A study in 247 untreated patients. *J Neurol.* 2012;**259**(8):1727–31.

4. de Tisi J, Bell GS, Peacock JL, et al. The long-term outcome of adult epilepsy surgery, patterns of seizure remission, and relapse: a cohort study. *Lancet.* 2011;**378**(9800):1388–95.

5. Wiebe S, Blume WT, Girvin JP, et al. A randomized, controlled trial of surgery for temporal-lobe epilepsy. *N Engl J Med.* 2001;**345**(5):311–8.

6. Duncan J. The current status of neuroimaging for epilepsy. *Curr Opin Neurol.* 2009;**22**:179–84.

7. Haag A, Knake S, Hamer HM, et al. The Wada test in Austrian, Dutch, German, and Swiss epilepsy centers from 2000 to 2005: a review of 1421 procedures. *Epilepsy Behav.* 2008;**13**(1):83–9.

8. Price CJ, Friston KJ. Scanning patients with tasks they can perform. *Hum Brain Mapp.* 1999;**8**(2–3):102–8.

9. Thornton R, Powell HW, Lemieux L. fMRI in epilepsy. fMRI techniques and protocols. *Humana.* 2009;**41**:681–730.

10. Jansen JF, Aldenkamp AP, Marian MHJ, et al. Functional MRI reveals declined prefrontal cortex activation in patients with epilepsy on topiramate therapy. *Epilepsy Behav.* 2006;**9**(1):181–5.

11. Duncan JS. Imaging in the surgical treatment of epilepsy. *Nat Rev Neurol.* 2010;**6**(10):537–50.

12. Fernandez G, Specht K, Weis S, et al. Intrasubject reproducibility of presurgical language lateralization and mapping using fMRI. *Neurology.* 2003;**60**(6):969–75.

13. Wilke M, Lidzba K. LI-tool: a new toolbox to assess lateralization in functional MR-data. *J Neurosci Methods.* 2007;**163**(1):128–36.

14. Wilke M, Schmithorst VJ. A combined bootstrap/histogram analysis approach for computing a lateralization index from neuroimaging data. *NeuroImage.* 2006;**33**(2):522–30.

15. Woermann FG, Labudda K. Clinical application of functional MRI for chronic epilepsy. *Radiologe.* 2010;**50**(2):123–30.

16. Dym RJ, Burns J, Freeman K, Lipton ML. Is functional MR imaging assessment of hemispheric language dominance as good as the Wada test? A meta-analysis. *Radiology.* 2011;**261**(2):446–55.

17. Janecek JK, Swanson SL, Sabsevitz DS, et al. Language lateralization by fMRI and Wada testing in 229 patients with epilepsy: rates and predictors of discordance. *Epilepsia*. 2013;**54**(2):314–22.

18. Benke T, Koylu B, Visani P, et al. Language lateralization in temporal lobe epilepsy: a comparison between fMRI and the Wada Test. *Epilepsia*. 2006;**47**(8):1308–19.

19. Lee D, Swanson SL, Sabsevitz DS, et al. Functional MRI and Wada studies in patients with interhemispheric dissociation of language functions. *Epilepsy Behav*. 2008;**13**(2):350–6.

20. Gaillard WD. Functional MR imaging of language, memory, and sensorimotor cortex. *Neuroimaging Clin N Am*. 2004;**14**(3):471–85.

21. Springer JA, Binder JR, Hammeke TA, et al. Language dominance in neurologically normal and epilepsy subjects: a functional MRI study. *Brain*. 1999;**122**:2033–46.

22. Berl MM, Balsamo LM, Xu B, et al. Seizure focus affects regional language networks assessed by fMRI. *Neurology*. 2005;**65**(10):1604–11.

23. Gaillard WD, Berl MM, Moore EN, et al. Atypical language in lesional and nonlesional complex partial epilepsy. *Neurology*. 2007;**69**(18):1761–71.

24. Duke ES, Tesfaye M, Berl MM, et al. The effect of seizure focus on regional language processing areas. *Epilepsia*. 2012;**53**(6):1044–50.

25. Voets NL, Adcock JE, Flitney DE, et al. Distinct right frontal lobe activation in language processing following left hemisphere injury. *Brain*. 2006;**129**:754–66.

26. Weber B, Wellmer J, Reuber M, et al. Left hippocampal pathology is associated with atypical language lateralization in patients with focal epilepsy. *Brain*. 2006;**129**:346–51.

27. Jensen EJ, Hargreaves IS, Pexman PM, et al. Abnormalities of lexical and semantic processing in left temporal lobe epilepsy: an fMRI study. *Epilepsia*. 2011;**52**(11):2013–21.

28. Bonelli SB, Powell R, Thompson PJ, et al. Hippocampal activation correlates with visual confrontation naming: fMRI findings in controls and patients with temporal lobe epilepsy. *Epilepsy Res*. 2011;**95**(3):246–54.

29. Bonelli SB, Thompson PJ, Yogarajah M, et al. Imaging language networks before and after anterior temporal lobe resection—results of a longitudinal fMRI study. *Epilepsia*. 2012;**53**:639–50.

30. Liegeois F, Connelly A, Cross JH, et al. Language reorganization in children with early-onset lesions of the left hemisphere: an fMRI study. *Brain*. 2004;**127**:1229–36.

31. Wellmer J, Weber B, Urbach H, et al. Cerebral lesions can impair fMRI-based language lateralization. *Epilepsia*. 2009;**50**(10):2213–24.

32. Janszky J, Mertens M, Janszky I, Ebner A, Woermann FG. Left-sided interictal epileptic activity induces shift of language lateralization in temporal lobe epilepsy: an fMRI study. *Epilepsia*. 2006;**47**(5):921–7.

33. Monjauze C, Broadbent H, Boyd SG, Neville BG, Baldeweg T. Language deficits and altered hemispheric lateralization in young people in remission from BECTS. *Epilepsia*. 2011;**52**(8):e79–83.

34. Wang G, Worrell G, Yang L, Wilke C, He B. Interictal spike analysis of high-density EEG in patients with partial epilepsy. *Clin Neurophysiol*. 2011;**122**(6):1098–105.

35. Szaflarski JP, Allendorfer JB. Topiramate and its effect on fMRI of language in patients with right or left temporal lobe epilepsy. *Epilepsy Behav*. 2012;**24**(1):74–80.

36. Yasuda CL, Centeno M, Vollmar C, et al. The effect of topiramate on cognitive fMRI. *Epilepsy Res*. 2013;**105**(1–2):250–5.

37. Noppeney U, Price CJ, Duncan JS, Koepp MJ. Reading skills after left anterior temporal lobe resection: an fMRI study. *Brain*. 2005;**128**:1377–85.

38. Helmstaedter C, Firtz NE, Gonzalez Perez PA, et al. Shift-back of right into left hemisphere language dominance after control of epileptic seizures: evidence for epilepsy driven functional cerebral organization. *Epilepsy Res*. 2006;**70**(2–3):257–62.

39. Wong SW, Jong L, Bandur D, et al. Cortical reorganization following anterior temporal lobectomy in patients with temporal lobe epilepsy. *Neurology*. 2009;**73**(7):518–25.

40. Gauffin H, van Ettinger-Veenstra H, Landtblom AM, et al. Impaired language function in generalized epilepsy: inadequate suppression of the default mode network. *Epilepsy Behav*. 2013;**28**(1):26–35.

41. Salek-Haddadi A, Mayer T, Hamandi K, et al. Imaging seizure activity: a combined EEG/EMG-fMRI study in reading epilepsy. *Epilepsia*. 2009;**50**(2):256–64.

42. Datta AN, Oser N, Bauder F, et al. Cognitive impairment and cortical reorganization in children with benign epilepsy with centrotemporal spikes. *Epilepsia*. 2013;**54**(3):487–94.

43. Yuan W, Szaflarski JP, Schmithorst VY, et al. fMRI shows atypical language lateralization in pediatric epilepsy patients. *Epilepsia*. 2006;**47**(3):593–600.

44. Lillywhite LM, Saling MM, Harvey AS, et al. Neuropsychological and functional MRI studies provide converging evidence of anterior language

dysfunction in BECTS. *Epilepsia.* 2009;**50**(10): 2276–84.

45. Pal DK. Epilepsy and neurodevelopmental disorders of language. *Curr Opin Neurol.* 2011;**24**(2):126–31.

46. Davies KG, Bell BD, Bush AJ, et al. Naming decline after left anterior temporal lobectomy correlates with pathological status of resected hippocampus. *Epilepsia.* 1998;**39**(4):407–19.

47. Hermann B, Davies K, Foley K, Bell B. Visual confrontation naming outcome after standard left anterior temporal lobectomy with sparing versus resection of the superior temporal gyrus: a randomized prospective clinical trial. *Epilepsia.* 1999;**40**(8):1070–6.

48. Croft LJ, Baldeweg T, Sepeta L, et al. Vulnerability of the ventral language network in children with focal epilepsy. *Brain.* 2014;**137**:2245–57.

49. Sabsevitz DS, Swanson SJ, Hammeke TA, et al. Use of preoperative functional neuroimaging to predict language deficits from epilepsy surgery. *Neurology.* 2003;**60**(11):1788–92.

50. Binder JR, Swanson SJ, Sabsevitz DS, et al. A comparison of two fMRI methods for predicting verbal memory decline after left temporal lobectomy: language lateralization versus hippocampal activation asymmetry. *Epilepsia.* 2010;**51**(4):618–26.

51. Labudda K, Mertens M, Janszky J, et al. Atypical language lateralisation associated with right fronto-temporal grey matter increases—a combined fMRI and VBM study in left-sided mesial temporal lobe epilepsy patients. *NeuroImage.* 2012;**59**(1):728–37.

52. Kunii N, Kamada K, Ota T, Kawai K, Saito N. A detailed analysis of functional magnetic resonance imaging in the frontal language area: a comparative study with extraoperative electrocortical stimulation. *Neurosurgery.* 2011;**69**(3):590–6; discussion 596–7.

53. Roux FE, Boulanouar K, Lotterie JA, et al. Language functional magnetic resonance imaging in preoperative assessment of language areas: correlation with direct cortical stimulation. *Neurosurgery.* 2003;**52** (6):1335–45; discussion 1345–7.

54. Rutten GJ, Ramsey NF, van Rijen PC, et al. Development of a functional magnetic resonance imaging protocol for intraoperative localization of critical temporoparietal language areas. *Ann Neurol.* 2002;**51**(3):350–60.

55. Addis DR, Moscovitch M, McAndrews MP. Consequences of hippocampal damage across the autobiographical memory network in left temporal lobe epilepsy. *Brain.* 2007;**130**:2327–42.

56. Milton F, Butler CR, Benattayallah A, Zeman AZ. The neural basis of autobiographical memory deficits in transient epileptic amnesia. *Neuropsychologia.* 2012;**50**(14):3528–41.

57. Baddeley A. The concept of episodic memory. *Philos Trans R Soc Lond B Biol Sci.* 2001;**356**(1413): 1345–50.

58. Squire LR, Zola-Morgan S. The medial temporal lobe memory system. *Science.* 1991;**253**(5026):1380–6.

59. Scoville WB, Milner B. Loss of recent memory after bilateral hippocampal lesions. *J Neuropsychiatry Clin Neurosci.* 2000;**12**(1):103–13.

60. Ivnik RJ, Sharbrough FW, Laws ER Jr. Effects of anterior temporal lobectomy on cognitive function. *J Clin Psychol.* 1987;**43**(1):128–37.

61. Spiers HJ, Burgess N, Maguire EA, et al. Unilateral temporal lobectomy patients show lateralized topographical and episodic memory deficits in a virtual town. *Brain.* 2001;**124**:2476–89.

62. Penfield W, Milner B. Memory deficit produced by bilateral lesions in the hippocampal zone. *AMA Arch Neurol Psychiatry.* 1958;**79**(5):475–97.

63. Warrington EK, Duchen LW. A re-appraisal of a case of persistent global amnesia following right temporal lobectomy: a clinico-pathological study. *Neuropsychologia.* 1992;**30**(5):437–50.

64. Loring DW, Hermann BP, Meador KJ, et al. Amnesia after unilateral temporal lobectomy: a case report. *Epilepsia.* 1994;**35**(4):757–63.

65. Bonelli SB, Powell RH, Yogarajah M, et al. Imaging memory in temporal lobe epilepsy: predicting the effects of temporal lobe resection. *Brain.* 2010;**133**: 1186–99.

66. Detre JA, Maccotta L, King D, et al. Functional MRI lateralization of memory in temporal lobe epilepsy. *Neurology.* 1998;**50**(4):926–32.

67. Golby AJ, Poldrack RA, Brewer JB, et al. Material-specific lateralization in the medial temporal lobe and prefrontal cortex during memory encoding. *Brain.* 2001;**124**:1841–54.

68. Powell HW, Koepp MJ, Symms MR, et al. Material-specific lateralization of memory encoding in the medial temporal lobe: blocked versus event-related design. *NeuroImage.* 2005;**27**(1):231–9.

69. Jokeit H, Okujava M, Woermann FG. Memory fMRI lateralizes temporal lobe epilepsy. *Neurology.* 2001;**57** (10):1786–93.

70. Powell HW, Richardson MP, Symms MR, et al. Reorganization of verbal and nonverbal memory in temporal lobe epilepsy due to unilateral hippocampal sclerosis. *Epilepsia.* 2007;**48**(8):1512–25.

71. Janszky J, Jokeit H, Kontopoulou K, et al. Functional MRI predicts memory performance after right

mesiotemporal epilepsy surgery. *Epilepsia*. 2005;**46**(2): 244–50.

72. Richardson MP, Strange BA, Duncan JS, Dolan RJ. Preserved verbal memory function in left medial temporal pathology involves reorganisation of function to right medial temporal lobe. *NeuroImage*. 2003;**20**(suppl 1):S112–9.

73. Alessio A, Pereira FR, Sercheli MS, et al. Brain plasticity for verbal and visual memories in patients with mesial temporal lobe epilepsy and hippocampal sclerosis: an fMRI study. *Hum Brain Mapp*. 2013;**34**(1): 186–99.

74. Guedj E, Bettus G, Barbeau EJ, et al. Hyperactivation of parahippocampal region and fusiform gyrus associated with successful encoding in medial temporal lobe epilepsy. *Epilepsia*. 2011;**52**(6):1100–9.

75. Sidhu MK, Stretton J, Winston GP, et al. A functional magnetic resonance imaging study mapping the episodic memory encoding network in temporal lobe epilepsy. *Brain*. 2013;**136**:1868–88.

76. Vannest J, Szaflarski JP, Privitera MD, et al. Medial temporal fMRI activation reflects memory lateralization and memory performance in patients with epilepsy. *Epilepsy Behav*. 2008;**12**(3):410–8.

77. Chelune GJ. Hippocampal adequacy versus functional reserve: predicting memory functions following temporal lobectomy. *Arch Clin Neuropsychol*. 1995;**10** (5):413–32.

78. Trenerry MR, Jack CR, Ivnik RJ, et al. MRI hippocampal volumes and memory function before and after temporal lobectomy. *Neurology*. 1993;**43**(9): 1800–5.

79. Baxendale S, Thompson P, Harkness W, Duncan J. Predicting memory decline following epilepsy surgery: a multivariate approach. *Epilepsia*. 2006;**47**(11): 1887–94.

80. Helmstaedter C, Elger CE. Cognitive consequences of two-thirds anterior temporal lobectomy on verbal memory in 144 patients: a three-month follow-up study. *Epilepsia*. 1996;**37**(2):171–80.

81. Baxendale S. The role of functional MRI in the presurgical investigation of temporal lobe epilepsy patients: a clinical perspective and review. *J Clin Exp Neuropsychol*. 2002;**24**(5):664–76.

82. Binder JR, Sabsevitz DS, Swanson SJ, et al. Use of preoperative functional MRI to predict verbal memory decline after temporal lobe epilepsy surgery. *Epilepsia*. 2008;**49**(8):1377–94.

83. Baxendale S, Thompson PJ, Duncan JS. Improvements in memory function following anterior temporal lobe resection for epilepsy. *Neurology*. 2008;**71**(17):1319–25.

84. Richardson MP, Strange BA, Duncan JS, Dolan RJ. Memory fMRI in left hippocampal sclerosis: optimizing the approach to predicting postsurgical memory. *Neurology*. 2006;**66**(5):699–705.

85. Richardson MP, Strange BA, Thompson PJ, et al. Pre-operative verbal memory fMRI predicts post-operative memory decline after left temporal lobe resection. *Brain*. 2004;**127**:2419–26.

86. Powell HW, Richardson MP, Symms MR, et al. Preoperative fMRI predicts memory decline following anterior temporal lobe resection. *J Neurol Neurosurg Psychiatry*. 2008;**79**(6):686–93.

87. Rabin ML, Narayn VM, Kimberg DY, et al. Functional MRI predicts post-surgical memory following temporal lobectomy. *Brain*. 2004;**127**:2286–98.

88. Dupont S, Duron E, Samson S, et al. Functional MR imaging or Wada test: which is the better predictor of individual postoperative memory outcome? *Radiology*. 2010;**255**(1):128–34.

89. Sidhu MK, Stretton J, Winston GP, et al. Memory fMRI predicts verbal memory decline after anterior temporal lobe resection. *Neurology*. 2015;**84**(15):1512–9.

90. Stretton J, Winston G, Sidhu M, et al. Neural correlates of working memory in temporal lobe epilepsy—an fMRI study. *NeuroImage*. 2012;**60**(3):1696–703.

91. Stretton J, Winston G, Sidhu M, et al. Disrupted segregation of working memory networks in temporal lobe epilepsy. *NeuroImage Clin*. 2013;**2**:273–81.

92. Centeno M, Thompson PJ, Koepp MJ, et al. Memory in frontal lobe epilepsy. *Epilepsy Res*. 2010;**91**(2–3): 123–32.

93. Braakman HM, Vaessen MJ, Jansen JF, et al. Frontal lobe connectivity and cognitive impairment in pediatric frontal lobe epilepsy. *Epilepsia*. 2013;**54**(3): 446–54.

94. Vlooswijk MC, Jansen JF, Jeukens CR, et al. Memory processes and prefrontal network dysfunction in cryptogenic epilepsy. *Epilepsia*. 2011;**52**(8):1467–75.

95. Vlooswijk MC, Jansen JF, Reijs RP, et al. Cognitive fMRI and neuropsychological assessment in patients with secondarily generalized seizures. *Clin Neurol Neurosurg*. 2008;**110**(5):441–50.

96. Alpherts WC, Vermeulen J, van Rijen PC, et al. Standard versus tailored left temporal lobe resections: differences in cognitive outcome? *Neuropsychologia*. 2008;**46**(2):455–60.

97. Helmstaedter C, Richter S, Roske S, et al. Differential effects of temporal pole resection with amygdalohippocampectomy versus selective amygdalohippocampectomy on material-specific memory in patients with mesial temporal lobe epilepsy. *Epilepsia*. 2008;**49**(1):88–97.

98. Helmstaedter C, Kurthen M, Lux S, et al. Chronic epilepsy and cognition: a longitudinal study in

temporal lobe epilepsy. *Ann Neurol.* 2003;**54**(4): 425–32.

99. Schramm J. Temporal lobe epilepsy surgery and the quest for optimal extent of resection: a review. *Epilepsia.* 2008;**49**(8):1296–307.

100. Cheung MC, Chan AS, Lam JM, Chan YL. Pre- and postoperative fMRI and clinical memory performance in temporal lobe epilepsy. *J Neurol Neurosurg Psychiatry.* 2009;**80**(10):1099–106.

101. Bonelli SB, Thompson PJ, Yogarajah M, et al. Memory reorganization following anterior temporal lobe resection: a longitudinal functional MRI study. *Brain.* 2013;**136**:1889–900.

102. Maccotta L, Buckner RL, Gilliam FG, Ojemann JG. Changing frontal contributions to memory before and after medial temporal lobectomy. *Cereb Cortex.* 2007;**17**(2):443–56.

103. Robinson S, Windischberger C, Rauscher A, Moser E. Optimized 3 T EPI of the amygdalae. *NeuroImage.* 2004;**22**(1):203–10.

104. Fernandez G, Effern A, Grunwald T, et al. Real-time tracking of memory formation in the human rhinal cortex and hippocampus. *Science.* 1999;**285**(5433): 1582–5.

105. Ojemann JG, Akbudak E, Snyder AZ, et al. Anatomic localization and quantitative analysis of gradient refocused echo-planar fMRI susceptibility artifacts. *NeuroImage.* 1997;**6**(3):156–67.

106. Greicius MD, Krasnow B, Boyett-Anderson JM, et al. Regional analysis of hippocampal activation during memory encoding and retrieval: fMRI study. *Hippocampus.* 2003;**13**(1):164–74.

107. Lipschutz B, Friston KJ, Ashburner J, Turner R, Price CJ. Assessing study-specific regional variations in fMRI signal. *NeuroImage.* 2001;**13**(2): 392–8.

108. Wagner AD, Koutstaal W, Schacter DL. When encoding yields remembering: insights from event-related neuroimaging. *Philos Trans R Soc Lond B Biol Sci.* 1999;**354**(1387):1307–24.

109. Axmacher N, Mormann F, Fernandez G, et al. Sustained neural activity patterns during working memory in the human medial temporal lobe. *J Neurosci.* 2007;**27**(29):7807–16.

110. Waites AB, Briellmann RS, Saling MM, et al. Functional connectivity networks are disrupted in left temporal lobe epilepsy. *Ann Neurol.* 2006;**59**(2): 335–43.

111. Pravata E, Sestieri C, Mantini D, et al. Functional connectivity MR imaging of the language network in patients with drug-resistant epilepsy. *AJNR Am J Neuroradiol.* 2011;**32**(3):532–40.

112. Vlooswijk MC, Jansen JF, Majoie HJ, et al. Functional connectivity and language impairment in cryptogenic localization-related epilepsy. *Neurology.* 2010;**75** (5):395–402.

113. Rodrigo S, Oppenheim C, Chassoux F, et al. Language lateralization in temporal lobe epilepsy using functional MRI and probabilistic tractography. *Epilepsia.* 2008;**49**(8):1367–76.

114. McCormick C, Quraan M, Cohn M, et al. Default mode network connectivity indicates episodic memory capacity in mesial temporal lobe epilepsy. *Epilepsia.* 2013;**54**(5):809–18.

115. Pereira FR, Alessio A, Sercheli MS, et al. Asymmetrical hippocampal connectivity in mesial temporal lobe epilepsy: evidence from resting state fMRI. *BMC Neurosci.* 2010;**11**:66.

116. Voets NL, Adcock JE, Stacey R, et al. Functional and structural changes in the memory network associated with left temporal lobe epilepsy. *Hum Brain Mapp.* 2009;**30**(12):4070–81.

117. Vollmar C, O'Muircheartaigh J, Barker GJ, et al. Motor system hyperconnectivity in juvenile myoclonic epilepsy: a cognitive functional magnetic resonance imaging study. *Brain.* 2011;**134**:1710–9.

118. Nimsky C, Ganslandt O, Merhof D, et al. Intraoperative visualization of the pyramidal tract by diffusion-tensor-imaging-based fiber tracking. *NeuroImage.* 2006;**30**(4):1219–29.

119. Beisteiner R, Robinson S, Wurnig M, et al. Clinical fMRI: evidence for a 7 T benefit over 3 T. *NeuroImage.* 2011;**57**(3):1015–21.

Index